A REVISION OF THE GENUS *LESPEDEZA*
SECTION *MACROLESPEDEZA* (LEGUMINOSAE)

# A REVISION OF THE GENUS *LESPEDEZA* SECTION *MACROLESPEDEZA* (LEGUMINOSAE)

Shinobu Akiyama

UNIVERSITY OF TOKYO PRESS

The University Museum, The University of Tokyo, Bulletin No. 33, 1988



ISBN  4-13-068145-1
ISBN  0-86008-429-9

a.  *Lespedeza bicolor* var. *nana*.

b.  *L. bicolor*.

**Plate 1.**

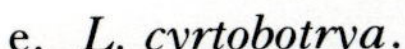

c.  *L. melanantha* f. *rosea*.

e.  *L. cyrtobotrya*.

d.  *L. melanantha*.

f.  *L. cyrtobotrya*.

a. *L. patens*.

b. *L. patens*.

c. *L. formosa* subsp. *formosa*.

**Plate 2.**

d. *L. formosa* subsp. *velutina* var. *satsumensis*.

e. *L. formosa* subsp. *velutina*.

f. *L. formosa* subsp. *elliptica*.

a. *L. homoloba*.

b. *L. formosa* subsp. *velutina*.

c. *L. Maximowiczii*.

d. *L. Davidii*.

e. *L. Buergeri*.

a.  *L. japonica* cv. Nipponica.

b.  *L. Thunbergii*.

d.  *L. × kagoshimensis*.

**Plate 4.**

e.  *L. japonica* cv. Versicolor.

c.  *L. Thunbergii*.

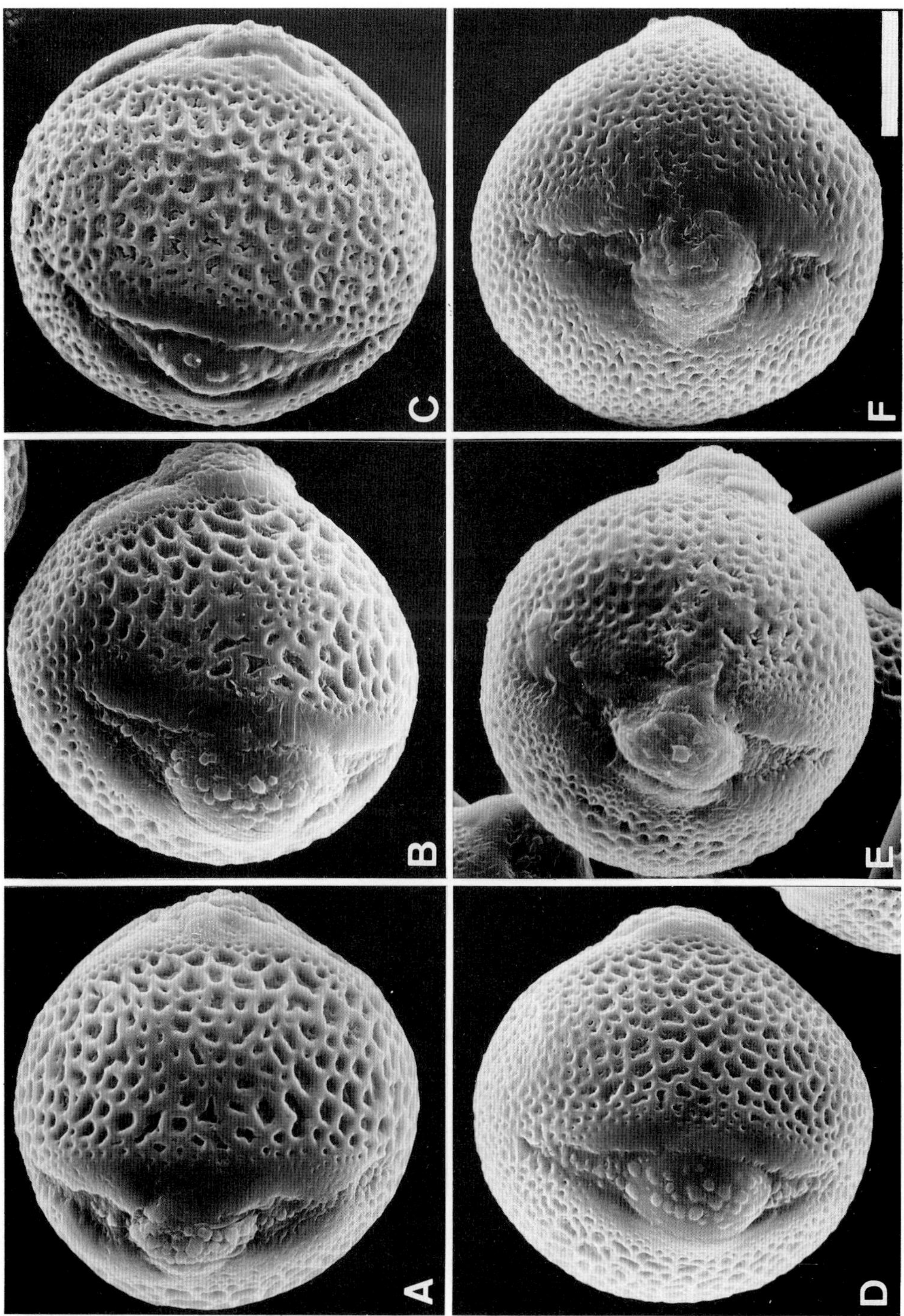

**Plate 5.** Pollen grains.   A: *L. formosa* subsp. *velutina* (Ohba & Akiyama 2284, TI). B: *L. patens* (Ohba & Akiyama 1650, TI).   C: *L. homoloba* (Ohba & Akiyama 578, TI). D: *L. Buergeri* (Akiyama 453, TI).   E: *L. bicolor* (Ohba & Akiyama 454, TI).   F: *L. cyrtobotrya* (Ohba & Akiyama 457, TI). Bar indicates 5μm.

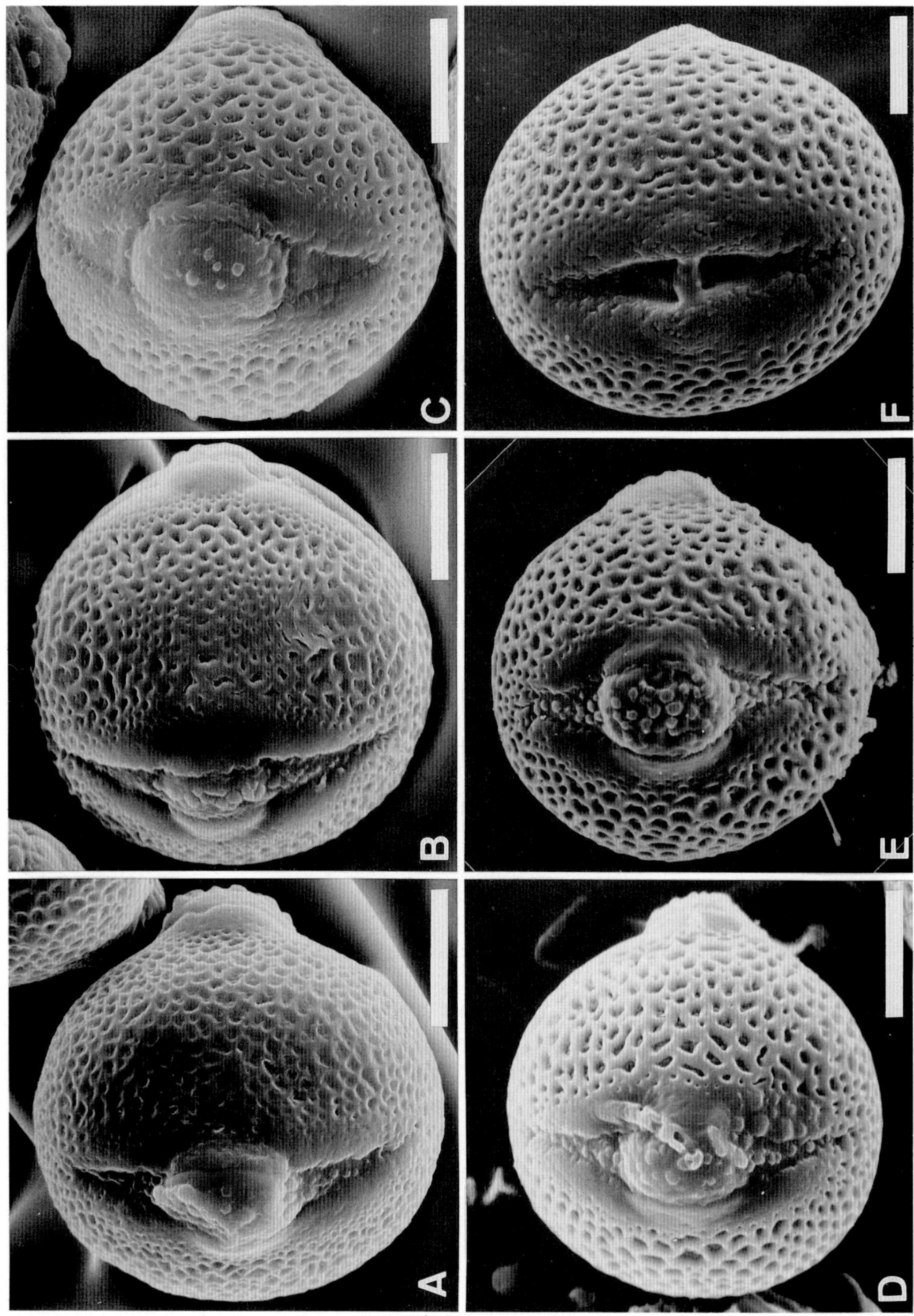

**Plate 6.** Pollen grains. A: *L. melanantha* (Akiyama, 1 Sept. 1984, TI). B: *L. Maximowiczii* (Murata 16098, TI). C: *L. formosa* subsp. *velutina* var. *satsumensis* (Ohba & Akiyama 807, TI). D: *L. Davidii* (Tsui 598, TI). E: *L. virgata* (Ohba & Akiyama 1036, TI). F: *Kummerowia striata* (Akiyama 285, TI). Bar indicates 5$\mu$m.

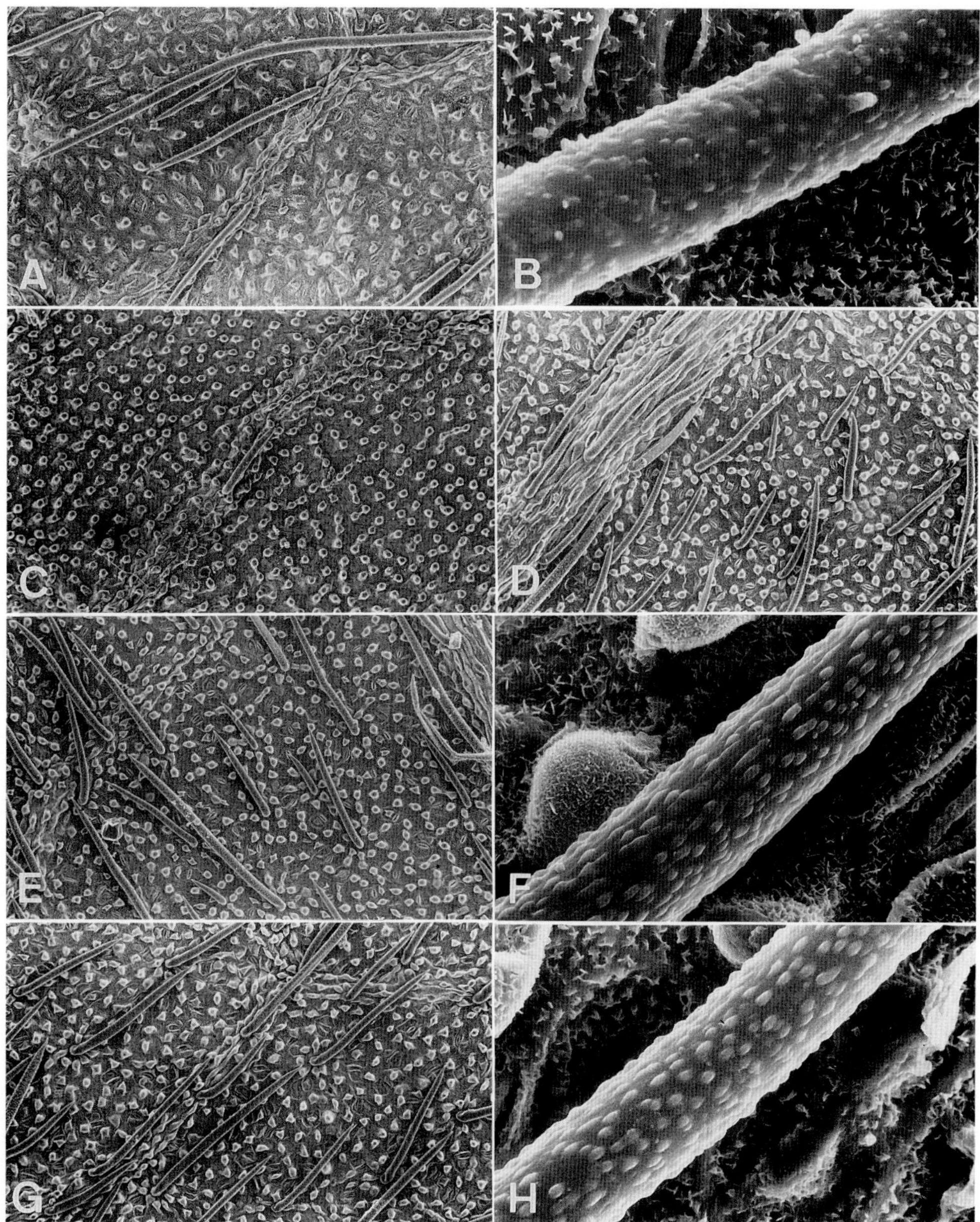

**Plate 7.** Leaf surface.　　A & B: Lower surface of *L. bicolor* (Ohba & Akiyama 648, TI) (A × 110, B × 1700).　C, D & H: Upper surface of *L. cyrtobotrya* (C: Ohba & Akiyama 621, TI; D & H: Ohba & Akiyama 624, TI) (C & D × 110, H × 1700). E–G: Lower surface of *L. cytobotrya* (E: Ohba & Akiyama 624, TI; F & G: Ohba & Akiyama 458c, TI) (E & G × 110, F × 1700).

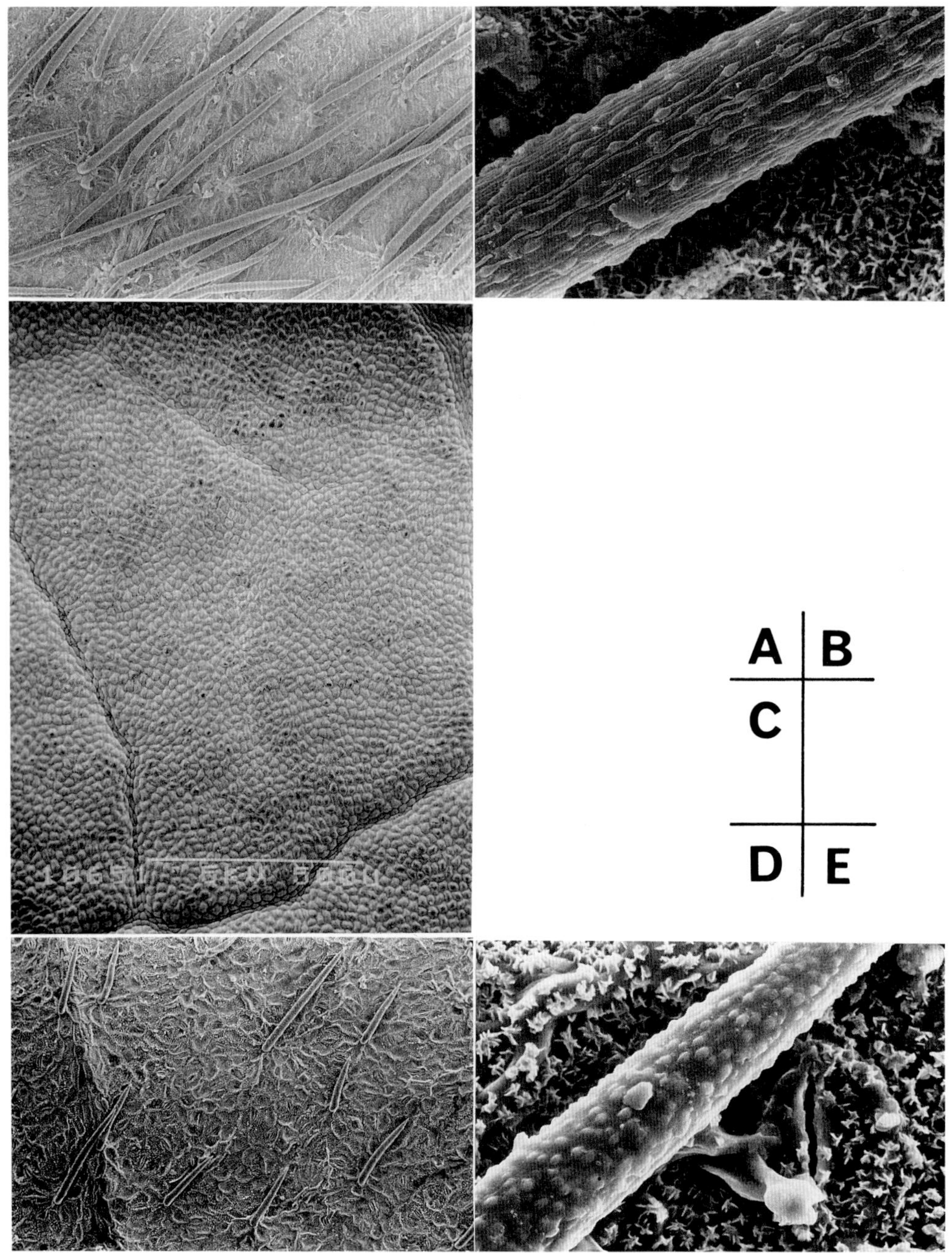

**Plate 8.** Leaf surface. A & B: Lower surface of *L. Buergeri* (Ohba & Akiyama 615, TI) (A × 110, B × 1700). C: Upper surface of *L. homoloba* (Akiyama 211, TI) (×60). D & E: Lower surface of *L. homoloba* (Ohba & Akiyama 756, TI) (D × 110, E × 1700).

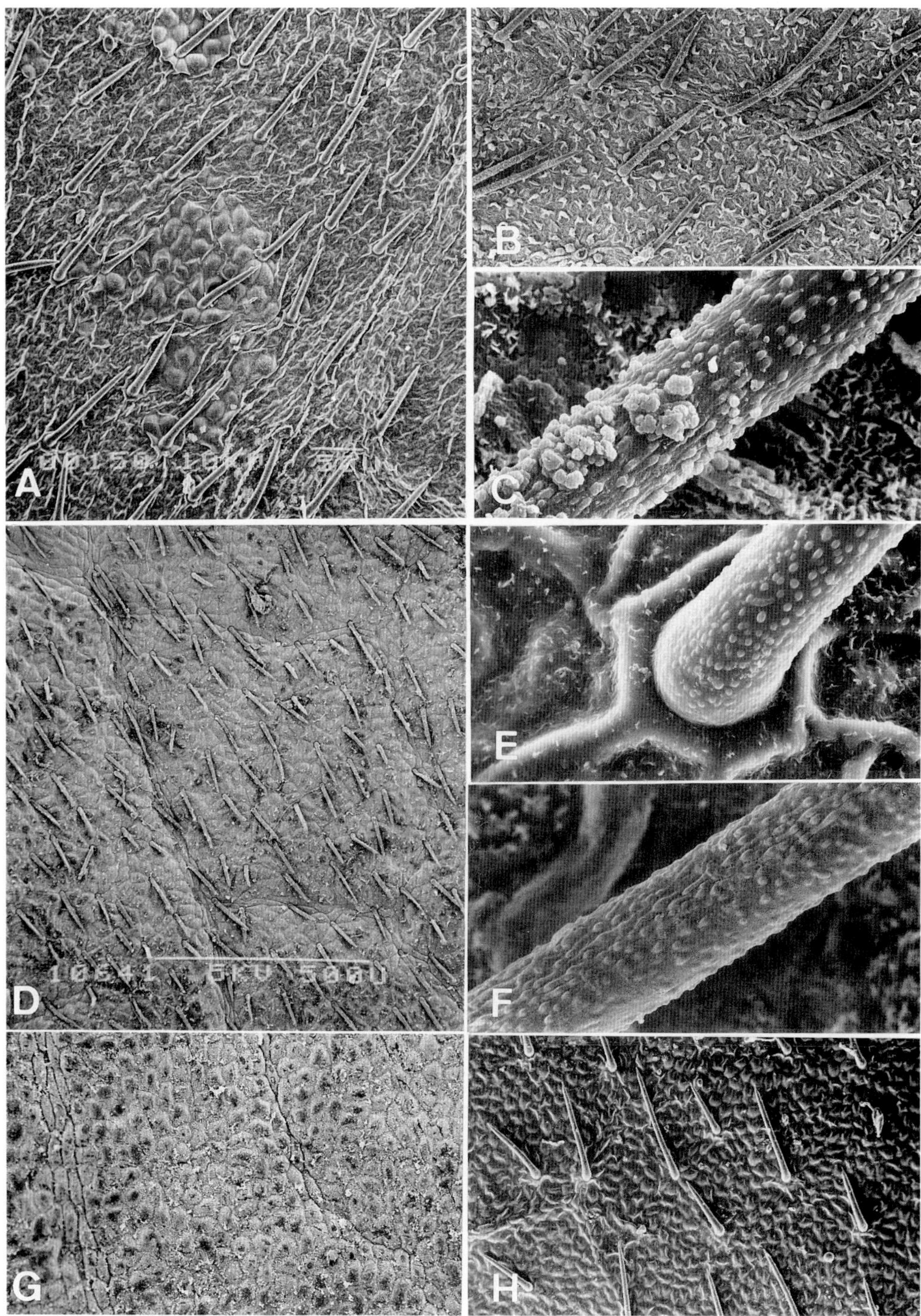

**Plate 9.** Leaf surface. A: Upper surface of *L. formosa* subsp. *velutina* (Ohba & Akiyama 2251, TI) (× 140). B & C: Lower surface of *L. formosa* subsp. *velutina* (Ohba & Akiyama 657, TI) (B × 110, C × 1600). D & F: Upper surface of *L. formosa* subsp. *velutina* var. *satsumensis* (D: Akiyama, 1 Sept. 1981, TI; F: Ohba & Akiyama 806, TI) (D × 60, F × 1700). E: Upper surface of *L. formosa* subsp. *velutina* (Ohba & Akiyama 2251, TI) (× 2000). G & H: Upper surface of *L. patens* (G: Akiyama, 1 Sept. 1981, TI; H: Ohba & Akiyama 2662, TI) (G × 160, H × 80).

**Plate 10.** *L. bicolor* (Tatarinow in 1856, LE—Syntype of *L bicolor β. intermedia* Maxim.).

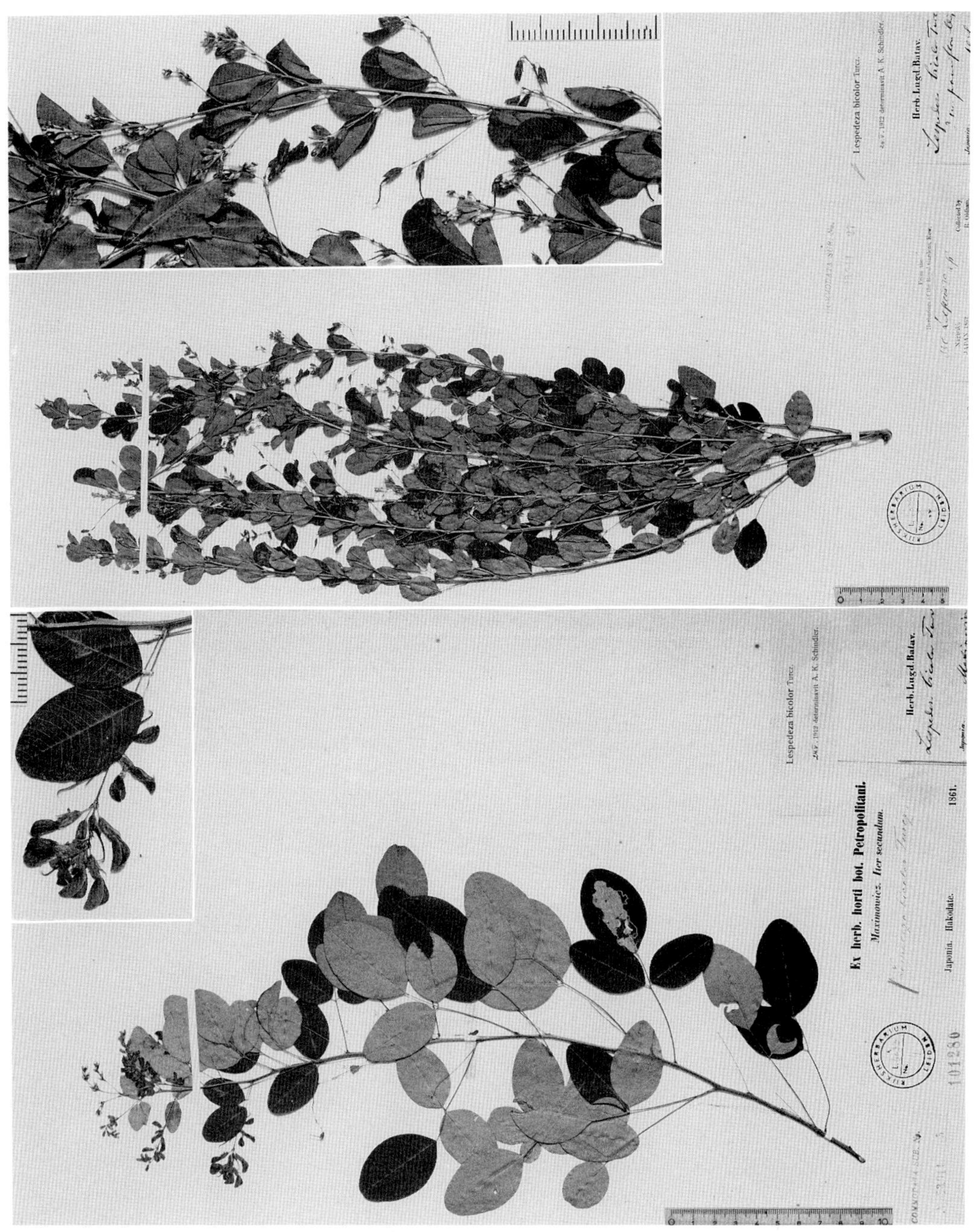

**Plate 11.**  *L. bicolor.*   Upper: Oldham 330, L 908.118-1055, cited by Miquel (1867).
Lower: Maximowicz in 1861, L 908.118-1089, cited by Miquel (1867).

**Plate 12.** *L. cyrtobotrya* (Buerger s.n., L 908.118-1365, determined as *L. cyrtobotrya* by Miquel).

**Plate 13.** *L. melanantha* (Nakai 7, TI–Lectotype).

**Plate 14.** A representative specimen of *L. homoloba* (Ohba & Akiyama 578, TI).

**Plate 15.** A representative specimen of *L. formosa* subsp. *velutina* (Ohba & Akiyama 2284, TI).

**Plate 16.** Holotype of *L. maritima* Nakai (Nakai 9821, TI).

**Plate 17.** Holotype of *L. Uekii* Nakai (Ueki 4497, TI).

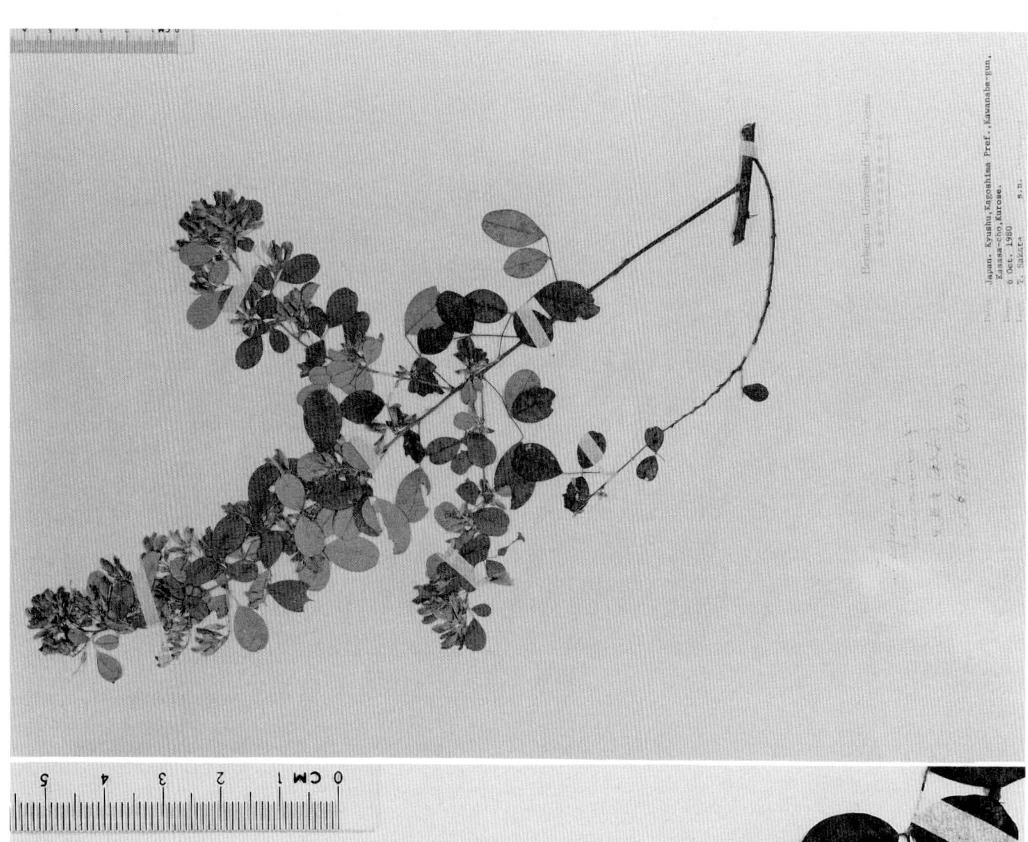

**Plate 18.**   A representative specimen of *L. formosa* subsp. *velutina* var. *satsumensis* (Sakata, 6 Oct. 1980, TI).

**Plate 19.**   A representative specimen of *L. patens* (Ohba & Akiyama 1636, TI).

**Plate 20.**  A representative specimen of *L. Davidii* (Takahashi, Sept. 1941, TI).

**Plate 21.** *L. Buergeri* Miq. (Siebold s.n., L 908.118-1368–Syntype).

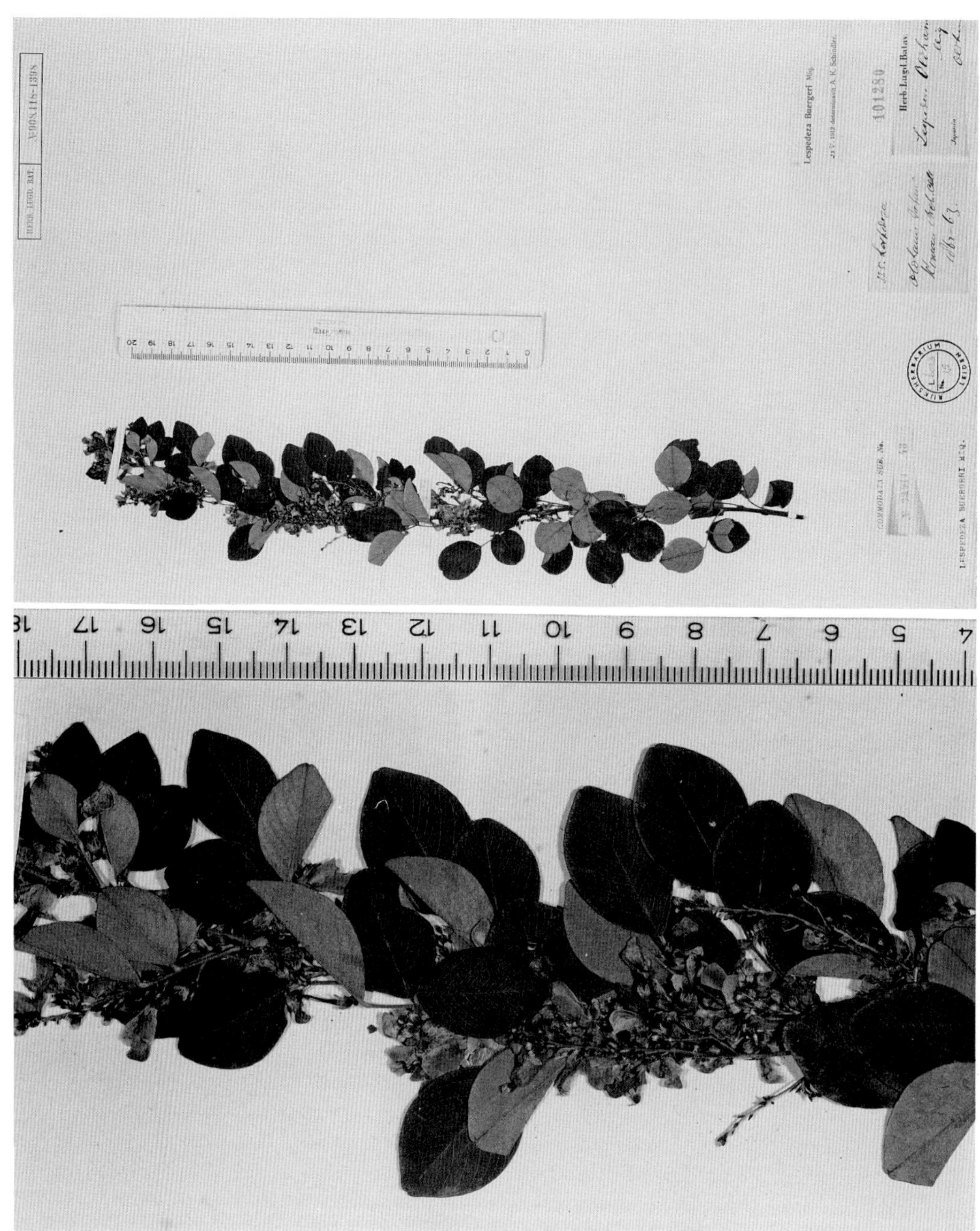

**Plate 22.** Holotype of *L. Oldhamii* Miq. (Oldham 335, L 908.118-1398).

**Plate 23.** *L. Maximowiczii* (Oldham 333, L 908.119-583—Syntype of *L. Sieboldii* Miq.).

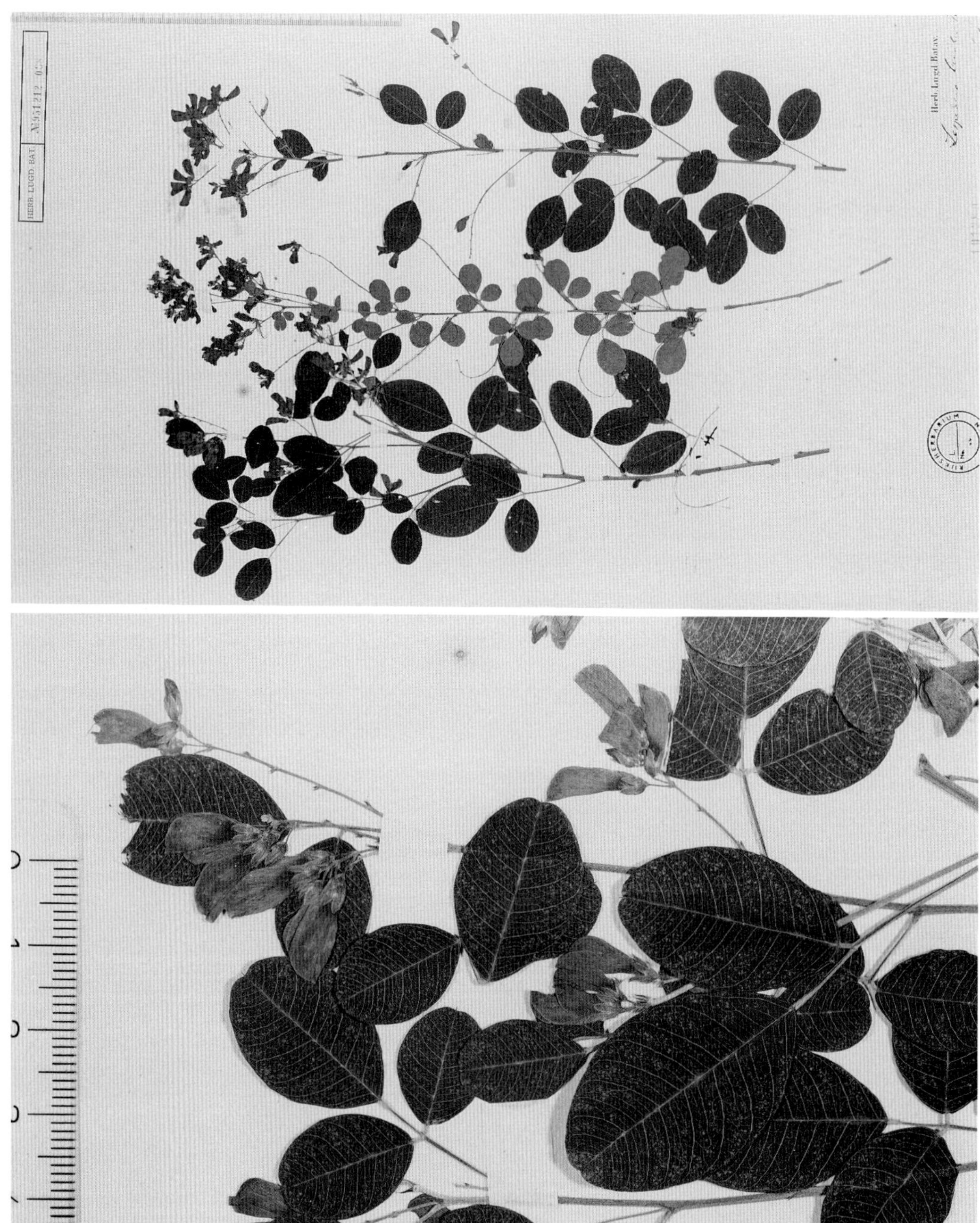

**Plate 24.** *L. japonica* (Textor s.n., L 951212 088, cited as *L. bicolor* by Miquel (1867)).

**Plate 25.** *L. japonica* (cv. Japonica ?) (Tschonoski in 1888, LE).

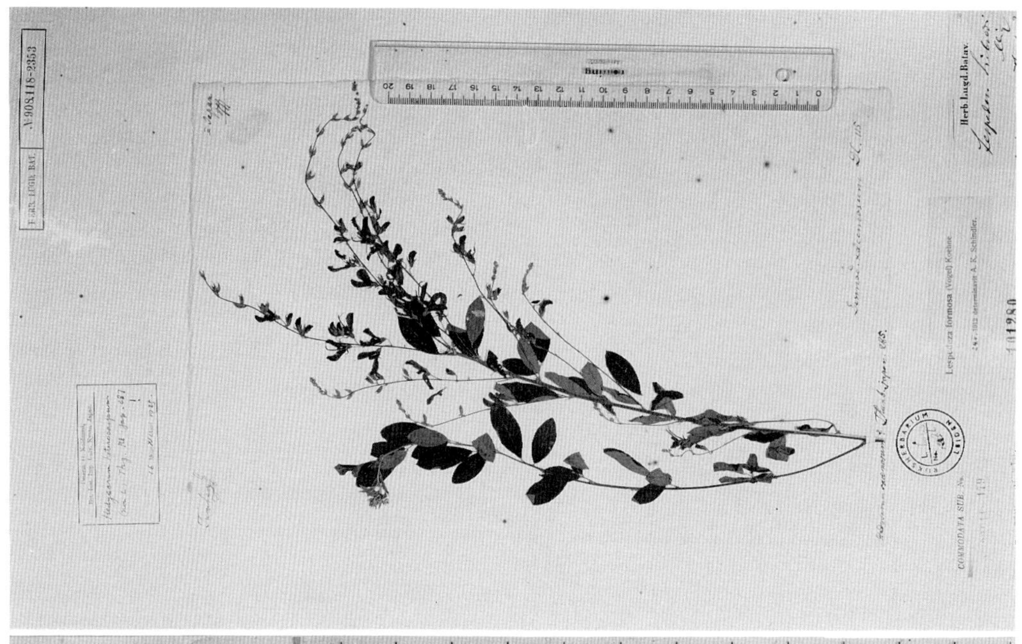

**Plate 26.** *L. Thunbergii* (Thunberg ? s.n., L 908.118-2353, determined as *L. Sieboldii* by Miquel).

**Plate 27.** Lectotype of *L. Sieboldii* Miq. (Siebold s.n., L 908.118-2391).

# CONTENTS

# Explanations of color plates

Plate 1. a. *L. bicolor* Turcz. var. *nana* Nakai.
Hokkaido, Hidaka, Mt. Apoi (Aug. 21, 1984).
b. *L. bicolor* Turcz.
Honshu, Nagano Pref., Oomachi-shi, Taira, nr. Lake Aoki (Aug. 17, 1979).
c. *L. melanantha* Nakai f. *rosea* Nakai.
Cultivated in Botanical Gardens, Nikko (July 12, 1980).
d. *L. melanantha* Nakai.
Cultivated in Tokyo (ex Tsushima Isl., Mt. Shiratake) (Nov. 12, 1987).
e. *L. cyrtobotrya* Miquel.
Kyushu, Kagoshima Pref., Kagoshima-shi, Kasamatsu (Oct. 7, 1979).
f. *L. cyrtobotrya* Miquel.
Honshu, Nagano Pref., Oomachi-shi, Taira, Kurosawa (Aug. 17, 1979).
Plate 2. a. *L. patens* Nakai.
Honshu, Nagano Pref., Iiyama-shi, Sekiya (Aug. 18, 1979).
b. *L. patens* Nakai.
Honshu, Fukushima Pref., Yama-gun, Yamato-machi, Aikawa—Ichinoki (Aug. 28, 1980).
c. *L. formosa* (Vogel) Koehne subsp. *formosa*.
Hong Kong, New Territories West, So Kwun Wat (near Tai Lam Chung Reservoir) (Oct. 29, 1985).
Photographed by H. Ohba.
d. *L. formosa* (Vogel) Koehne subsp. *velutina* (Nakai) S. Akiyama et H. Ohba var. *satsumensis* (Nakai) S. Akiyama et H. Ohba.
Cultivated in Tokyo (ex Kagoshima Pref., Kawanabe-gun, Kasasa-machi) (Nov. 12, 1987).
e. *L. formosa* (Vogel) Koehne subsp. *velutina* (Nakai) S. Akiyama et H. Ohba.
Honshu, Tottori Pref., Nichinan-machi, near Kasagi (Sept. 9, 1980).
f. *L. formosa* (Vogel) Koehne subsp. *elliptica* (Benth. ex Maxim.) S. Akiyama et H. Ohba.
China, Yunnan, Lijiang, Dongshan (Sept. 29, 1987).
Photographed by H. Ohba.
Plate 3. a. *L. homoloba* Nakai.
Honshu, Tochigi Pref., Nikko-shi, Hanaishi-cho (Aug. 31, 1979).
b. *L. formosa* (Vogel) Koehne subsp. *velutina* (Nakai) S. Akiyama et H. Ohba.
Cultivated in Tokyo (ex Hyogo Pref., Shikama-gun, Yumesaki-machi) (Nov. 12, 1987).
c. *L. Maximowiczii* Schneider.
Cultivated in Botanical Gardens, Koishikawa, Tokyo (May 6, 1978).
Photographed by Y. Fukuda.
d. *L. Davidii* Franch.
Cultivated in Tokyo (ex China, Kiangsi, Lushan Botanical Garden) (Nov. 12, 1987).
e. *L. Buergeri* Miquel.
Cultivated in Botanical Gardens, Koishikawa, Tokyo (ex Saitama Pref., Mt. Izugadake) (July 20, 1981).

# Introduction

The genus *Lespedeza* is disjunctively distributed in Asia and North America, and consists of about 40 species (Ohashi, Polhill, & Schubert, 1981). Since Maximowicz (1873) this genus has been divided into two sections: sect. *Lespedeza* and sect. *Macrolespedeza* (see p. 92). The section *Macrolespedeza* is distributed only in East Asia, ranging from the Himalayas to Japan and Siberia through mainland China, Taiwan, and Korea.

In Japan sect. *Macrolespedeza* is represented by several species and infraspecific taxa. Most grow abundantly on open habitat throughout Japan, and some even extend into thickets. The Japanese species have been studied by many taxonomists, although they have not yet been successfully circumscribed or clearly distinguished from each other due to their lack of significant characters and also to the presence of interspecific hybrids. Since 1979, we have been studying *Lespedeza* and its allied genera through field surveys in various localities in Japan and Hong Kong (Akiyama & Ohba, 1982, 1983a, b, c, 1985, 1988).

The present paper aims to provide a detailed revision of the species of sect. *Macrolespedeza* that occur in Japan and neighboring regions. We have discovered significant characters, especially in the flowers, which we think are sufficient to circumscribe the species of sect. *Macrolespedeza*. These characters were primarily investigated in the wild populations in Japan. In order to clarify the variation and distribution range of each species, herbarium specimens from various herbaria were also examined. In this paper the specific and infraspecific taxa of sect. *Macrolespedeza* will be revised based on a synthesis of present knowledge.

In the text, the following abbreviations of the herbaria (according to *Index Herbariorum*, Part 1, Ed. 7 (Holmgren, Keuken, & Schofield, 1981)) are used, except for that with an asterisk.

A: Arnold Arboretum of Harvard University, Cambridge, U.S.A. CAL: Central National Herbarium, Botanical Survey of India, Howrah, India. GH: Gray Herbarium of Harvard University, Cambridge U.S.A. HK: Herbarium, Agricultural and Fisheries Department, Hong Kong. K: The Herbarium Royal Botanic Gardens, Kew, Richmond, England, U.K. KAG: Herbarium, Faculty of Agriculture, Kagoshima University, Kagoshima, Japan. KANA: Herbarium, Faculty of Science, University of Kanazawa, Kanazawa, Japan. KIEL: Botanisches Institut der Universität Kiel, Federal Republic of Germany. KYO: Herbarium, Department of Botany, Faculty of Science, Kyoto University, Kyoto, Japan. L: Rijksherbarium, Leiden, the Netherlands. LE: Herbarium of Department of Higher Plants, V. L. Komarov Botanical Institute of the Academy of Sciences of U.S.S.R., Leningrad, U.S.S.R. MAK: Makino Herbarium, Faculty of Science, Tokyo Metropolitan University, Tokyo, Japan. NA: Herbarium, U.S. National Arboretum, Washington, D.C., U.S.A. SAP: Laboratory of Plant Taxonomy and Ecology, Botanical Institute, Faculty of Agriculture, Hokkaido University, Sapporo, Japan. SHO*: Herbarium, Biological Laboratory, Shoei

2

Junior College, Kobe, Japan. TI: Department of Botany, University Museum, University of Tokyo, Tokyo, Japan. TNS: Department of Botany, National Science Museum, Tokyo, Japan. TOFO: Department of Forest Botany, University Museum, University of Tokyo, Tokyo, Japan. TUSG: Herbarium, Botanical Garden, Faculty of Science, Tohoku University, Sendai, Japan.


## History

The genus *Lespedeza* Michaux was described by Richard in 1803 (Nakai, 1925; Ricker, 1934; Hochreutiner, 1934) based on a North American species, *L. sessiliflora* Michx. (now regarded as a synonym of *L. virginaica* (L.) Britton). A detailed account of the taxonomic history of the genus *Lespedeza* covering the period up to 1873 is given in a monograph on this genus by Maximowicz (1873).

The generic delimitation of *Lespedeza* was controversial. Maximowicz (1873) regarded *Campylotropis*, *Lespedeza*, and *Microlespedeza* as subgenera of the genus *Lespedeza*. Schindler (1912) raised the subgenus *Microlespedeza* of Maximowicz to a new genus, *Kummerowia*. Based on its inflorescence and other characters Schindler (1913) recognized the three subgenera of Maximowicz as distinct genera and also recognized two sections in the genus *Lespedeza*, sect. *Macrolespedeza* and *Eulespedeza* (=*Lespedeza*), as did Maximowicz (1873).

Although Nakai (1927) regarded *Campylotropis* as congeneric with *Lespedeza*, he recognized no subgenera and only three sections in *Lespedeza*: *Campylotropis*, *Macrolespedeza*, and *Eulespedeza*. *Kummerowia* was regarded as a distinct genus by him. Nakai (1939) established the section *Heterolespedeza* based on Momiyama's observation of bud phyllotaxy in *L. Buergeri* (Momiyama, 1933), and *L. Maximowiczii* is also included in this section.

Ohashi (1971) recognized *Lespedeza* and *Kummerowia* as distinct genera and *Campylotropis* as congeneric with *Lespedeza* in his study of the tribe Coronilleae. But later, he regarded *Kummerowia* as congeneric with *Lespedeza*, while *Campylotropis* as a distinct genus (Ohashi, 1982b). Ohashi (1982a) also recognized *Macrolespedeza* as a subgenus in the genus *Lespedeza*. This subgenus, according to him, consists of two sections, *Macrolespedeza* and *Heterolespedeza*. We studied the branching pattern of the vegetative shoots and the inflorescences of *Kummerowia* and gave evidence for the distinguishing between *Lespedeza* and *Kummerowia* (Akiyama & Ohba, 1985).

In the section *Macrolespedeza* Maximowicz recognized five species three of them reported from Japan. Schindler (1913) examined almost all the type and authentic specimens, and then revised the taxonomy. He recognized eight species in sect. *Macrolespedeza*, four of them from Japan.

Nakai studied Japanese and Korean species of *Lespedeza*, chiefly herbarium specimens, and used various combinations of vegetative characters to distinguish the species but paid little attention to floral features. In 1927 he published "*Lespedeza* of Japan and Korea." This is regarded as a standard work on species growing in the two countries. He recognized 18 species in sect. *Macrolespedeza* of which 14 are from Japan, including seven new species. He also regarded ten species as endemic.

Hatusima (1967) studied the species and the infraspecific taxa under sect. *Macro-*

*lespedeza* and *Heterolespedeza* from Japan, Korea, and Taiwan. He partially evaluated the importance of the characters used by Nakai (1927) and others including Ohwi (1953, 1965a, b) and Kitamura & Murata (1961). His evaluation, however, depended on personal impressions, and are unsatisfactory. He did not attempt to find significant characters within flowers and did not mention the floral features except for the calyx. No revision has been published after that, although some of the Japanese species belonging to sect. *Macrolespedeza* were discussed by Murata (1978) and Ohashi (1981, 1986).

Interspecific hybridization is known in both sections of *Lespedeza*, sect. *Lespedeza* (Clewell, 1964, 1966) and sect. *Macrolespedeza* (Lee, 1965). Clewell revealed the intermediate appearance of morphological characters as well as cytological evidence and pollen stainability for detecting interspecific hybridization. However, Lee did not mention any concept or criteria to distinguish the hybrids from the species recognized by him. We studied the interspecific hybridization of sect. *Macrolespedeza* in Japan (Akiyama & Ohba, 1982, 1983a, c), and judged the occurrence of hybrids from the intermediate appearance of morphological characters and some reduction of pollen stainability (Akiyama & Ohba, 1982). We also noticed the occurrence of introgressive hybridization which shows a morphological continuity and irregularity in the reduction of pollen stainability (Akiyama & Ohba, 1983a).

# Morphological Diversity and Variation

In the section *Macrolespedeza*, approximately 60 taxa have been described as species, a large number of which still remain uncertain or confusing. The minor morphological differences that are found in only one or a few specimens have been exhaustively used for delimitation of species. The uncertainty or confusion in specific delimitation is mainly due to poor understanding of such differences and the absence of infallible distinguishing characters. For the purpose of the species delimitation in sect. *Macrolespedeza* it is necessary to determine characteristics with constant appearances and narrow ranges of variation in each species.

## Floral Characters

Generally floral features are regarded as the most reliable characters for species delimitation, although in sect. *Macrolespedeza* the floral features have not been fully investigated. Variation is due to genetically inherited differences or to a particular range of environmental factors which produce largely reversible modifications from the same genotype. It is not possible to determine the nature of the variation—whether genotypic or environmentally induced—in the herbarium or even in the field (Davis & Heywood, 1973). However, the variation seen among many flowers from a single individual is environmentally induced and regarded as phenotypic plasticity. Variation in terms of populations can be divided into individual and collective variation. Individual variation, including phenotypic plasticity, occurs among individuals within a population, while collective variation occurs between populations.

We have attempted to clarify phenotypic plasticity and individual and collective variations of various floral features to determine reliable floral characters for taxonomic purposes.

*Materials and Methods*

The field survey was carried out in Japan and Hong Kong. Collections were made from one or more populations at each locality. All specimens collected were deposited in the Herbarium, Department of Botany, University Museum, University of Tokyo (TI). This study is also based on herbarium specimens deposited in various herbaria, as mentioned in the Introduction.

Measurements and sketches were based on fully developed flowers. Flowers taken from herbarium specimens were examined under a binocular microscope with camera lucida after boiling in water. Measurements were made of the morphological features shown in Fig. 1.

Both phenotypic plasticity and individual variation of each species were investigated based on six populations of six species, which are widely distributed in Japan (Table 1). Ten fully developed flowers were taken from a single individual within a population and dissected. To investigate individual variation, a single flower was taken from

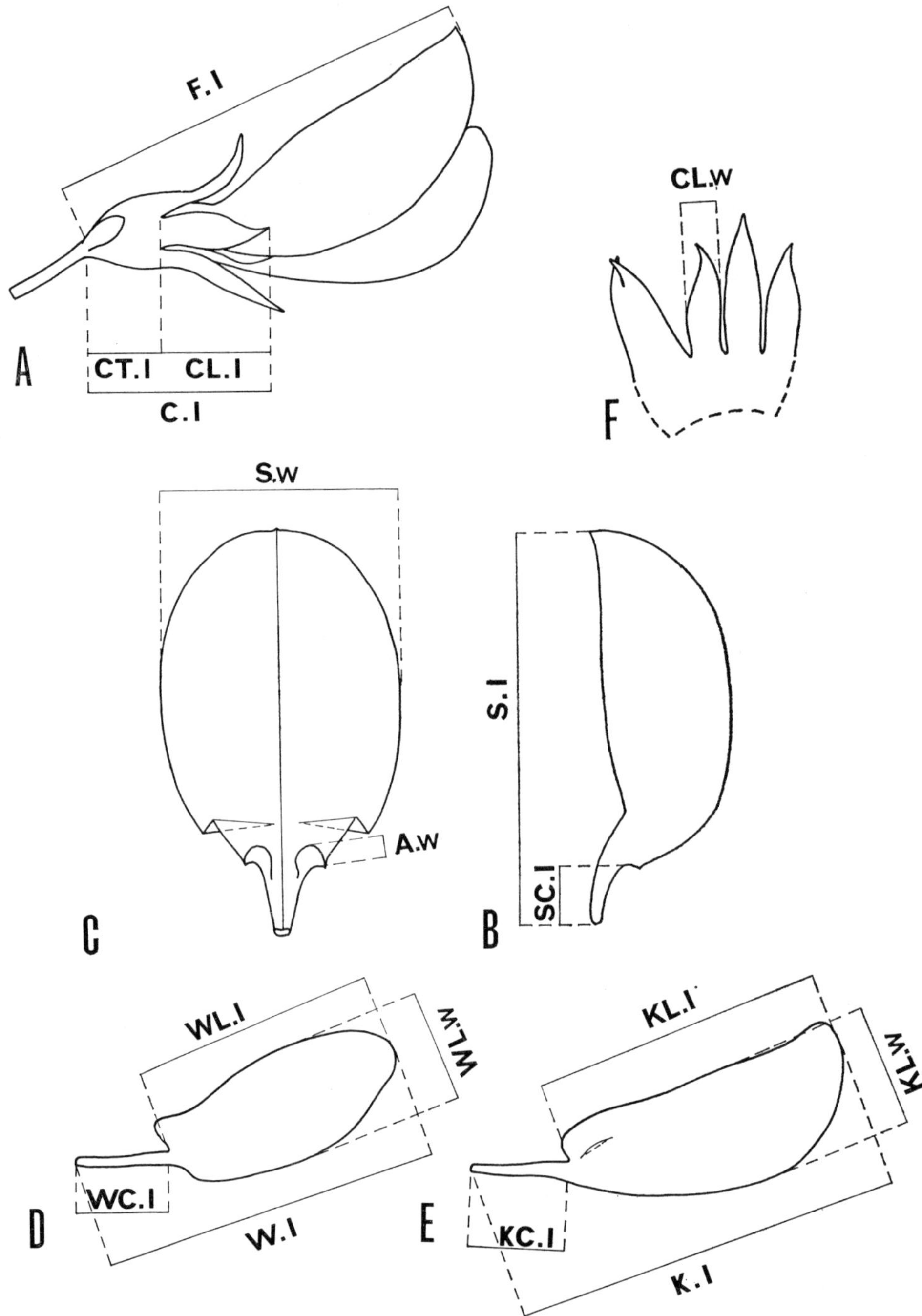


Fig. 1.   Flower of the section *Macrolespedeza*.
A, flower;   B, standard, lateral view;   C, standard, opened;   D, wing;
E, keel-petal;   F, calyx, dissected;   F.l, length of flower;   S.l, length of
standard;   SC.l, length of standard-claw;   S.w, width of standard;   A.w,→

Table 1.   List of voucher specimens corresponding to the phenotypic plasticity and individual variation examined.

| Species | Specimens |
| --- | --- |
| *L. bicolor* | Gunma Pref., Tone-gun, Katashina-mura, nr. Lake Marunuma (Ohba & Akiyama 1496, 1499–1507, TI) |
| *L. cyrtobotrya* | Kanagawa Pref., Ashigarashimo-gun, Hakone-machi, Miyagino (Ohba & Akiyama 621–629, TI) |
| *L. homoloba* | Tochigi Pref., Nikko-shi, Hanaishi-machi (Ohba & Akiyama 1422–1429, TI) |
| *L. formosa* ssp. *velutina* | Okayama Pref., Mitsu-gun, Mitsu-machi, Kubo (Ohba & Akiyama 650–657, TI) |
| *L. patens* | Ishikawa Pref., Nomi-gun, Tatsunokuchi-machi, Tedori-gawara (Akiyama 1513–1515, 1519–1522, 1524–1525, TI) |
| *L. Buergeri* | Kanagawa Pref., Ashigarashimo-gun, Hakone-machi, Tounosawa (Ohba & Akiyama 614–615, TI) |

all individuals constituting a population. To study variation between populations (i.e., collective variation), ten populations of *L. bicolor* Turcz. from various localities were investigated. Moreover, in five other species one flower from each individual in various localities was investigated.

*Observation*

### *Lespedeza bicolor* Turcz.

Phenotypic plasticity: Ten flowers are illustrated in Fig. 2 and sizes of each floral character are summarized in Table 2. The standard is apparently longer than the wing and the keel-petal (Fig. 2A). The wing is shorter than the keel-petal (Fig. 2D). The ratio of the length of the standard to the keel-petal is $1.06\pm0.02$, wing to standard $0.88\pm0.01$, and wing to keel-petal $0.93\pm0.02$.

The shape of the standard is obovate with an attenuate base and a retuse apex; the shape of the auricle is lunate (Fig. 2B & C). The shape of the lamina of the wing is narrowly obovate with an auriculate upper base and tapering lower base (Fig. 2E). The ratio of the length of the wing-claw to the wing is $0.44\pm0.01$. The shape of the lamina of the keel-petal is obovate with a cordate upper base and tapering lower base (Fig. 2F). The ratio of the length of the keel-petal-claw to the keel-petal is $0.46\pm0.01$.

The shape of the lateral calyx-lobe is triangular with an acute apex (Fig. 2G). Two upper calyx-lobes are connated 4/5–7/8 of the way along its length. The calyx-lobes are apparently shorter than the tube, and the ratio of the length of the lateral calyx-lobe to the calyx-tube is $0.64\pm0.09$.

width of auricle; W.l, length of wing; WC.l, length of wing-claw; WL.l, length of wing-lamina; WL.w, width of wing-lamina; K.l, length of keel-petal; KC.l, length of keel-petal-claw; KL.l, length of keel-petal-lamina; KL.w, width of keel-petal-lamina; C.l, length of calyx; CT.l, length of calyx tube; CL.l, length of lateral calyx-lobe; CL.w, width of lateral calyx-lobe.

8

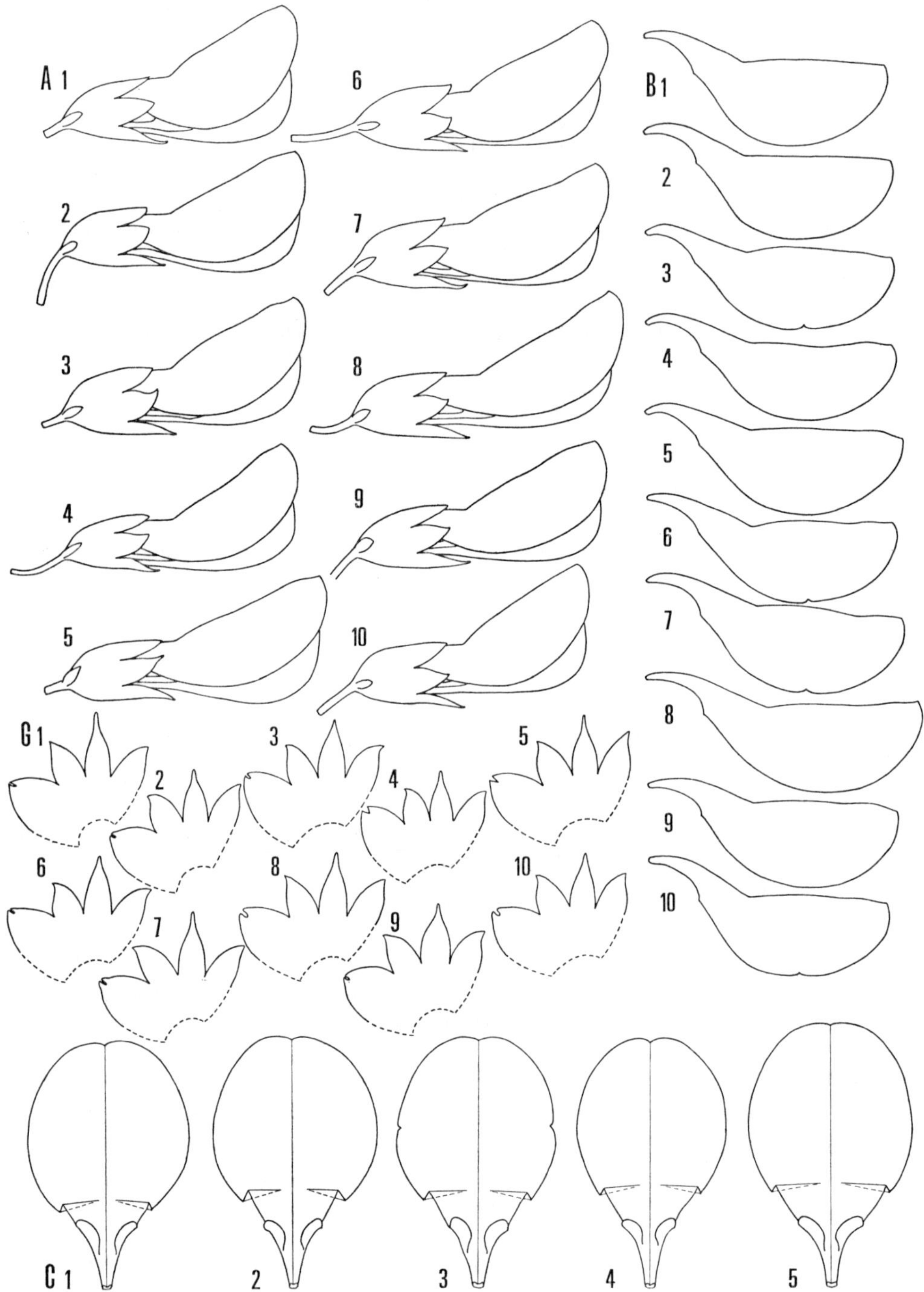

Fig. 2–1.  Phenotypic plasticity of *Lespedeza bicolor*.
A, flowers, lateral view;  B, standards, lateral view;  C, standards, opened;
G, calyces, dissected.   All×3.3.

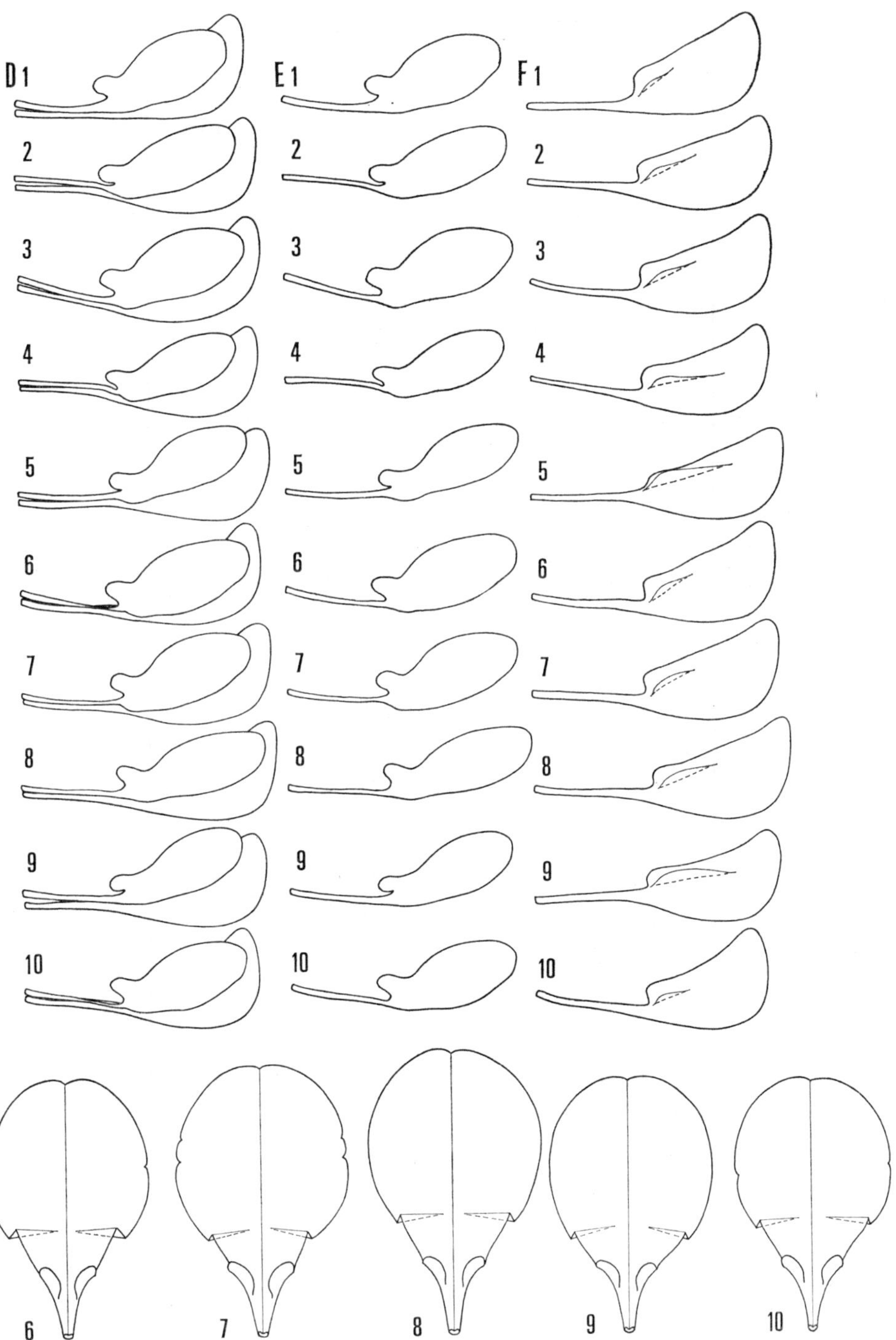

Fig. 2–2.  Phenotypic plasticity of *L. bicolor*.
D, wings and keel-petals;  E, wings;  F, keel-petals.  All×3.3.

Table 2.   Phenotypic plasticity of *L. bicolor*.

| | F.1 | C.1 | CT.1 | CL.1/CT.1 | S.1 | W.1 | WC.1 | WC.1/W.1 | K.1 | KC.1 | KC.1/K.1 | W.1/K.1 | S.1/K.1 | W.1/S.1 |
|---|---|---|---|---|---|---|---|---|---|---|---|---|---|---|
| 1 | 10.8 | 3.9 | 2.2 | 0.77 | 10.5 | 9.2 | 3.9 | 0.42 | 10.0 | 4.5 | 0.45 | 0.92 | 1.05 | 0.88 |
| 2 | 10.8 | 3.7 | 2.3 | 0.61 | 10.7 | 9.3 | 4.2 | 0.45 | 10.1 | 4.6 | 0.46 | 0.92 | 1.06 | 0.87 |
| 3 | 11.2 | 4.1 | 2.5 | 0.64 | 10.6 | 9.4 | 4.0 | 0.43 | 10.1 | 4.5 | 0.45 | 0.93 | 1.05 | 0.89 |
| 4 | 10.7 | 3.6 | 2.2 | 0.64 | 10.7 | 9.1 | 4.1 | 0.45 | 9.9 | 4.6 | 0.46 | 0.92 | 1.08 | 0.85 |
| 5 | 11.8 | 4.2 | 2.4 | 0.75 | 11.0 | 9.7 | 4.3 | 0.44 | 10.5 | 4.6 | 0.44 | 0.92 | 1.05 | 0.88 |
| 6 | 11.2 | 4.1 | 2.8 | 0.46 | 10.7 | 9.6 | 4.1 | 0.43 | 10.1 | 4.6 | 0.46 | 0.95 | 1.06 | 0.90 |
| 7 | 11.4 | 4.1 | 2.4 | 0.71 | 11.1 | 9.7 | 4.2 | 0.43 | 10.5 | 4.7 | 0.45 | 0.92 | 1.06 | 0.87 |
| 8 | 12.3 | 4.0 | 2.6 | 0.54 | 11.8 | 10.2 | 4.2 | 0.41 | 10.8 | 4.9 | 0.45 | 0.94 | 1.09 | 0.86 |
| 9 | 11.2 | 4.1 | 2.5 | 0.64 | 10.7 | 9.3 | 4.2 | 0.45 | 10.3 | 4.7 | 0.46 | 0.90 | 1.04 | 0.87 |
| 10 | 10.8 | 4.0 | 2.5 | 0.60 | 10.5 | 9.3 | 4.1 | 0.44 | 9.7 | 4.7 | 0.48 | 0.96 | 1.08 | 0.89 |
| Min. | 10.7 | 3.6 | 2.2 | 0.46 | 10.5 | 9.1 | 3.9 | 0.41 | 9.7 | 4.5 | 0.44 | 0.90 | 1.04 | 0.85 |
| Max. | 12.3 | 4.2 | 2.8 | 0.77 | 11.8 | 10.2 | 4.3 | 0.45 | 10.8 | 4.9 | 0.48 | 0.96 | 1.09 | 0.90 |
| Average | 11.2 | 4.0 | 2.4 | 0.64 | 10.8 | 9.5 | 4.1 | 0.44 | 10.2 | 4.6 | 0.46 | 0.93 | 1.06 | 0.88 |
| S.D. | 0.51 | 0.19 | 0.18 | 0.09 | 0.39 | 0.33 | 0.12 | 0.01 | 0.33 | 0.12 | 0.01 | 0.02 | 0.02 | 0.01 |

The abbreviations in this and following tables are as given in Fig. 1.

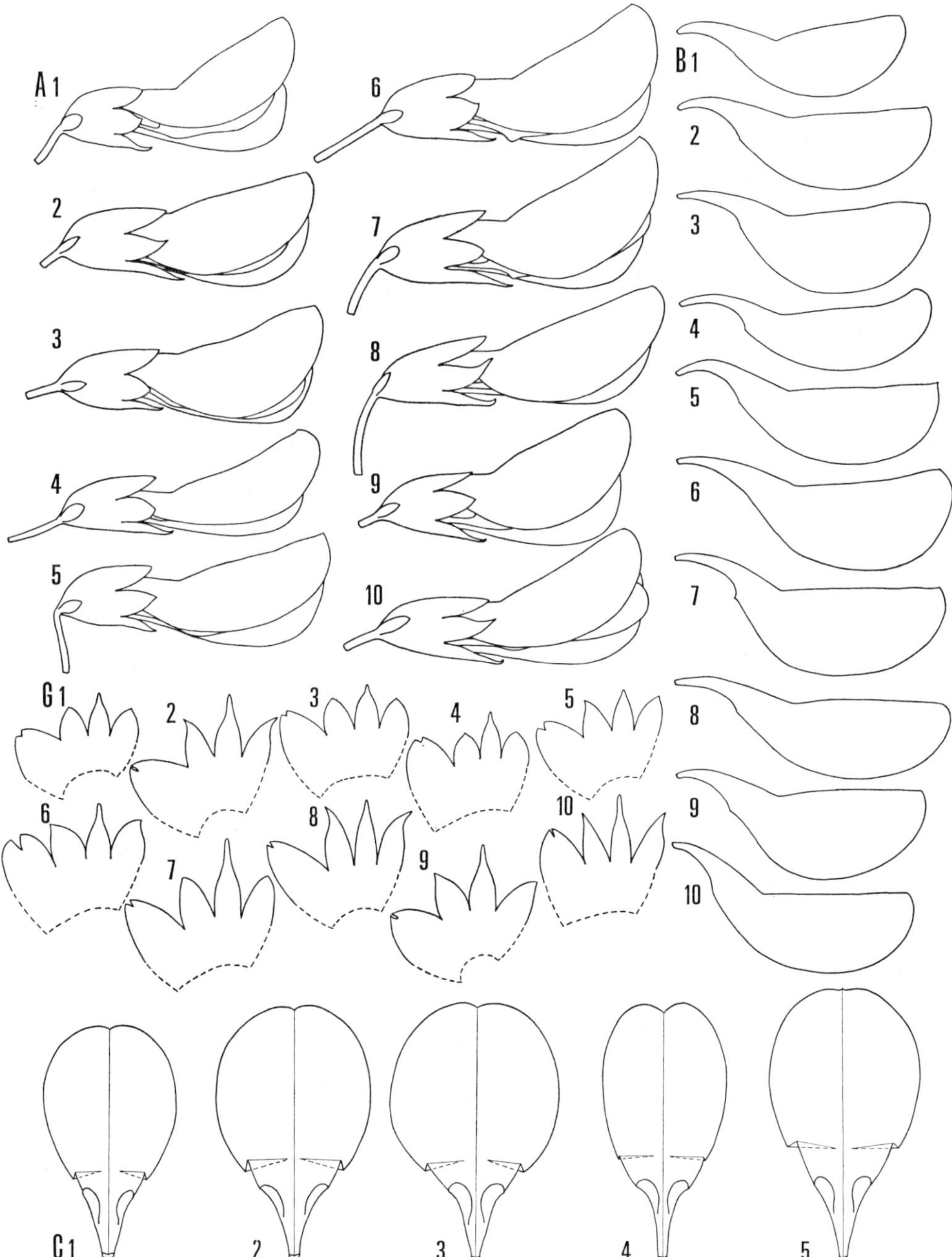

Fig. 3–1.  Individual variation in a population of *L. bicolor*.
A, flowers, lateral view;  B, standards, lateral view;  C, standards, opened;
G, calyces, dissected.  All × 3.3.

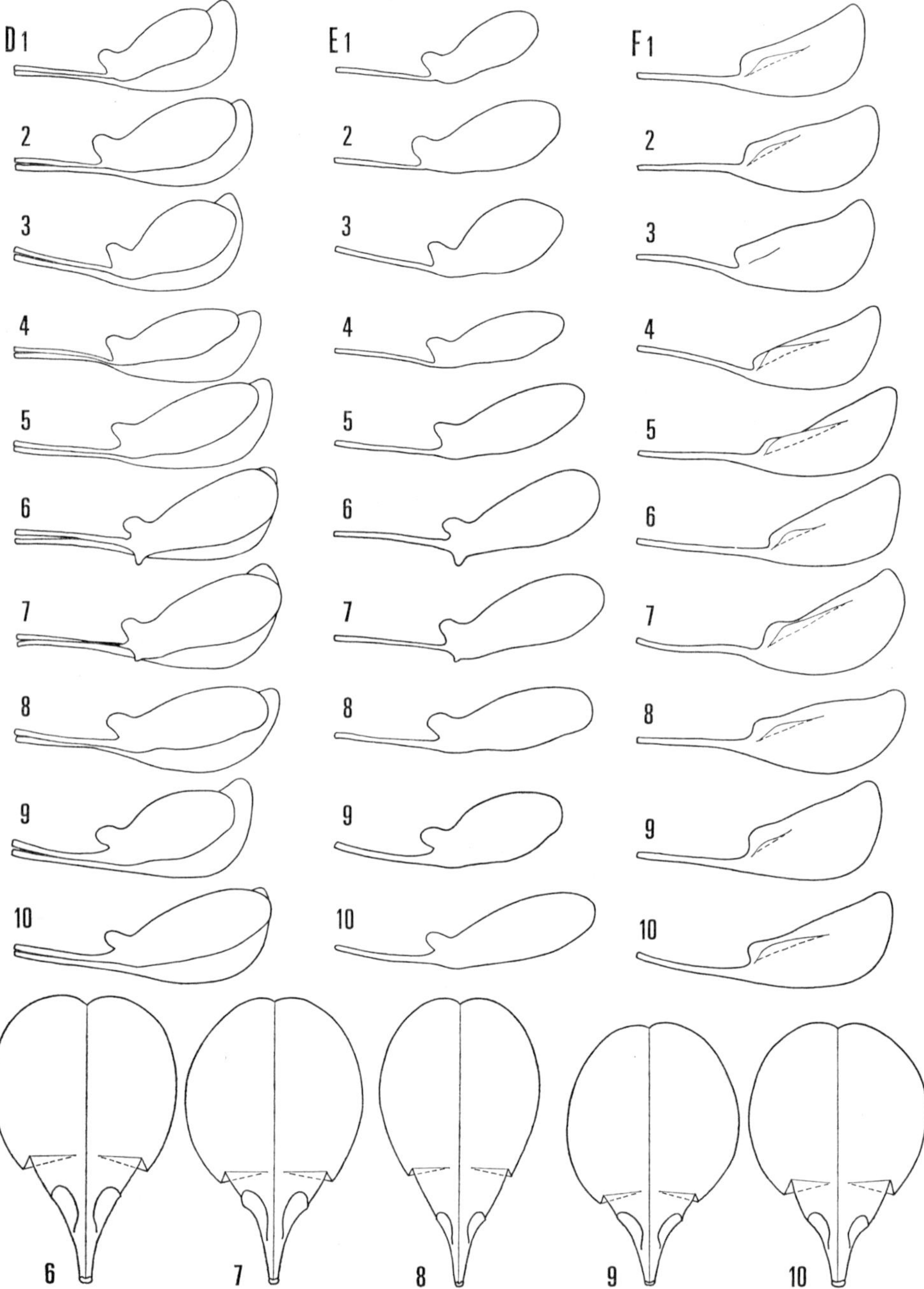

Fig. 3–2.  Individual variation in a population of *L. bicolor*.
D, wings and keel-petals;  E, wings;  F, keel-petals.  All×3.3.

Table 3.  Individual variation of *L. bicolor*.

| | F.1 | C.1 | CT.1 | CL.1/CT.1 | S.1 | W.1 | WC.1 | WC.1/W.1 | K.1 | KC.1 | KC.1/K.1 | W.1/K.1 | S.1/K.1 | W.1/S.1 |
|---|---|---|---|---|---|---|---|---|---|---|---|---|---|---|
| 1 | 10.4 | 3.3 | 2.1 | 0.57 | 9.4 | 8.2 | 3.8 | 0.46 | 9.4 | 4.2 | 0.45 | 0.87 | 1.00 | 0.87 |
| 2 | 10.5 | 4.6 | 2.7 | 0.70 | 10.5 | 9.1 | 3.5 | 0.38 | 9.8 | 4.2 | 0.43 | 0.93 | 1.07 | 0.87 |
| 3 | 11.1 | 3.7 | 2.6 | 0.42 | 10.5 | 9.1 | 4.0 | 0.44 | 9.6 | 4.1 | 0.43 | 0.95 | 1.09 | 0.87 |
| 4 | 10.9 | 3.8 | 2.2 | 0.73 | 10.4 | 9.2 | 4.0 | 0.43 | 9.8 | 4.8 | 0.49 | 0.94 | 1.06 | 0.88 |
| 5 | 11.4 | 4.0 | 2.5 | 0.60 | 10.9 | 10.1 | 4.3 | 0.43 | 10.7 | 4.8 | 0.45 | 0.94 | 1.02 | 0.93 |
| 6 | 11.8 | 4.0 | 2.7 | 0.48 | 11.6 | 10.7 | 4.7 | 0.44 | 10.8 | 5.4 | 0.50 | 0.99 | 1.07 | 0.92 |
| 7 | 12.1 | 4.3 | 2.6 | 0.65 | 11.4 | 10.9 | 4.5 | 0.41 | 10.9 | 5.1 | 0.47 | 1.00 | 1.05 | 0.96 |
| 8 | 12.3 | 5.0 | 2.6 | 0.92 | 11.6 | 10.3 | 4.2 | 0.41 | 10.9 | 4.6 | 0.42 | 0.94 | 1.06 | 0.89 |
| 9 | 10.8 | 3.9 | 2.2 | 0.77 | 10.5 | 9.2 | 3.9 | 0.42 | 10.0 | 4.5 | 0.45 | 0.92 | 1.05 | 0.88 |
| 10 | 11.3 | 4.9 | 2.7 | 0.81 | 10.5 | 10.5 | 4.0 | 0.38 | 10.5 | 4.5 | 0.43 | 1.00 | 1.00 | 1.00 |
| Min. | 10.4 | 3.3 | 2.1 | 0.42 | 9.4 | 8.2 | 3.5 | 0.38 | 9.4 | 4.1 | 0.42 | 0.87 | 1.00 | 0.87 |
| Max. | 12.3 | 5.0 | 2.7 | 0.92 | 11.6 | 10.9 | 4.7 | 0.46 | 10.9 | 5.4 | 0.50 | 1.00 | 1.09 | 1.00 |
| Average | 11.3 | 4.2 | 2.5 | 0.67 | 10.7 | 9.7 | 4.1 | 0.42 | 10.2 | 4.62 | 0.45 | 0.95 | 1.05 | 0.91 |
| S.D. | 0.65 | 0.54 | 0.23 | 0.15 | 0.67 | 0.89 | 0.35 | 0.03 | 0.58 | 0.42 | 0.03 | 0.04 | 0.03 | 0.04 |

14

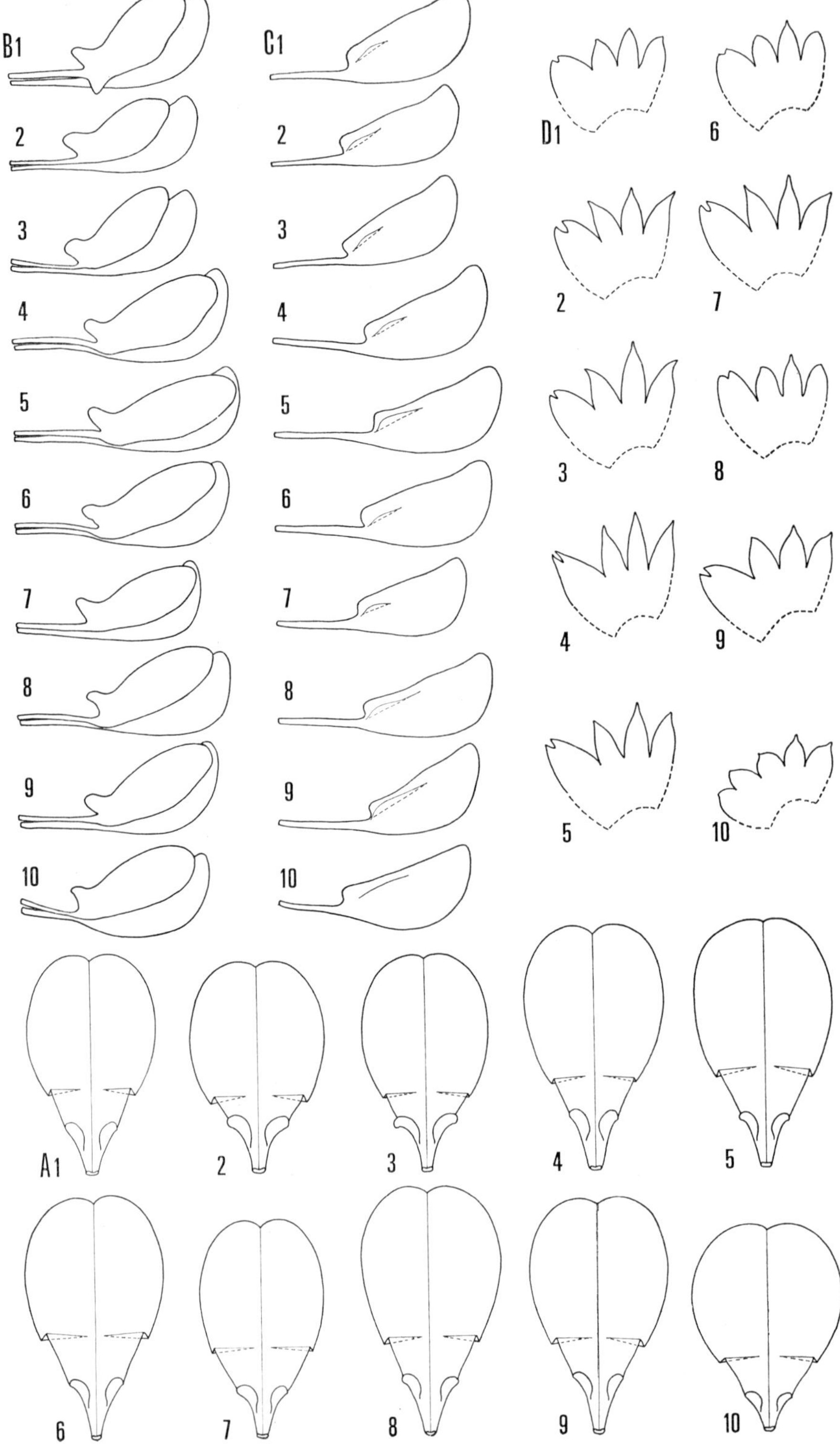

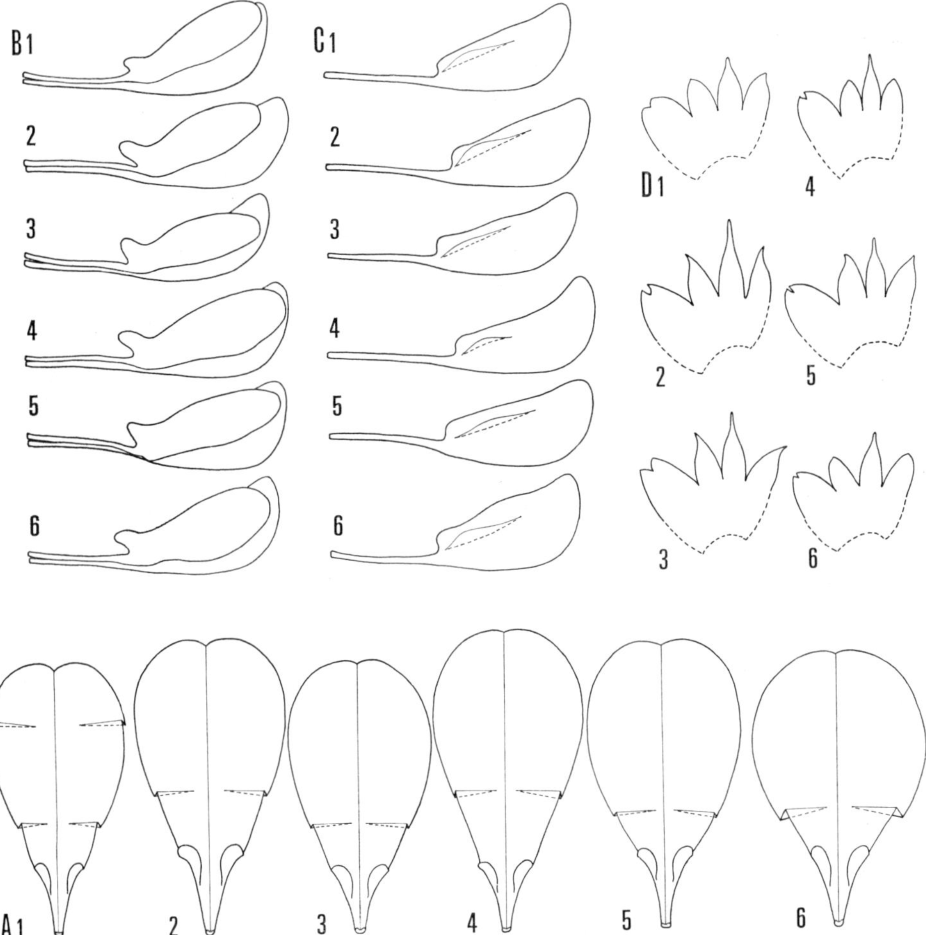

Fig. 5.　Individual variation in a population of *L. bicolor* from Hokkaido (Kuramoto 8108101–8108106, TI).
　　　A, standards, opened;　B, wings and keel-petals;　C, keel-petals;　D, calyces, dissected.　All×3.3.

←Fig. 4.　Individual variation in a population of *L. bicolor* from Hokkaido (Ohba & Akiyama 3501–3510, TI).
　　　A, standards, opened;　B, wings and keel-petals;　C, keel-petals;　D, calyces, dissected.　All×3.3.

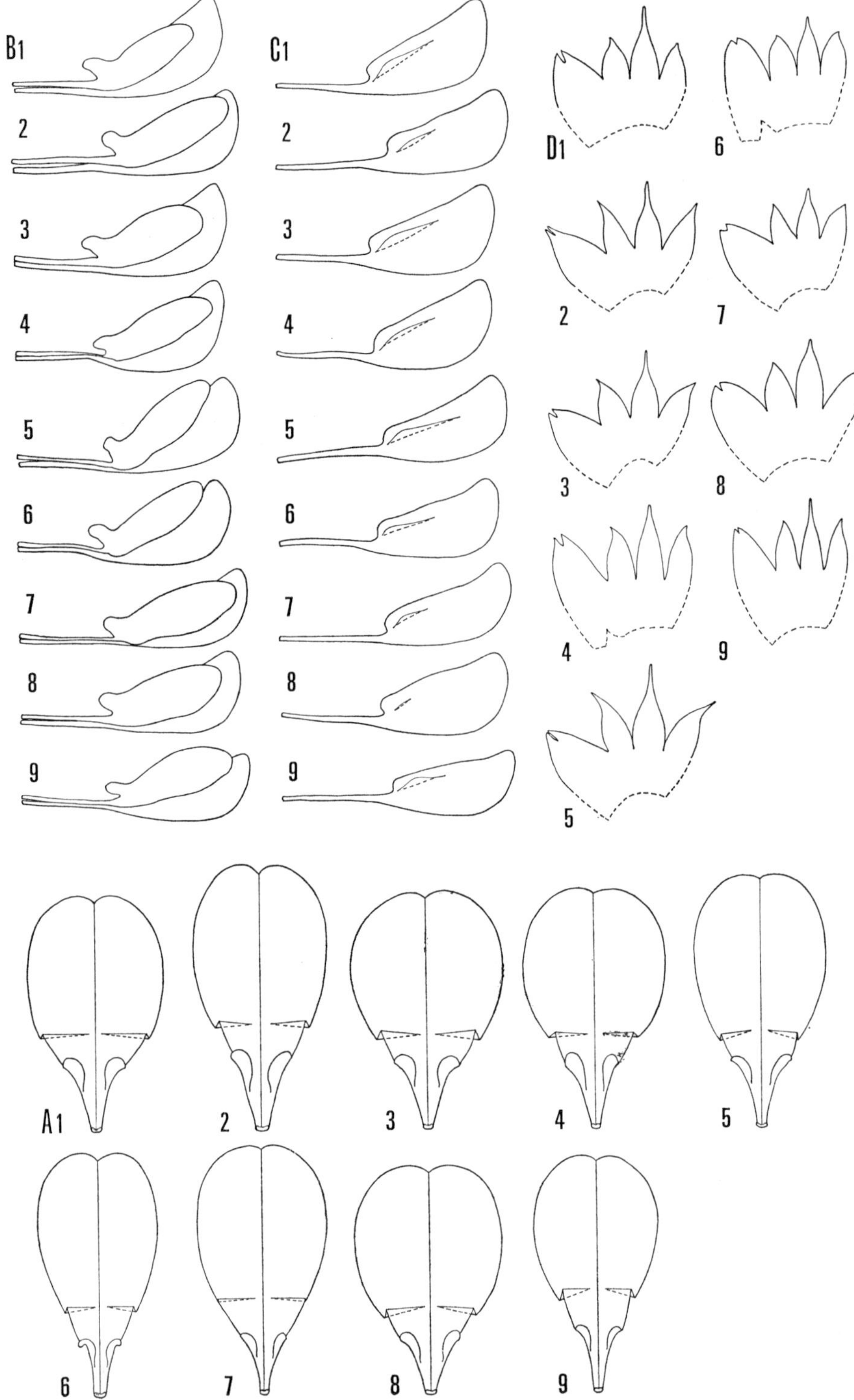

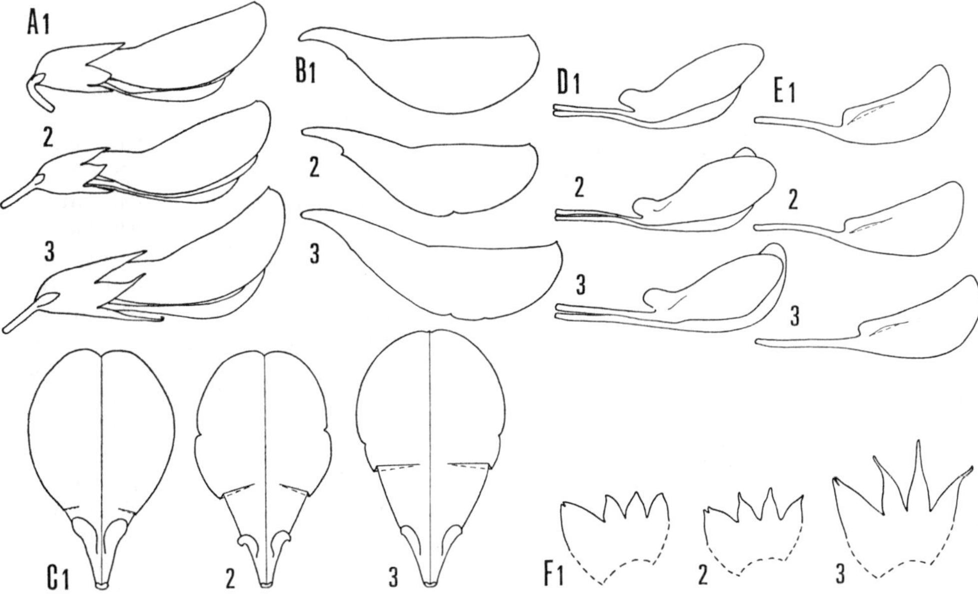

Fig. 7. Individual variation in a population of *L. bicolor* var. *nana* from Mt. Apoi (Ohba & Akiyama, 22 Aug. 1984, nos. 1–3, TI).
A, flowers, lateral view; B, standards, lateral view; C, standards, opened; D, wings and keel-petals; E, keel-petals; F, calyces, dissected. All×3.3.

←Fig. 6. Individual variation in a population of *L. bicolor* from Fukushima (Ohba & Akiyama 1820–1828, TI).
A, standards, opened; B, wings and keel-petals; C, keel-petals; D, calyces, dissected. All×3.3.

18

Table 4.  Phenotypic plasticity and individual variation of L. bicolor.

| | Character | Phenotypic plasticity | Individual variation |
|---|---|---|---|
| Calyx | shape of lobe | triangular | triangular |
| | shape of apex | acute | acute |
| | length ratio of lateral lobe to tube | 0.64±0.09 | 0.67±0.15 |
| | ratio of connated part of upper lobes | 4/5–7/8 | 2/3–7/8 |
| Corolla | color | red-purple | red-purple |
| Length ratio | wing to keel-petal | 0.93±0.02 | 0.95±0.04 |
| | standard to keel-petal | 1.06±0.02 | 1.05±0.03 |
| | wing to standard | 0.88±0.01 | 0.91±0.04 |
| Standard | shape | obovate with attenuate base | obovate with attenuate base |
| | shape of apex | retuse | retuse with a point or without a point |
| | shape of auricle | lunate | lunate |
| | length ratio of claw to standard | — | — |
| Wing | shape of lamina | narrowly obovate | narrowly obovate |
| | shape of upper base | auriculate | auriculate |
| | shape of lower base | tapering | minutely auriculate-tapering |
| | length ratio of claw to wing | 0.44±0.01 | 0.42±0.03 |
| Keel-petal | shape of lamina | obovate | obovate |
| | shape of upper base | cordate | cordate-truncate |
| | shape of lower base | tapering | tapering |
| | length ratio of claw to keel-petal | 0.46±0.01 | 0.45±0.03 |

Individual variation: The population studied consisted of nine blooming individuals (Fig. 3; Table 3). Almost all characters that have constant expressions in phenotypic plasticity are stable within this population (Table 4). However, the lower base of the lamina of the wing varies from tapering to minutely auriculate and the upper base of the lamina of the keel-petal also varies from cordate to truncate. Two upper calyx-lobes are connated 2/3–7/8 of the way along its length. The length ratios between different petals or parts of petals show a slightly wider range of variation than those in phenotypic plasticity (Table 4). That is, the ratio of the length of the standard to the keel-petal is 1.05±0.03, wing to standard 0.91±0.04, wing to keel-petal 0.95± 0.04, wing-claw to wing 0.42±0.03, keel-petal-claw to keel-petal 0.45±0.03, and lateral calyx-lobe to calyx-tube 0.67±0.15.

Collective variation: The flowers from three of ten representative populations from various localities are illustrated in Figs. 4–6 and the individual variations of characters within each population are summarized in Table 5. Most of the characters with constant expression in both phenotypic plasticity and individual variation also exhibit constant expression in these ten populations. A slight difference in expression of some characters was observed in the population from Mt. Apoi, Hokkaido (Fig. 7).* In this population the wing is slightly longer than or nearly equal to the keel-petal. Most

---

* *Lespedeza bicolor* Turcz. var. *nana* Nakai.
Further discussion of this will be made in the note on taxonomic treatment (p. 104).

Table 5.   Individual variation of the populations of *L. bicolor*.

| | Character | Population 1 | Population 2 | Population 3 | Population 4 | Population 5 |
|---|---|---|---|---|---|---|
| Calyx | shape of lobe | triangular | triangular | triangular | triangular | triangular |
| | shape of apex | acute | acute | acute | acute | acute—obtuse |
| | length ratio of lateral lobe to tube | 0.67±0.15 | 0.63±0.16 | 0.70±0.08 | 0.69±0.19 | 0.54±0.08 |
| Corolla | color | red-purple | red-purple | red-purple | red-purple | red-purple |
| Length ratio | wing to keel-petal | 0.95±0.04 | 0.94±0.04 | 0.95±0.03 | 0.90±0.04 | 0.93±0.03 |
| | standard to keel-petal | 1.05±0.03 | 1.10±0.03 | 1.09±0.01 | 1.07±0.03 | 1.07±0.03 |
| | wing to standard | 0.91±0.04 | 0.86±0.04 | 0.87±0.03 | 0.85±0.04 | 0.86±0.03 |
| Standard | shape | obovate with attenuate base | obovate with attenuate base | obovate with attenuate base | obovate with attenuate base | obovate with attenuate base |
| | shape of apex | retuse with a point or without a point | retuse with a point or without a point | retuse with a point or without a point | retuse with a point or without a point | retuse with a point or without a point |
| | shape of auricle | lunate | lunate | lunate | lunate | lunate |
| | length ratio of claw to standard | — | — | — | — | — |
| Wing | shape of lamina | narrowly obovate | narrowly obovate | narrowly obovate | narrowly obovate | narrowly obovate |
| | shape of upper base | auriculate | auriculate | auriculate | auriculate | auriculate |
| | shape of lower base | minutely auriculate -tapering | minutely auriculate -tapering | minutely auriculate -tapering | minutely auriculate -tapering | minutely auriculate -tapering |
| | length ratio of claw to wing | 0.42±0.03 | 0.39±0.03 | 0.43±0.02 | 0.44±0.01 | 0.42±0.02 |
| Keel-petal | shape of lamina | obovate | obovate | obovate | obovate | obovate |
| | shape of upper base | cordate-truncate | cordate-truncate | cordate-truncate | cordate-truncate | cordate-truncate |
| | shape of lower base | tapering | tapering | tapering | tapering | tapering |
| | length ratio of claw to keel-petal | 0.45±0.03 | 0.39±0.04 | 0.44±0.02 | 0.45±0.02 | 0.45±0.02 |

Population   1:   Gunma Pref., Tone-gun, Katashina-mura, nr. Lake Maru-numa, alt. ca. 1500 m (Ohba & Akiyama 1496, 1499–1507, TI).
Population   2:   Hokkaido: Iburi, Mukawa-machi, near Shiomi Station (Ohba & Akiyama 3501–3510, TI).
Population   3:   Nagano Pref., Oomachi-shi, Taira, nr. Lake Aoki, alt. ca. 750 m (Ohba & Akiyama 453–456, TI).
Population   4:   Fukushima Pref., Kawanuma-gun, Aidzubange-machi, Toudera-Ketanomiya (Ohba & Akiyama 1820–1828, TI).
Population   5:   Fukushima Pref., Minamiaidzu-gun, Shimogou-machi, Ono, alt. 400–500 m (Ohba & Akiyama 1929–1935, TI).

Table 5.  (Continued)

| Character | | Population 6 | Population 7 | Population 8 | Population 9 | Population 10 |
|---|---|---|---|---|---|---|
| Calyx | shape of lobe | triangular | triangular | triangular | triangular | triangular |
| | shape of apex | acute | acute | acute | acute | acute |
| | length ratio of lateral lobe to tube | 0.60±0.08 | 0.53±0.17 | 0.83±0.05 | 0.59±0.18 | 0.70±0.11 |
| Corolla | color | red-purple | red-purple | red-purple | red-purple | red-purple |
| Length ratio | wing to keel-petal | 0.96±0.04 | 0.95±0.02 | 0.95±0.04 | 0.94±0.03 | 0.92±0.04 |
| | standard to keel-petal | 1.11±0.05 | 1.10±0.04 | 1.08±0.03 | 1.05±0.03 | 1.10±0.03 |
| | wing to standard | 0.87±0.04 | 0.86±0.03 | 0.88±0.04 | 0.89±0.02 | 0.84±0.03 |
| Standard | shape | obovate with attenuate base | obovate with attenuate base | obovate with attenuate base | obovate with attenuate base | obovate with attenuate base |
| | shape of apex | retuse with a point or without a point | retuse with a point or without a point | retuse with a point or without a point | retuse with a point or without a point | retuse with a point or without a point |
| | shape of auricle | lunate | lunate | lunate | lunate | lunate |
| | length ratio of claw to standard | — | — | — | — | — |
| Wing | shape of lamina | narrowly obovate | narrowly obovate | narrowly obovate | narrowly obovate | narrowly obovate |
| | shape of upper base | auriculate | auriculate | auriculate | auriculate | auriculate |
| | shape of lower base | minutely auriculate -tapering | minutely auriculate -tapering | minutely auriculate -tapering | minutely auriculate -tapering | minutely auriculate -tapering |
| | length ratio of claw to wing | 0.40±0.03 | 0.42±0.03 | 0.40±0.01 | 0.42±0.00 | 0.39±0.02 |
| Keel-petal | shape of lamina | obovate | obovate | obovate | obovate | obovate |
| | shape of upper base | cordate-truncate | cordate-truncate | cordate-truncate | cordate-truncate | cordate-truncate |
| | shape of lower base | tapering | tapering | tapering | tapering | tapering |
| | length ratio of claw to keel-petal | 0.43±0.03 | 0.45±0.02 | 0.42±0.01 | 0.43±0.03 | 0.40±0.03 |

Population  6:  Niigata Pref., Minamikanbara-gun, Shitada-mura, Taya (Ohba & Akiyama 1747-1756, TI).
Population  7:  Niigata Pref., Kitauonuma-gun, Kawaguchi-machi, alt. ca. 100 m (Ohba & Akiyama 1594-1599, 1601-1604, TI).
Population  8:  Niigata Pref., Yuzawa-machi, near Asagai, alt. ca. 900 m (Ohba & Akiyama 1579, 1581-1585, TI).
Population  9:  Shimane Pref., Nita-gun, Yokota-machi, near Nakamura, alt. ca. 400 m (Ohba & Akiyama 2131-2134, TI).
Population 10:  Hokkaido:  Sorachi, Yubari-shi, Oyubari (Kuramoto 8108101-8108106, TI).

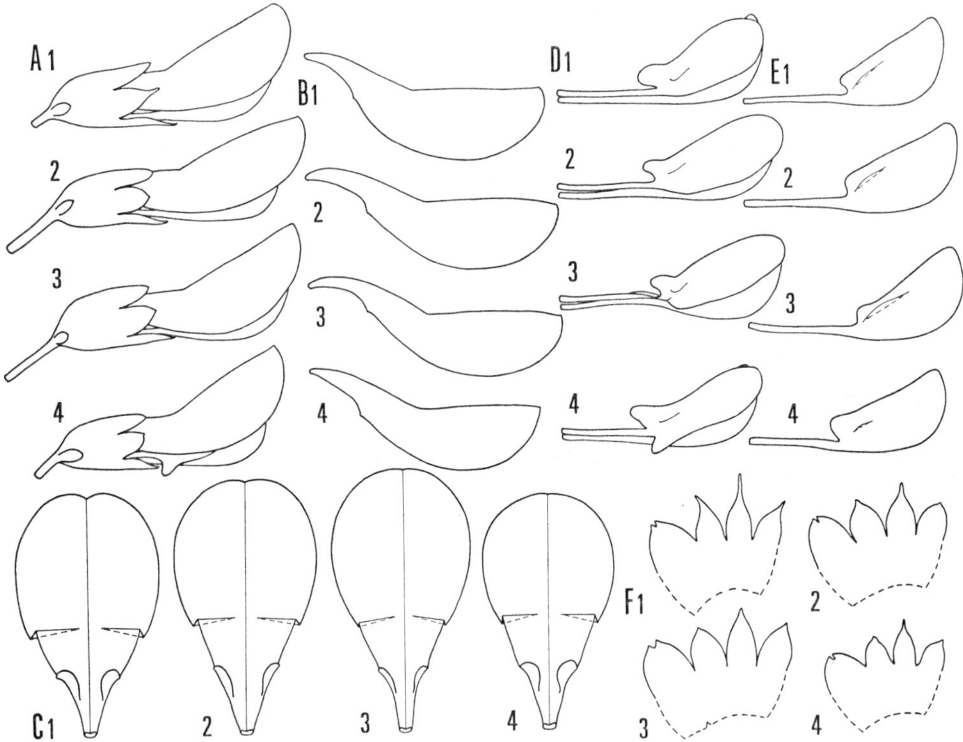

Fig. 8. Individual variation in a population of *L. bicolor* from Tokushima, Shikoku
(Ohba & Akiyama 3764–3767, TI).
A, flowers, lateral view;   B, standards, lateral view;   C, standards, opened;
D, wings and keel-petals;   E, keel-petals;   F, calyces, dissected.   All×3.3.

populations from Shikoku also exhibit a slight difference in expression of the relative
length between the wing and the keel-petal (Fig. 8). The wing is slightly longer than
or nearly equal to the keel-petal. Therefore var. *nana* and the population from Shikoku are excluded from the further observations described below.

The wing is usually shorter than the keel-petal but rarely is the same as the keel-
petal in length. The auricles of standard vary from lunate to narrowly lunate. The
shape of the base of the wing lamina is more variable. The apices of the lateral calyx-
lobes are variable from acute to obtuse. The ratio of the length of the wing to
the keel-petal is 0.90(±0.04)–0.96(±0.04), standard to keel-petal 1.05(±0.03)–1.11
(±0.05), wing to keel-petal 0.84(±0.03)–0.91(±0.04), wing-claw to wing 0.39
(±0.03)–0.44(±0.01), keel-petal-claw to keel-claw 0.39(±0.04)–0.45(±0.03), and
lateral calyx-lobe to calyx-tube 0.53(±0.17)–0.83(±0.05).

Flowers from various localities are shown in Fig. 9 and Table 6. The expression
of each character does not exceed the range of variation of those among the ten populations mentioned above.

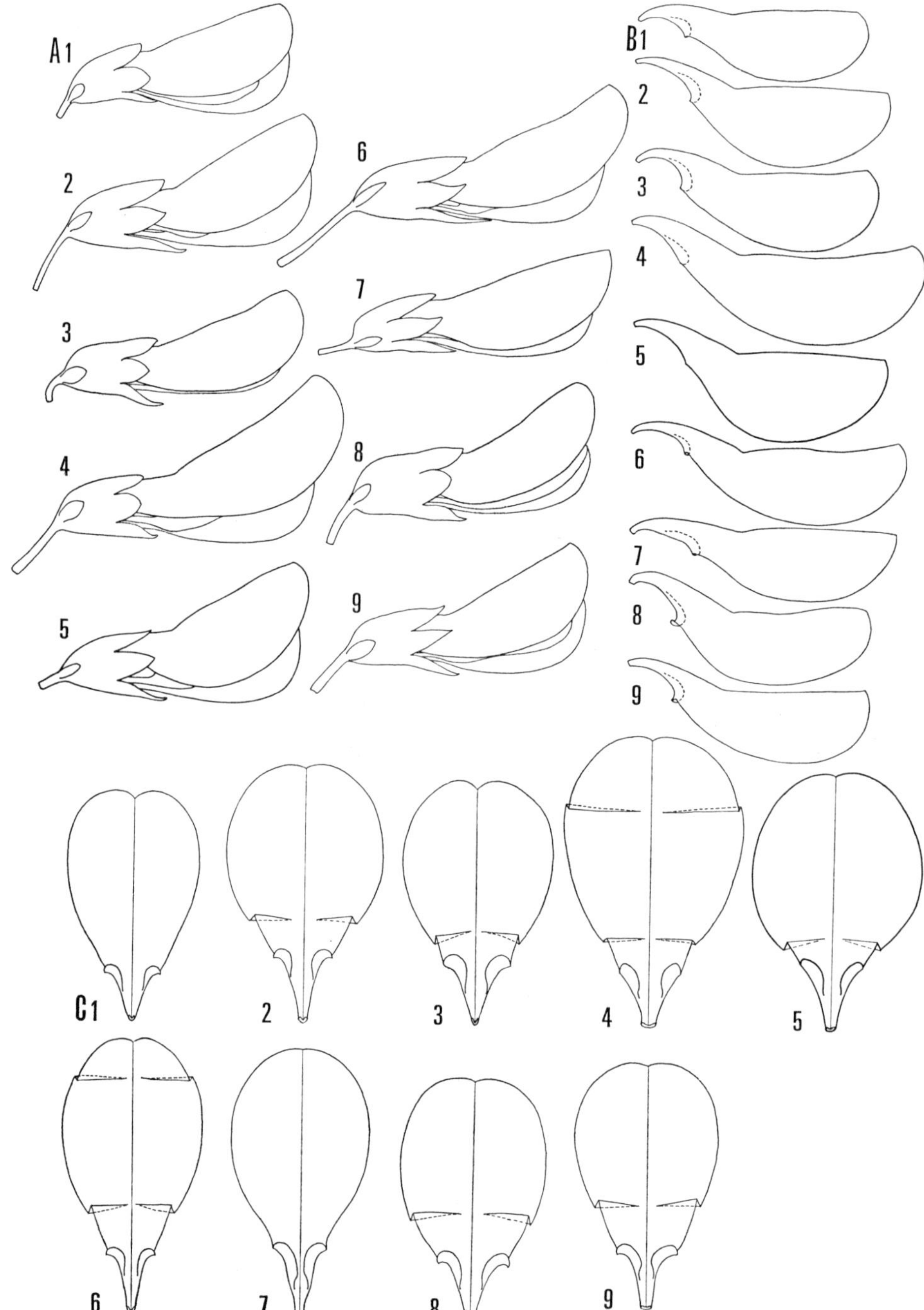

Fig. 9–1.   Variation of *L. bicolor*.
A, flowers, lateral view;   B, standards, lateral view;   C, standards, opened.
All×3.3.
1. Hokkaido (Koidzumi, Aug. 1917, KYO).   2. Miyagi (Ohba & Akiyama 589, TI).   3. Chiba (Hisauchi, 23 Sept. 1933, TI).   4. Tochigi (Ohba &→

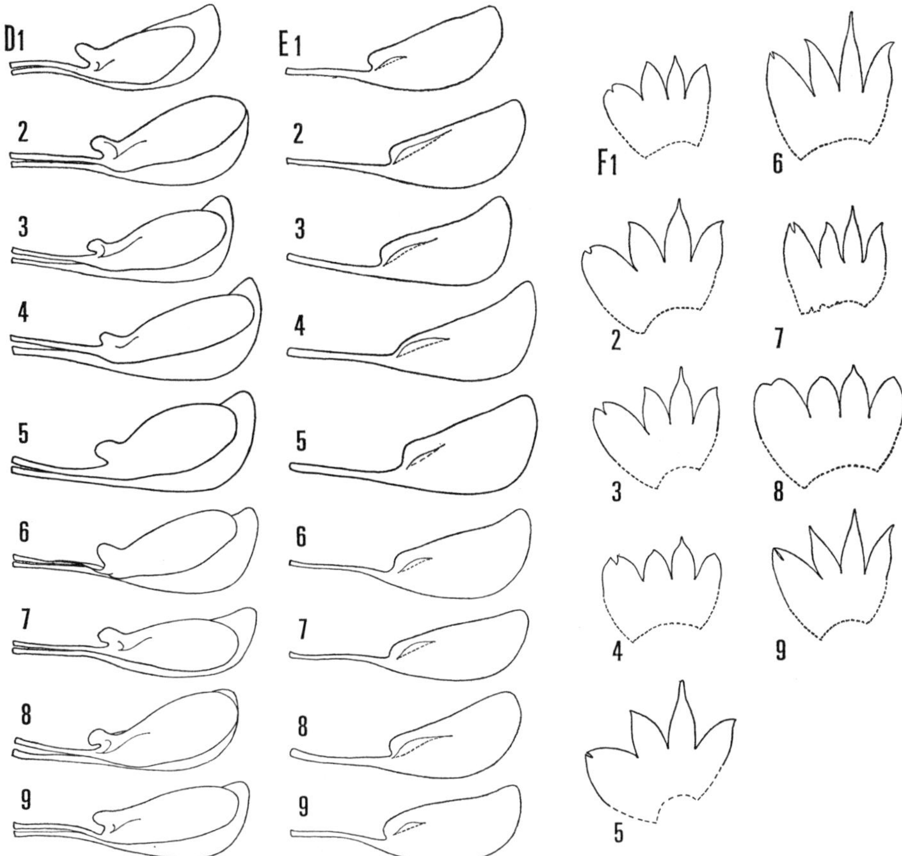

Fig. 9–2.   Variation *of L. bicolor*.
D, wings;   E, keel-petals;   F, calyces, dissected.   All×3.3.

Akiyama 608, TI).   5. Gunma (Ohba & Akiyama 1506, TI).   6. Nagano (Ohba & Akiyama 525, TI).   7. Kyoto (Togashi, 21 Sept. 1949, TI).   8. Nagasaki (Ohashi, Tateishi & Ohba 67, TI).   9. Kagoshima (Ohba & Akiyama 2615, TI).

Table 6. Collective variation of *L. bicolor*.

| | F.1 | C.1 | CT.1 | CL.1/CT.1 | S.1 | W.1 | WC.1 | WC.1/W.1 | K.1 | KC.1 | KC.1/K.1 | W.1/K.1 | S.1/K.1 | W.1/S.1 |
|---|---|---|---|---|---|---|---|---|---|---|---|---|---|---|
| 1 | 9.6 | 3.6 | 2.1 | 0.71 | 9.3 | 7.3 | 3.2 | 0.44 | 8.6 | 3.4 | 0.40 | 0.85 | 1.08 | 0.78 |
| 2 | 11.1 | 4.2 | 2.2 | 0.91 | 10.4 | 9.2 | 3.8 | 0.41 | 9.4 | 4.1 | 0.44 | 0.98 | 1.11 | 0.88 |
| 3 | 10.2 | 4.0 | 2.2 | 0.82 | 9.9 | 8.4 | 3.3 | 0.39 | 8.7 | 3.8 | 0.44 | 0.97 | 1.14 | 0.85 |
| 4 | 12.6 | 3.5 | 2.3 | 0.52 | 11.8 | 9.5 | 3.6 | 0.38 | 9.8 | 4.3 | 0.44 | 0.97 | 1.20 | 0.81 |
| 5 | 11.8 | 4.7 | 2.4 | 0.96 | 11.1 | 8.9 | 3.9 | 0.44 | 9.4 | 4.2 | 0.45 | 0.95 | 1.18 | 0.80 |
| 6 | 11.0 | 3.1 | 1.5 | 1.07 | 10.8 | 9.1 | 3.8 | 0.42 | 9.4 | 3.9 | 0.41 | 0.97 | 1.15 | 0.84 |
| 7 | 10.6 | 3.6 | 2.0 | 0.80 | 9.8 | 8.9 | 3.5 | 0.39 | 9.1 | 4.0 | 0.44 | 0.98 | 1.08 | 0.91 |
| 8 | 10.3 | 4.7 | 2.5 | 0.88 | 10.0 | 8.8 | 3.6 | 0.41 | 9.0 | 3.8 | 0.42 | 0.98 | 1.11 | 0.88 |
| Min. | 9.6 | 3.1 | 1.5 | 0.52 | 9.3 | 7.3 | 3.2 | 0.38 | 8.6 | 3.4 | 0.40 | 0.85 | 1.08 | 0.78 |
| Max. | 12.6 | 4.7 | 2.5 | 1.07 | 11.8 | 9.5 | 3.9 | 0.44 | 9.8 | 4.3 | 0.45 | 0.98 | 1.20 | 0.91 |
| Average | 10.9 | 3.9 | 2.2 | 0.83 | 10.4 | 8.8 | 3.6 | 0.41 | 9.2 | 3.9 | 0.43 | 0.95 | 1.13 | 0.84 |
| S.D. | 0.95 | 0.58 | 0.31 | 0.17 | 0.81 | 0.67 | 0.25 | 0.02 | 0.40 | 0.28 | 0.02 | 0.04 | 0.05 | 0.04 |

### *Lespedeza cyrtobotrya* Miquel

Phenotypic plasticity: Ten flowers are illustrated (Fig. 10) and measured (Table 7). The standard is significantly longer than the wing and the keel-petal (Fig. 10A), and the wing itself is apparently longer than the keel-petal (Fig. 10D). The ratio of the length of the standard to the keel-petal is $1.21\pm0.03$, wing to standard $0.87\pm0.02$, and wing to keel-petal $1.06\pm0.02$ (Table 7).

The shape of the standard is obovate with an attenuated base like that of *L. bicolor*, with or without lateral notches. The apex is retuse with a point (Fig. 10B & C). The shape of the auricle near the base of the standard is lunate. The shape of the lamina of the wing is narrowly obovate with an auriculated upper base and tapering lower base (Fig. 10D). The ratio of the length of the wing-claw to the wing is $0.41\pm0.02$. The shape of the lamina of the keel-petal is obovate with a cordate upper base and tapering lower base (Fig. 10E). The ratio of the length of the keel-petal-claw to the keelpetal is $0.50\pm0.02$.

The shape of the lateral calyx-lobe is triangular with a fine acuminate apex (Fig. 10F). Two upper calyx-lobes are connated 2/3–3/4 of the way along its length. The calyx-lobes are apparently longer than the tube and the ratio of the length of the lateral calyx-lobe to the calyx-tube is $1.19\pm0.10$.

Individual variation: A population consisting of nine blooming individuals was investigated (Fig. 11 and Table 8). As in the case of *L. bicolor* almost all of the characters with constant expressions phenotypically are stable within this population (Table 9). However, some characters become slightly more variable in expression: in the standard, the auricle varies from lunate to narrowly lunate, and the apex also varies from retuse with a point to retuse without a point (Fig. 11C). The upper base of the lamina of the keel-petal is cordate or truncate (Fig. 11F).

The ratios of the length of each petal show a wider range of variation, but the standard is always longer than the other petals and the wing is longer than the keel-petals (Table 9). That is, the ratio of the length of the wing to the keel-petal is $1.10\pm0.04$, standard to keel-petal $1.22\pm0.03$, wing to standard $0.91\pm0.03$, wing-claw to wing $0.41\pm0.01$, keel-petal-claw to keel-petal $0.50\pm0.02$, and lateral calyx-lobe to calyx-tube $1.25\pm0.17$. Two upper calyx-lobes are connated 2/3 to the entire way along its length (Fig. 11F).

Collective variation: Flowers from various localities were illustrated (Fig. 12) and measured (Table 10). The expressions of the characters investigated do not exceed the variation range of those within the single population mentioned above. The relative length among various characters is slightly different from that in the single population.

### *Lespedeza homoloba* Nakai

Phenotypic plasticity: Ten flowers are illustrated in Fig. 13 and the measurement of each character is summarized in Table 11. The standard is nearly equal to or slightly shorter than the keel-petal (Fig. 13A), while the keel-petal is longer than the wing (Fig. 13D). The ratios of the length of each petal are summarized in Table 11. That of the standard to the keel-petal is $0.99\pm0.03$, wing to standard $0.80\pm0.02$, and wing to keel-petal $0.79\pm0.03$.

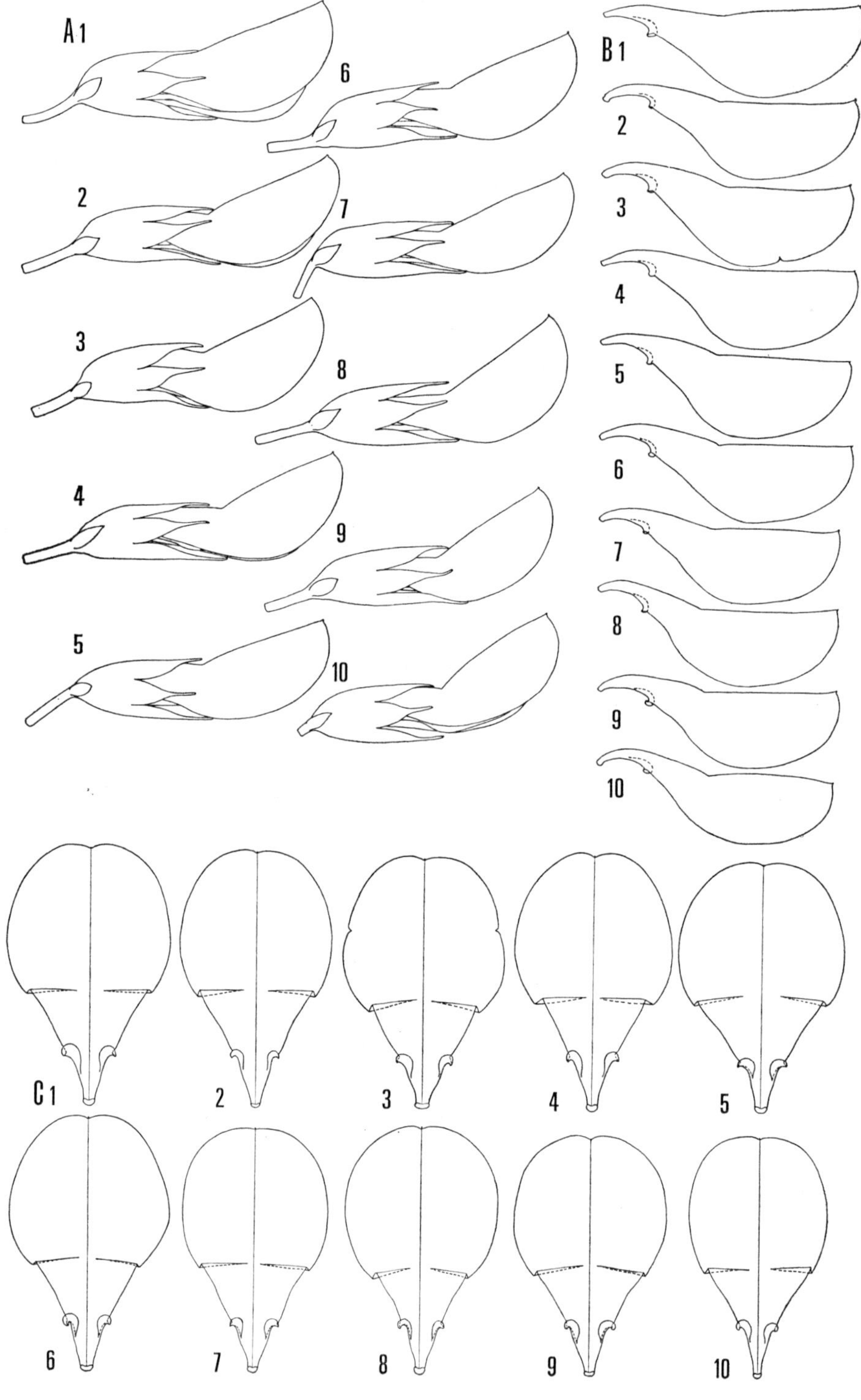

A 1
2
3
4
5
6
7
8
9
10
B 1
2
3
4
5
6
7
8
9
10
C 1
2
3
4
5
6
7
8
9
10

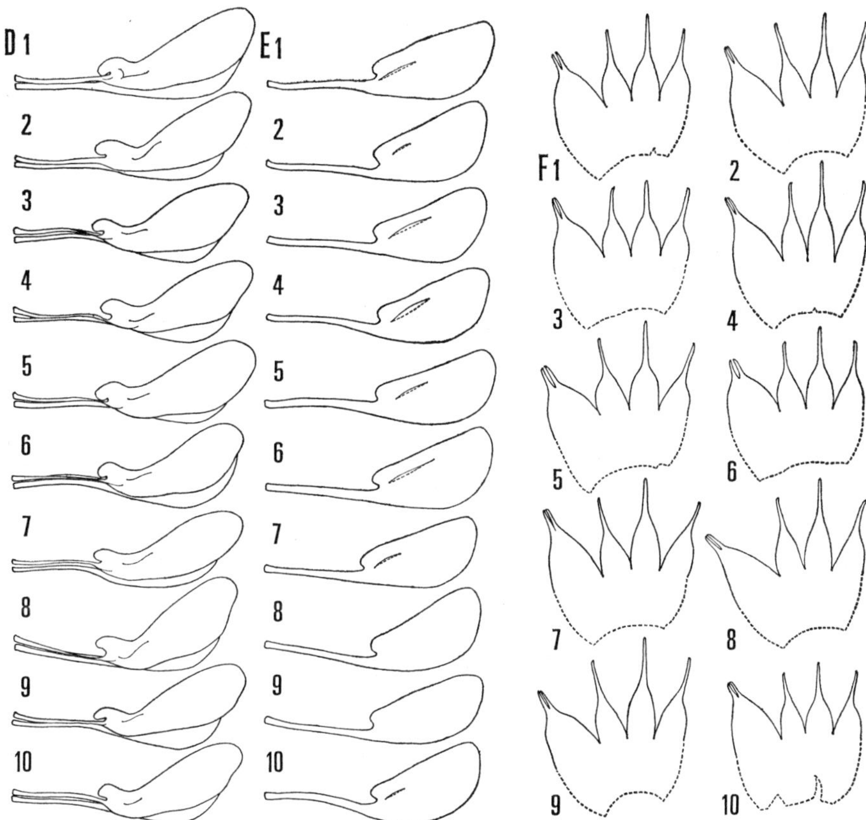

Fig. 10–2. Phenotypic plasticity of *L. cyrtobotrya*.
D, wings; E, keel-petals; F, calyces, dissected. All×3.3.

←Fig. 10–1. Phenotypic plasticity of *L. cyrtobotrya*.
A, flowers, lateral view; B, standards, lateral view; C, standards, opened. All×3.3.

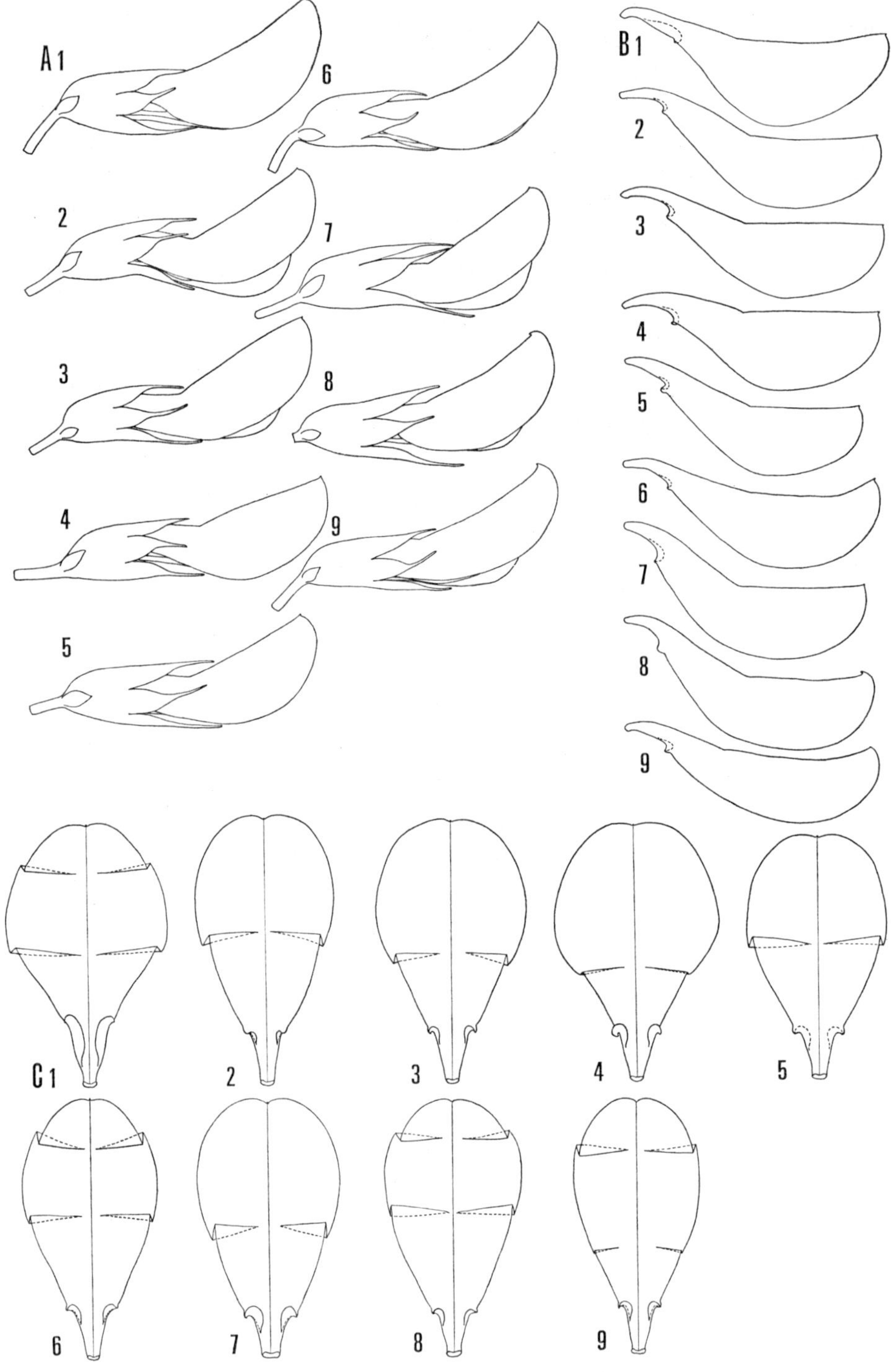

A 1
2
3
4
5
6
7
8
9
B 1
2
3
4
5
6
7
8
9
C 1
2
3
4
5
6
7
8
9

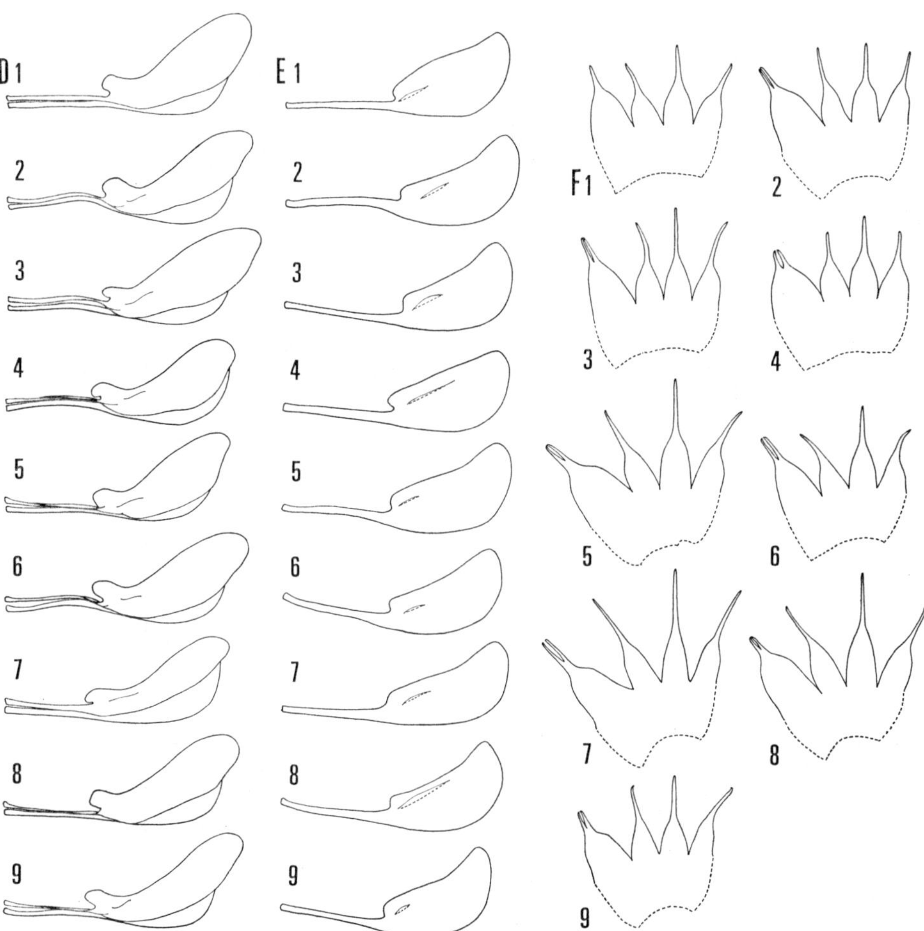

Fig. 11–2.   Individual variation in a population of *L. cyrtobotrya.*
D, wings;   E, keel-petals;   F, calyces, dissected.   All×3.3.


←Fig. 11–1.   Individual variation in a population of *L. cyrtobotrya.*
A, flowers, lateral view;   B, standards, lateral view;   C, standards,
opened.   All×3.3.

Table 7.  Phenotypic plasticity of *L. cyrtobotrya*.

| | F.1 | C.1 | CT.1 | CL.1/CT.1 | S.1 | W.1 | WC.1 | WC.1/W.1 | K.1 | KC.1 | KC.1/K.1 | W.1/K.1 | S.1/K.1 | W.1/S.1 |
|---|---|---|---|---|---|---|---|---|---|---|---|---|---|---|
| 1 | 11.2 | 5.3 | 2.3 | 1.30 | 10.9 | 9.4 | 3.8 | 0.40 | 9.0 | 4.3 | 0.48 | 1.04 | 1.21 | 0.86 |
| 2 | 11.5 | 6.0 | 2.9 | 1.07 | 10.7 | 9.3 | 3.8 | 0.41 | 8.5 | 4.4 | 0.52 | 1.09 | 1.26 | 0.87 |
| 3 | 10.8 | 5.3 | 2.5 | 1.12 | 10.5 | 9.0 | 3.7 | 0.41 | 8.6 | 4.3 | 0.50 | 1.05 | 1.22 | 0.86 |
| 4 | 11.8 | 5.6 | 2.5 | 1.24 | 10.7 | 9.2 | 4.1 | 0.45 | 8.6 | 4.4 | 0.51 | 1.07 | 1.24 | 0.86 |
| 5 | 11.1 | 5.3 | 2.3 | 1.30 | 10.4 | 9.4 | 3.7 | 0.39 | 9.0 | 4.4 | 0.49 | 1.04 | 1.16 | 0.90 |
| 6 | 11.4 | 5.2 | 2.5 | 1.08 | 10.7 | 9.5 | 3.9 | 0.41 | 8.8 | 4.4 | 0.50 | 1.08 | 1.22 | 0.89 |
| 7 | 11.1 | 5.3 | 2.3 | 1.30 | 10.1 | 8.7 | 3.6 | 0.41 | 8.3 | 3.9 | 0.47 | 1.05 | 1.22 | 0.86 |
| 8 | 11.1 | 5.5 | 2.5 | 1.20 | 10.3 | 8.8 | 3.5 | 0.40 | 8.4 | 4.4 | 0.52 | 1.05 | 1.23 | 0.85 |
| 9 | 10.7 | 5.8 | 2.6 | 1.23 | 10.0 | 9.1 | 3.8 | 0.42 | 8.5 | 4.3 | 0.51 | 1.07 | 1.18 | 0.91 |
| 10 | 10.7 | 5.1 | 2.5 | 1.04 | 10.0 | 8.8 | 3.9 | 0.44 | 8.4 | 4.4 | 0.52 | 1.05 | 1.19 | 0.88 |
| Min. | 10.7 | 5.1 | 2.3 | 1.04 | 10.0 | 8.7 | 3.5 | 0.39 | 8.3 | 3.9 | 0.47 | 1.04 | 1.16 | 0.85 |
| Max. | 11.8 | 6.0 | 2.9 | 1.30 | 10.9 | 9.5 | 4.1 | 0.45 | 9.0 | 4.4 | 0.52 | 1.09 | 1.26 | 0.91 |
| Average | 11.1 | 5.4 | 2.5 | 1.19 | 10.4 | 9.1 | 3.8 | 0.41 | 8.6 | 4.3 | 0.50 | 1.06 | 1.21 | 0.87 |
| S.D. | 0.36 | 0.28 | 0.18 | 0.10 | 0.32 | 0.29 | 0.17 | 0.02 | 0.25 | 0.15 | 0.02 | 0.02 | 0.03 | 0.02 |

Table 8.  Individual variation of *L. cyrtobotrya*.

| | F.1 | C.1 | CT.1 | CL.1/CT.1 | S.1 | W.1 | WC.1 | WC.1/W.1 | K.1 | KC.1 | KC.1/K.1 | W.1/K.1 | S.1/K.1 | W.1/S.1 |
|---|---|---|---|---|---|---|---|---|---|---|---|---|---|---|
| 1 | 11.5 | 4.9 | 2.2 | 1.23 | 10.9 | 10.0 | 4.2 | 0.42 | 8.9 | 4.4 | 0.49 | 1.12 | 1.22 | 0.92 |
| 2 | 11.0 | 5.8 | 2.7 | 1.15 | 11.1 | 10.0 | 4.1 | 0.41 | 9.2 | 4.6 | 0.50 | 1.09 | 1.21 | 0.90 |
| 3 | 11.2 | 5.5 | 2.3 | 1.39 | 10.8 | 10.2 | 4.3 | 0.42 | 8.9 | 4.7 | 0.53 | 1.15 | 1.21 | 0.94 |
| 4 | 11.4 | 5.2 | 2.5 | 1.08 | 10.7 | 9.5 | 3.9 | 0.41 | 8.8 | 4.4 | 0.50 | 1.08 | 1.22 | 0.89 |
| 5 | 10.9 | 6.1 | 2.6 | 1.35 | 10.0 | 9.4 | 3.9 | 0.41 | 8.4 | 4.3 | 0.51 | 1.12 | 1.19 | 0.94 |
| 6 | 11.2 | 5.1 | 2.4 | 1.13 | 10.8 | 10.0 | 4.1 | 0.41 | 9.0 | 4.3 | 0.48 | 1.11 | 1.20 | 0.93 |
| 7 | 10.9 | 6.6 | 2.7 | 1.44 | 10.7 | 9.0 | 3.7 | 0.41 | 8.9 | 4.1 | 0.46 | 1.01 | 1.20 | 0.84 |
| 8 | 10.9 | 6.2 | 2.5 | 1.48 | 10.6 | 9.5 | 3.9 | 0.41 | 8.7 | 4.4 | 0.51 | 1.09 | 1.22 | 0.90 |
| 9 | 11.0 | 5.5 | 2.7 | 1.04 | 10.5 | 9.4 | 3.7 | 0.39 | 8.2 | 4.1 | 0.50 | 1.15 | 1.28 | 0.90 |
| Min. | 10.9 | 4.9 | 2.2 | 1.04 | 10.0 | 9.0 | 3.7 | 0.39 | 8.2 | 4.1 | 0.46 | 1.01 | 1.19 | 0.84 |
| Max. | 11.5 | 6.6 | 2.7 | 1.48 | 11.1 | 10.2 | 4.3 | 0.42 | 9.2 | 4.7 | 0.53 | 1.15 | 1.28 | 0.94 |
| Average | 11.1 | 5.7 | 2.5 | 1.25 | 10.7 | 9.7 | 4.0 | 0.41 | 8.8 | 4.4 | 0.50 | 1.10 | 1.22 | 0.91 |
| S.D. | 0.23 | 0.56 | 0.18 | 0.17 | 0.31 | 0.40 | 0.21 | 0.01 | 0.31 | 0.20 | 0.02 | 0.04 | 0.03 | 0.03 |

Table 9. Phenotypic plasticity and individual variation of *L. cyrtobotrya*.

| Character | | Phenotypic plasticity | Individual variation |
|---|---|---|---|
| Calyx | shape of lobe | triangular | triangular |
| | shape of apex | acuminate | acuminate |
| | length ratio of lateral lobe to tube | 1.19±0.10 | 1.25±0.17 |
| | ratio of connated part of upper lobes | 2/3–3/4 | 2/3–1 |
| Corolla | color | red-purple | red-purple |
| Length ratio | wing to keel-petal | 1.06±0.02 | 1.10±0.04 |
| | standard to keel-petal | 1.21±0.03 | 1.22±0.03 |
| | wing to standard | 0.87±0.02 | 0.91±0.03 |
| Standard | shape | obovate with attenuate base | obovate with attenuate base |
| | shape of apex | retuse with a point | retuse with a point or without a point |
| | shape of auricle | lunate | lunate–narrowly lunate |
| | length ratio of claw to standard | — | — |
| Wing | shape of lamina | narrowly obovate | narrowly obovate |
| | shape of upper base | auriculate | auriculate |
| | shape of lower base | tapering | tapering |
| | length ratio of claw to wing | 0.41±0.02 | 0.41±0.01 |
| Keel-petal | shape of lamina | obovate | obovate |
| | shape of upper base | cordate | cordate–truncate |
| | shape of lower base | tapering | tapering |
| | length ratio of claw to keel-petal | 0.50±0.02 | 0.50±0.02 |

The standard has a distinct claw; the shape of the lamina of the standard is elliptic with a retuse apex, with or without lateral notches; the auricle at the base of the lamina is well developed and its shape is reniform (Fig. 13B & C). The ratio of the length of the standard-claw to the standard itself is 0.21±0.03. The shape of the lamina of the wing is narrowly oblong with a truncate or auriculate upper base and a minutely auriculate or cordate lower base (Fig. 13D). The ratio of the length of the wing-claw to the wing itself is 0.35±0.03. The shape of the lamina of the keel-petal is narrowly obovate with a cordate upper base and a tapering lower base (Fig. 13E). The ratio of the length of the keel-petal-claw to the keel-petal is 0.28±0.02.

The shape of the lateral calyx-lobe is elliptic; the apex is obtuse with a point (Fig. 13F). Two upper calyx-lobes are connated 1/2–9/10 of the way along its length. The calyx-lobes are shorter than the tube and the ratio of the length of the lateral calyx-lobe to the tube is 0.77±0.09.

Individual variation: A population consisting of eight blooming individuals was investigated (Fig. 14 and Table 12). As in the former two cases almost all of the characters with constant expressions phenotypically are stable within this population, although some become slightly more variable (Table 13). The ratio of the length of the standard to the keel-petal is 0.99±0.03, wing to standard 0.83±0.03, and wing to keel-petal 0.82±0.05.

In the standard, the shape of the lamina is elliptic or broadly obovate with a retuse apex (the apex itself can be with or without a point) (Fig. 14C). The ratio of the length of the claw to the standard itself is 0.23±0.03. The lower base of the wing lamina is

Table 10.  Collective variation of *L. cyrtobotrya*.

| | F.1 | C.1 | CT.1 | CL.1/CT.1 | S.1 | W.1 | WC.1 | WC.1/W.1 | K.1 | KC.1 | KC.1/K.1 | W.1/K.1 | S.1/K.1 | W.1/S.1 |
|---|---|---|---|---|---|---|---|---|---|---|---|---|---|---|
| 1 | 11.1 | 4.5 | 2.2 | 1.05 | 10.5 | 9.7 | 3.5 | 0.36 | 8.7 | 4.0 | 0.46 | 1.11 | 1.21 | 0.92 |
| 2 | 12.6 | 5.5 | 2.6 | 1.12 | 12.4 | 10.7 | 4.3 | 0.40 | 9.6 | 5.0 | 0.52 | 1.11 | 1.29 | 0.86 |
| 3 | 12.2 | 5.7 | 2.4 | 1.38 | 11.3 | 10.3 | 4.5 | 0.44 | 9.5 | 5.0 | 0.53 | 1.08 | 1.19 | 0.91 |
| 4 | 8.9 | 5.7 | 2.5 | 1.28 | 8.1 | 8.1 | 3.5 | 0.43 | 7.7 | 3.6 | 0.47 | 1.05 | 1.05 | 1.00 |
| 5 | 11.6 | 6.2 | 2.7 | 1.30 | 11.4 | 9.0 | 4.0 | 0.44 | 8.7 | 4.4 | 0.51 | 1.03 | 1.31 | 0.79 |
| 6 | 11.2 | 5.1 | 2.4 | 1.13 | 10.8 | 10.0 | 4.1 | 0.41 | 9.0 | 4.3 | 0.48 | 1.11 | 1.20 | 0.93 |
| 7 | 10.8 | 5.8 | 2.4 | 1.42 | 10.4 | 9.0 | 4.1 | 0.46 | 8.7 | 4.5 | 0.52 | 1.03 | 1.20 | 0.87 |
| 8 | 10.5 | 6.2 | 2.7 | 1.30 | 10.1 | 9.3 | 3.9 | 0.42 | 9.1 | 4.6 | 0.51 | 1.02 | 1.11 | 0.92 |
| 9 | 9.1 | 5.5 | 2.3 | 1.39 | 8.8 | 8.2 | 3.6 | 0.44 | 7.6 | 4.1 | 0.54 | 1.08 | 1.16 | 0.93 |
| 10 | 10.8 | 5.4 | 2.4 | 1.25 | 10.2 | 9.4 | 4.2 | 0.45 | 9.1 | 4.6 | 0.51 | 1.03 | 1.12 | 0.92 |
| Min. | 8.9 | 4.5 | 2.2 | 1.05 | 8.1 | 8.1 | 3.5 | 0.36 | 7.6 | 3.6 | 0.46 | 1.02 | 1.05 | 0.79 |
| Max. | 12.6 | 6.2 | 2.7 | 1.42 | 12.4 | 10.7 | 4.5 | 0.46 | 9.6 | 5.0 | 0.54 | 1.11 | 1.31 | 1.00 |
| Average | 10.9 | 5.6 | 2.5 | 1.26 | 10.4 | 9.4 | 4.0 | 0.42 | 8.8 | 4.4 | 0.50 | 1.07 | 1.18 | 0.91 |
| S.D. | 1.18 | 0.50 | 0.16 | 0.13 | 1.25 | 0.84 | 0.34 | 0.03 | 0.67 | 0.44 | 0.03 | 0.04 | 0.08 | 0.06 |

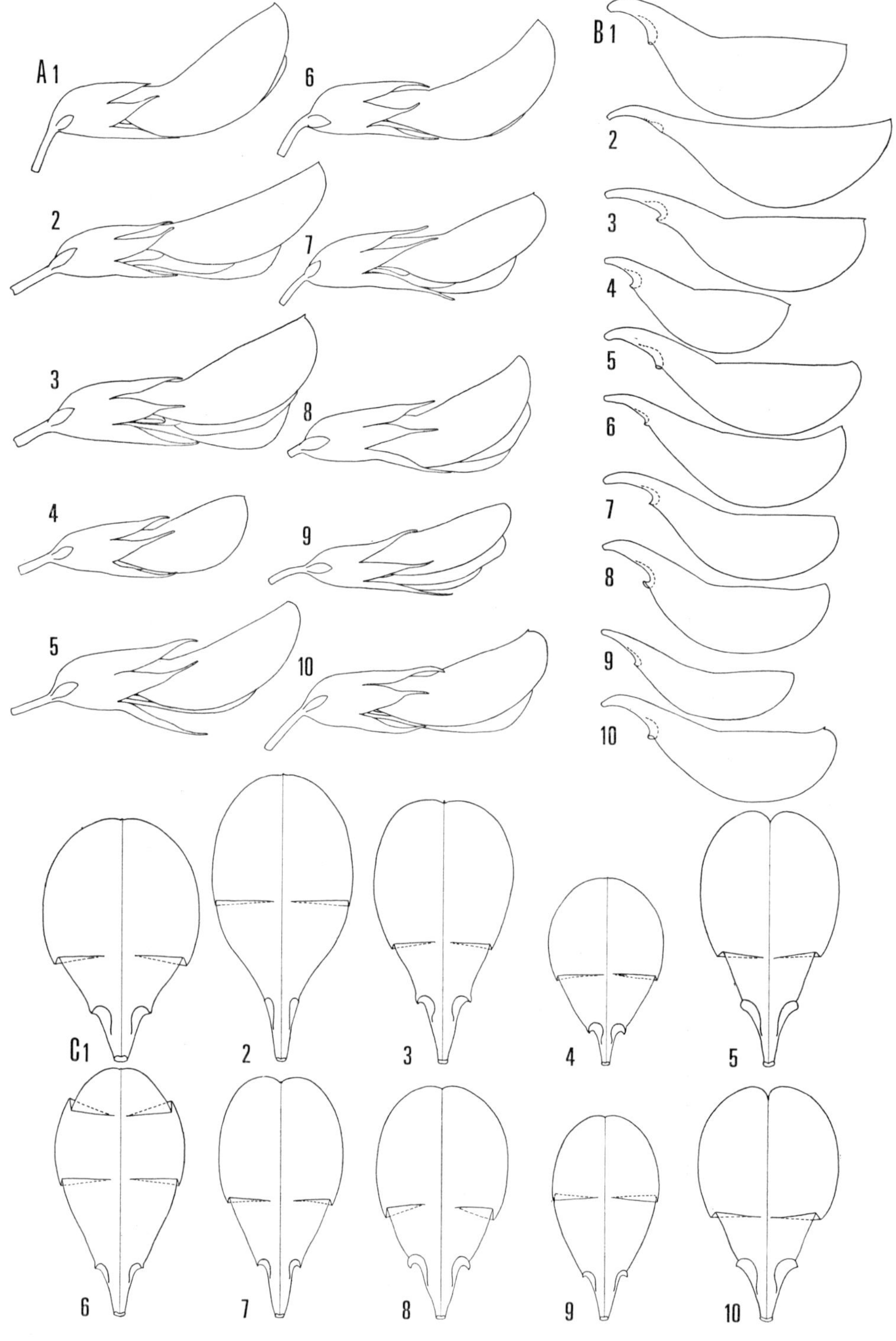

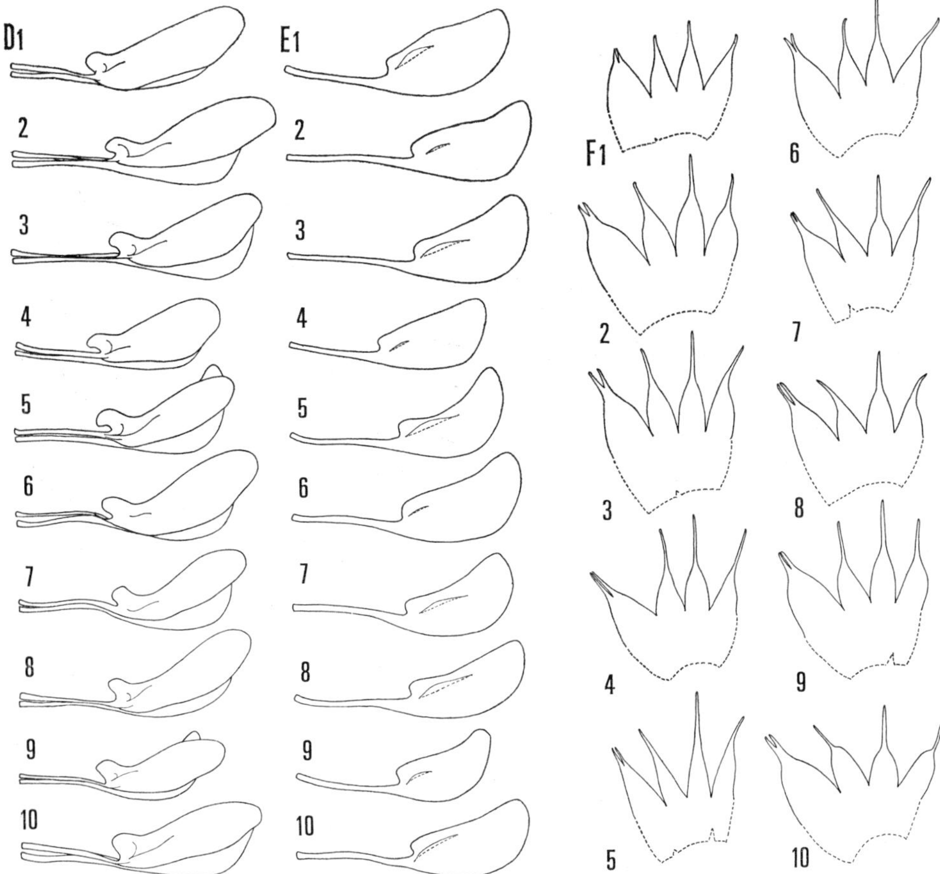

Fig. 12–2.   Variation of *L. cyrtobotrya*.
D, wings;   E, keel-petals;   F, calyces, dissected.   All × 3.3.

←Fig. 12–1.   Variation of *L. cyrtobotrya*.
A, flowers, lateral view;   B, standards, lateral view;   C, standards, opened.   All × 3.3.
1. Yamagata (Ohba & Akiyama 2700, TI).   2. Yamagata (Koidzumi, 23 Aug. 1927, KYO).   3. Fukushima (Ohba & Akiyama 1943, TI).   4. Ibaraki (Tsurumachi 17, TI).   5. Nagano (Ohba & Akiyama 458c, TI).   6. Kanagawa (Ohba & Akiyama 624, TI).   7. Shimane (Ohba & Akiyama 2121, TI).   8. Yamaguchi (Tateishi 3638, TI).   9. Kouchi (Okuyama 16019, TNS).   10. Kagoshima (Ohba & Akiyama 776, TI).

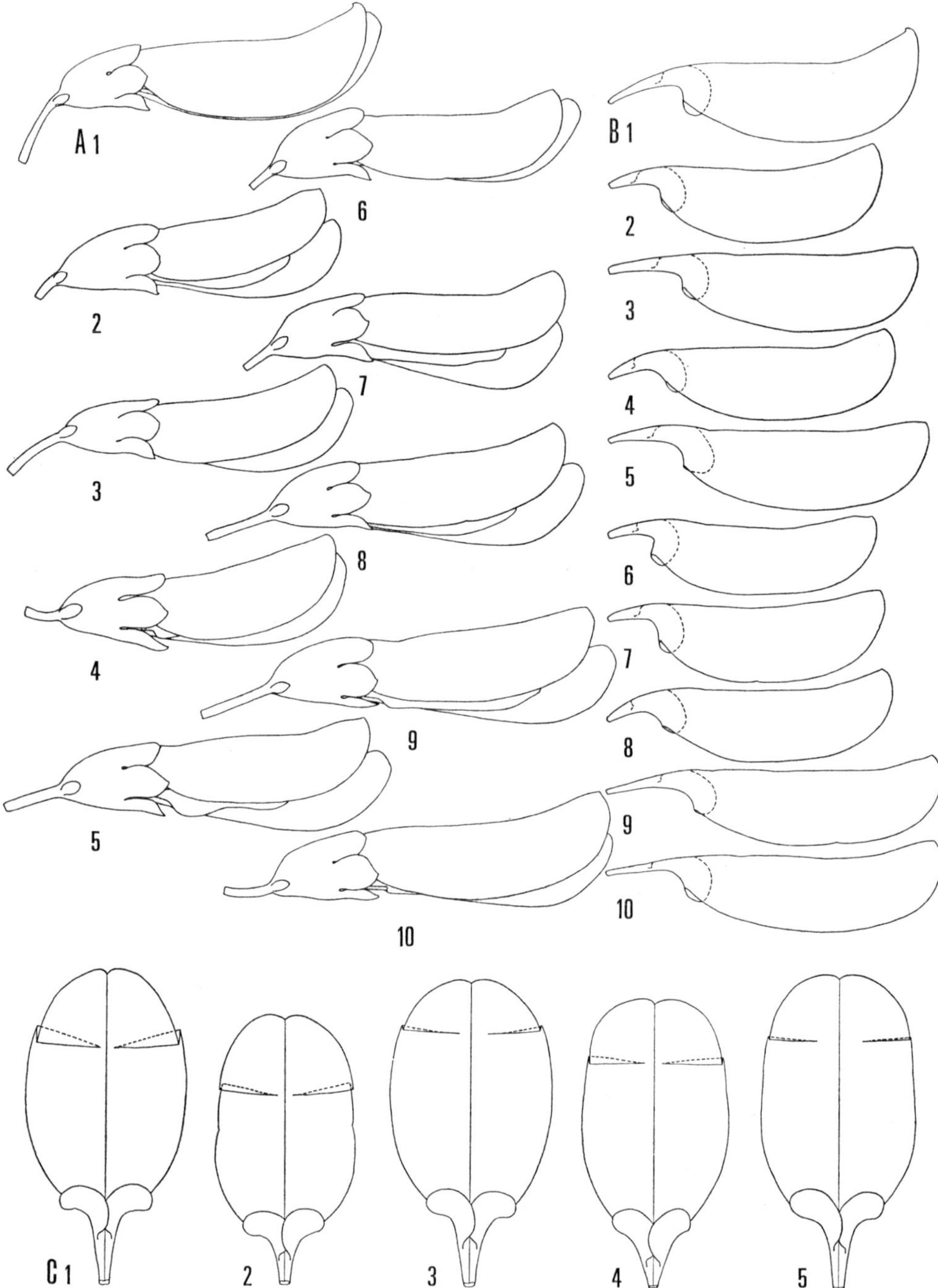

Fig. 13–1.  Phenotypic plasticity of *L. homoloba*.
A, flowers, lateral view;  B, standards, lateral view;  C, standards, opened.  All×3.3.

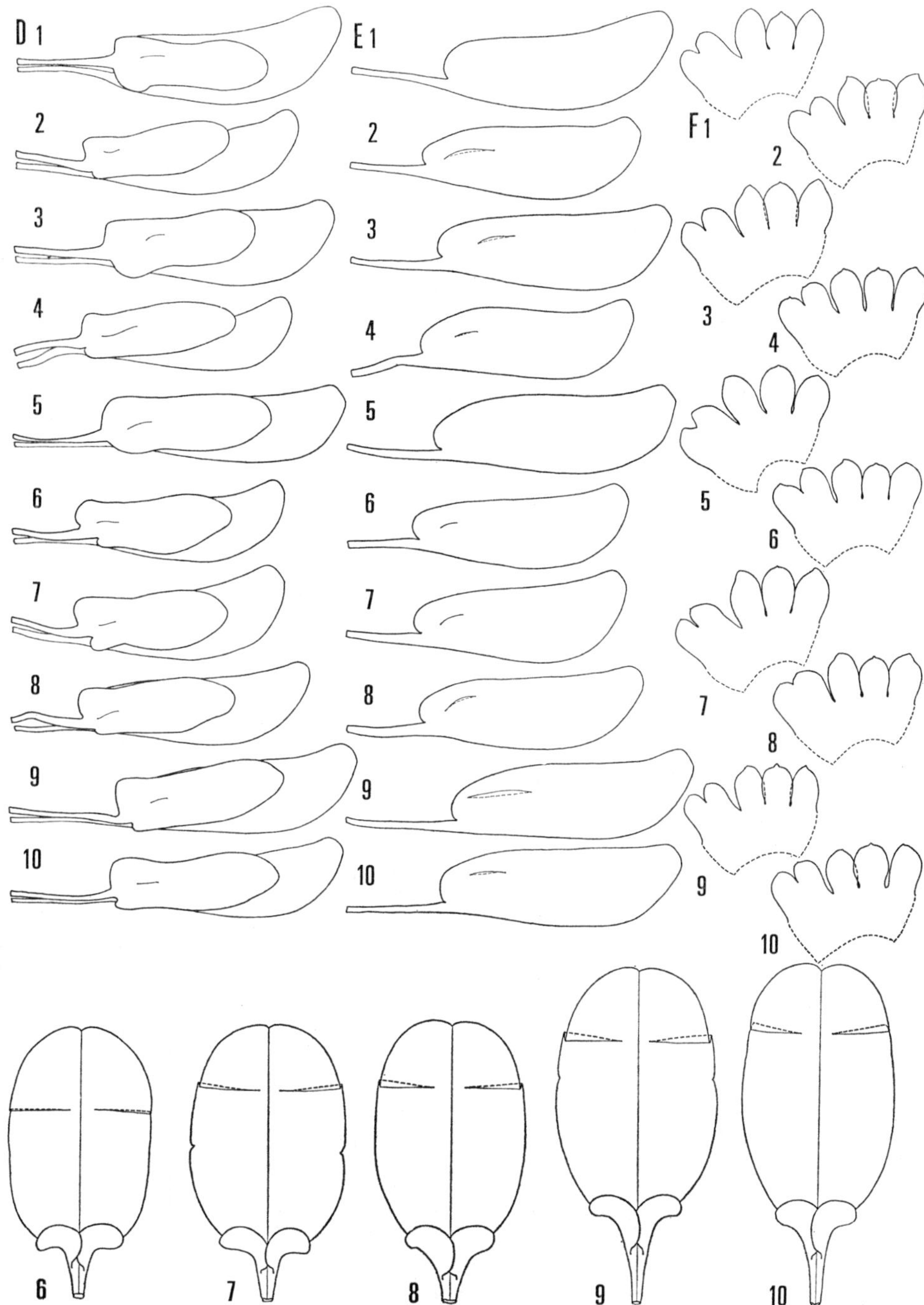

Fig. 13-2.  Phenotypic plasticity of *L. homoloba*.
D, wings;   E, keel-petals;   F, calyces, dissected.   All ×3.3.

Table 11.  Phenotypic plasticity of *L. homoloba*.

| | F.1 | C.1 | CT.1 | CL.1/CT.1 | S.1 | SC.1 | SC.1/S.1 | W.1 | WC.1 | WC.1/W.1 | K.1 | KC.1 | KC.1/K.1 | W.1/K.1 | S.1/K.1 | W.1/S.1 |
|---|---|---|---|---|---|---|---|---|---|---|---|---|---|---|---|---|
| 1 | 13.6 | 3.9 | 2.1 | 0.86 | 12.9 | 3.0 | 0.23 | 10.3 | 4.0 | 0.39 | 12.7 | 3.9 | 0.31 | 0.81 | 1.02 | 0.80 |
| 2 | 12.2 | 4.1 | 2.2 | 0.86 | 11.1 | 2.0 | 0.18 | 8.7 | 2.9 | 0.33 | 11.6 | 3.1 | 0.27 | 0.75 | 0.96 | 0.78 |
| 3 | 13.7 | 4.5 | 2.6 | 0.73 | 12.6 | 3.0 | 0.24 | 9.9 | 3.6 | 0.36 | 12.9 | 3.7 | 0.29 | 0.77 | 0.98 | 0.79 |
| 4 | 12.2 | 4.1 | 2.3 | 0.78 | 11.8 | 2.3 | 0.19 | 9.3 | 2.9 | 0.31 | 11.6 | 3.0 | 0.26 | 0.80 | 1.02 | 0.79 |
| 5 | 14.1 | 4.8 | 2.7 | 0.78 | 13.0 | 3.0 | 0.23 | 10.8 | 3.7 | 0.34 | 13.2 | 3.8 | 0.29 | 0.82 | 0.98 | 0.83 |
| 6 | 11.9 | 4.4 | 2.7 | 0.63 | 11.0 | 2.0 | 0.18 | 9.0 | 2.8 | 0.31 | 11.3 | 2.9 | 0.26 | 0.80 | 0.97 | 0.82 |
| 7 | 12.0 | 4.5 | 2.4 | 0.88 | 11.3 | 2.0 | 0.18 | 8.9 | 2.8 | 0.31 | 11.3 | 3.1 | 0.27 | 0.79 | 1.00 | 0.79 |
| 8 | 12.7 | 4.3 | 2.4 | 0.79 | 11.5 | 2.3 | 0.20 | 9.1 | 3.0 | 0.33 | 12.0 | 3.2 | 0.27 | 0.76 | 0.96 | 0.79 |
| 9 | 14.5 | 4.1 | 2.4 | 0.71 | 13.7 | 3.6 | 0.26 | 11.5 | 4.6 | 0.40 | 14.0 | 4.5 | 0.32 | 0.82 | 0.98 | 0.84 |
| 10 | 14.1 | 4.3 | 2.6 | 0.65 | 13.9 | 3.2 | 0.23 | 11.2 | 4.2 | 0.38 | 13.5 | 4.0 | 0.30 | 0.83 | 1.03 | 0.81 |
| Min. | 11.9 | 3.9 | 2.1 | 0.63 | 11.0 | 2.0 | 0.18 | 8.7 | 2.8 | 0.31 | 11.3 | 2.9 | 0.26 | 0.75 | 0.96 | 0.78 |
| Max. | 14.5 | 4.8 | 2.7 | 0.88 | 13.9 | 3.6 | 0.26 | 11.5 | 4.6 | 0.40 | 14.0 | 4.5 | 0.32 | 0.83 | 1.03 | 0.84 |
| Average | 13.1 | 4.3 | 2.4 | 0.77 | 12.3 | 2.6 | 0.21 | 9.9 | 3.5 | 0.35 | 12.4 | 3.5 | 0.28 | 0.79 | 0.99 | 0.80 |
| S.D. | 1.00 | 0.26 | 0.21 | 0.09 | 1.08 | 0.59 | 0.03 | 1.03 | 0.66 | 0.03 | 0.98 | 0.53 | 0.02 | 0.03 | 0.03 | 0.02 |

Table 12. Individual variation of *L. homoloba*.

| | F.1 | C.1 | CT.1 | CL.1/CT.1 | S.1 | SC.1 | SC.1/S.1 | W.1 | WC.1 | WC.1/W.1 | K.1 | KC.1 | KC.1/K.1 | W.1/K.1 | S.1/K.1 | W.1/S.1 |
|---|---|---|---|---|---|---|---|---|---|---|---|---|---|---|---|---|
| 1 | 11.5 | 3.9 | 1.9 | 1.05 | 10.7 | 2.6 | 0.24 | 8.4 | 3.4 | 0.40 | 11.0 | 3.4 | 0.31 | 0.76 | 0.97 | 0.79 |
| 2 | 12.7 | 4.3 | 2.4 | 0.79 | 11.5 | 3.0 | 0.26 | 9.1 | 3.0 | 0.33 | 12.0 | 4.1 | 0.34 | 0.76 | 0.96 | 0.79 |
| 3 | 14.4 | 4.8 | 2.6 | 0.85 | 12.6 | 2.7 | 0.21 | 10.5 | 3.7 | 0.35 | 13.2 | 3.8 | 0.29 | 0.80 | 0.95 | 0.83 |
| 4 | 13.1 | 4.6 | 2.3 | 1.00 | 12.4 | 2.8 | 0.23 | 10.3 | 3.7 | 0.36 | 12.2 | 4.3 | 0.35 | 0.84 | 1.02 | 0.83 |
| 5 | 14.4 | 4.4 | 2.4 | 0.83 | 13.1 | 3.1 | 0.24 | 11.0 | 3.9 | 0.35 | 13.5 | 4.2 | 0.31 | 0.81 | 0.97 | 0.84 |
| 6 | 10.8 | 4.3 | 2.1 | 1.05 | 10.6 | 2.1 | 0.20 | 9.1 | 3.4 | 0.37 | 10.2 | 3.9 | 0.38 | 0.89 | 1.04 | 0.86 |
| 7 | 13.9 | 4.9 | 2.4 | 1.04 | 13.4 | 3.7 | 0.28 | 11.6 | 4.5 | 0.39 | 13.5 | 4.0 | 0.30 | 0.86 | 0.99 | 0.87 |
| 8 | 11.3 | 4.7 | 2.3 | 1.04 | 10.6 | 2.2 | 0.21 | 9.0 | 3.3 | 0.37 | 10.5 | 3.8 | 0.36 | 0.86 | 1.01 | 0.85 |
| Min. | 10.8 | 3.9 | 1.9 | 0.79 | 10.6 | 2.1 | 0.20 | 8.4 | 3.0 | 0.33 | 10.2 | 3.4 | 0.29 | 0.76 | 0.95 | 0.79 |
| Max. | 14.4 | 4.9 | 2.6 | 1.05 | 13.4 | 3.7 | 0.28 | 11.6 | 4.5 | 0.40 | 13.5 | 4.3 | 0.38 | 0.89 | 1.04 | 0.87 |
| Average | 12.8 | 4.5 | 2.3 | 0.96 | 11.9 | 2.8 | 0.23 | 9.9 | 3.6 | 0.37 | 12.0 | 3.9 | 0.33 | 0.82 | 0.99 | 0.83 |
| S.D. | 1.43 | 0.33 | 0.21 | 0.11 | 1.16 | 0.51 | 0.03 | 1.13 | 0.45 | 0.02 | 1.33 | 0.28 | 0.03 | 0.05 | 0.03 | 0.03 |

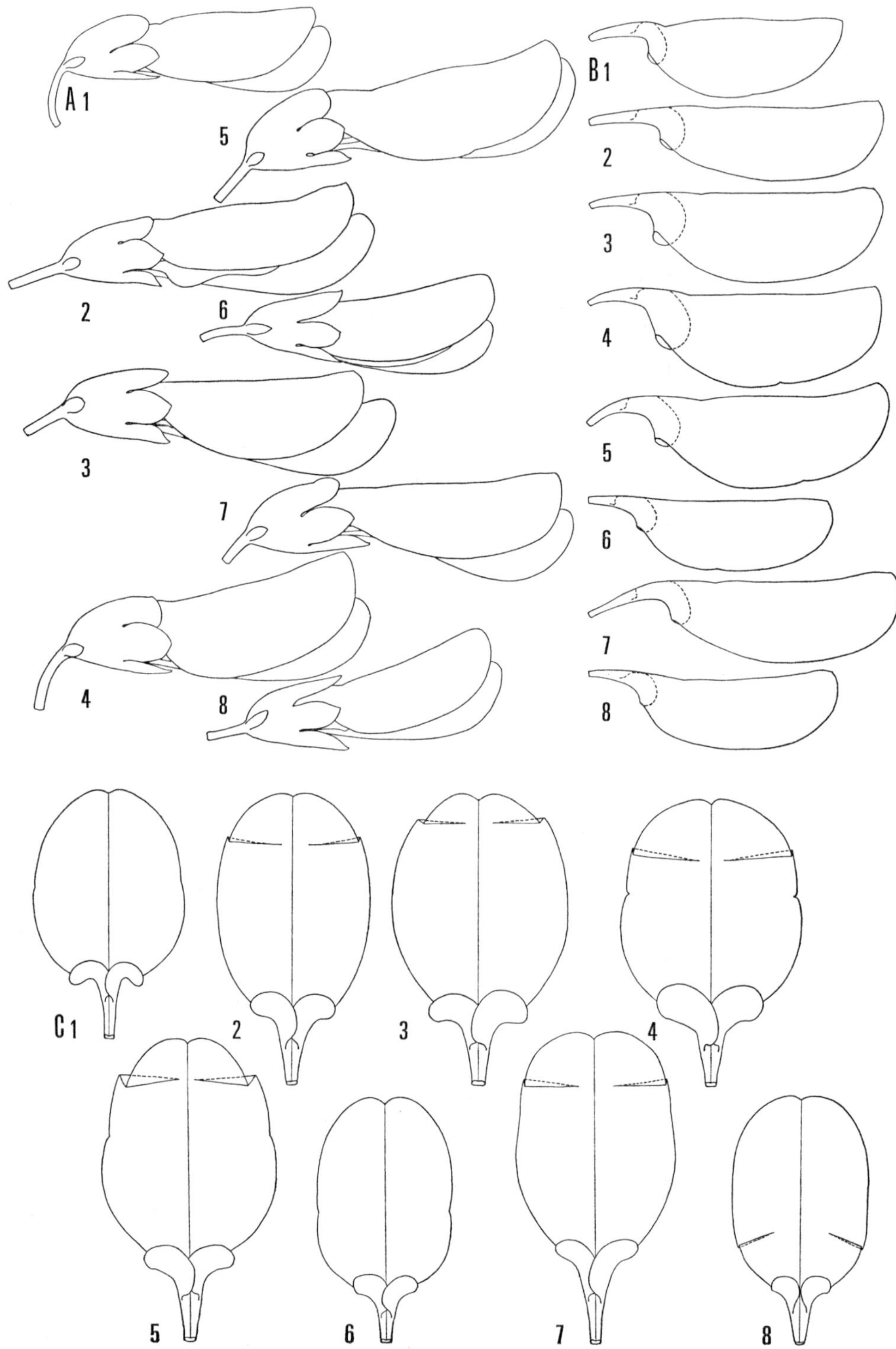

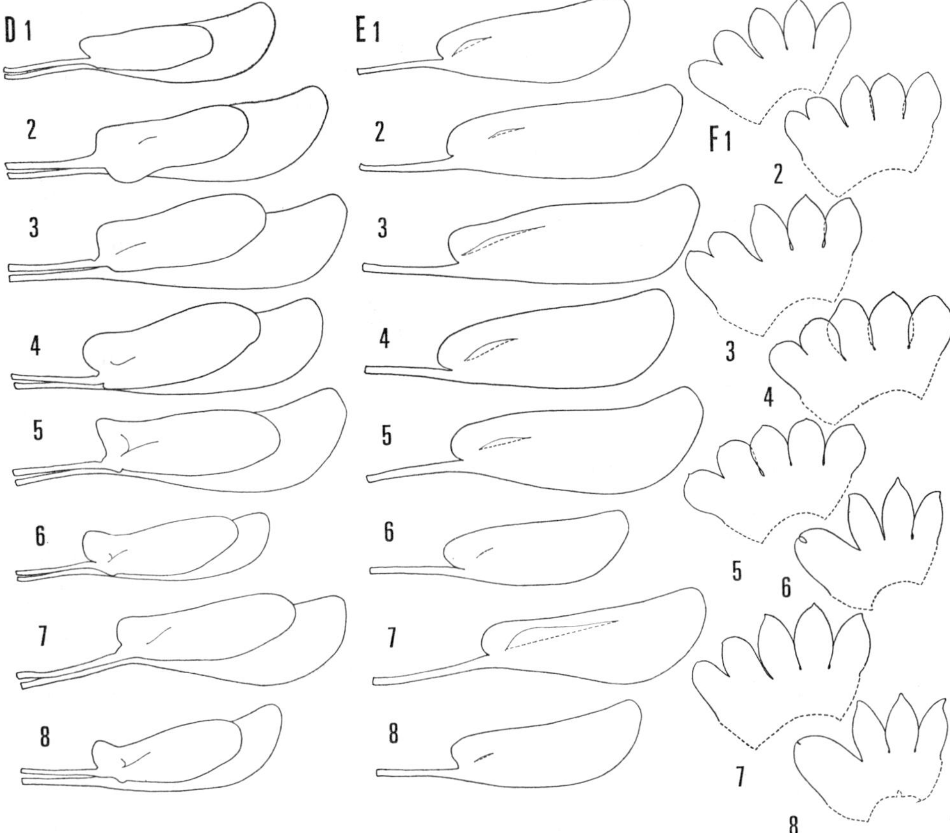

Fig. 14–2.  Individual variation in a population of *L. homoloba*.
D, wings;   E, keel-petals;   F, calyces, dissected.   All×3.3.

←Fig. 14–1.   Individual variation in a population of *L. homoloba*.
A, flowers, lateral view;   B, standards, lateral view;   C, standards,
opened.   All×3.3.

Table 13. Phenotypic plasticity and individual variation of *L. homoloba*.

| | Character | Phenotypic plasticity | Individual variation |
|---|---|---|---|
| Calyx | shape of lobe | broadly elliptic | broadly elliptic |
| | shape of apex | obtuse with a point | obtuse with a point |
| | length ratio of lateral lobe to tube | $0.77 \pm 0.09$ | $0.96 \pm 0.11$ |
| | ratio of connated part of upper lobes | 1/2–9/10 | 1/2–9/10 |
| Corolla | color | red-purple | red-purple |
| Length ratio | wing to keel-petal | $0.79 \pm 0.03$ | $0.82 \pm 0.05$ |
| | standard to keel-petal | $0.99 \pm 0.03$ | $0.99 \pm 0.03$ |
| | wing to standard | $0.80 \pm 0.02$ | $0.83 \pm 0.03$ |
| Standard | shape | elliptic with distinct claw | elliptic–broadly obovate with distinct claw |
| | shape of apex | retuse | retuse with a point or without a point |
| | shape of auricle | reniform | reniform |
| | length ratio of claw to standard | $0.21 \pm 0.03$ | $0.23 \pm 0.03$ |
| Wing | shape of lamina | narrowly oblong | narrowly oblong |
| | shape of upper base | truncate-auriculate | truncate-auriculate |
| | shape of lower base | minutely auriculate-cordate | minutely auriculate-cordate-tapering |
| | length ratio of claw to wing | $0.35 \pm 0.03$ | $0.37 \pm 0.02$ |
| Keel-petal | shape of lamina | narrowly obovate | narrowly obovate |
| | shape of upper base | cordate-truncate | cordate-truncate |
| | shape of lower base | tapering | tapering |
| | length ratio of claw to keel-petal | $0.28 \pm 0.02$ | $0.33 \pm 0.03$ |

minutely auriculate, cordate, or tapering (Fig. 14D). The ratio of the length of the wing-claw to the wing itself is $0.37 \pm 0.02$. The upper base of the keel-petal-lamina is cordate or truncate (Fig. 14E). The ratio of the length of the keel-petal-claw to the keel-petal is $0.33 \pm 0.03$.

The shape of the lateral calyx-lobe is broadly elliptic to elliptic (Fig. 14C). Two upper calyx-lobes are connated 1/2–9/10 of the way along its length. The length of the calyx-lobes varies from shorter than the tube to nearly the same. The ratio of the length of the lateral calyx-lobe to the tube is $0.96 \pm 0.11$.

Collective variation: Flowers from various localities were illustrated (Fig. 15) and measured (Table 14). The expressions of the characters investigated scarcely exceed the variation range of those in the single population mentioned above. The relative length among various characters is slightly different from that in the single population. However, the shape of the standard was much variable, from elliptic to obovate to broadly obovate (to ovate) (Fig. 15C). The upper and lower bases of wing also have increased variation in shape (Fig. 15D). The shape of the lateral calyx-lobe varies from broadly elliptic to elliptic. The apex of the lateral calyx-lobes is obtuse with or without a point (Fig. 15F).

*Lespedeza formosa* (Vogel) Koehne subsp. *velutina* (Nakai)
S. Akiyama et H. Ohba

Phenotypic plasticity: Ten flowers were illustrated (Fig. 16) and measured (Table

Table 14.  Collective variation of *L. homoloba*.

| | F.1 | C.1 | CT.1 | CL.1/CT.1 | S.1 | SC.1 | SC.1/S.1 | W.1 | WC.1 | WC.1/W.1 | K.1 | KC.1 | KC.1/K.1 | W.1/K.1 | S.1/K.1 | W.1/S.1 |
|---|---|---|---|---|---|---|---|---|---|---|---|---|---|---|---|---|
| 1 | 11.2 | 3.6 | 1.9 | 0.89 | 9.9 | 2.0 | 0.20 | 8.1 | 3.1 | 0.38 | 10.4 | 3.2 | 0.31 | 0.78 | 0.95 | 0.82 |
| 2 | 12.2 | 4.0 | 2.1 | 0.90 | 11.7 | 2.2 | 0.19 | 9.9 | 3.5 | 0.35 | 11.8 | 3.6 | 0.31 | 0.84 | 0.99 | 0.85 |
| 3 | 10.6 | 3.9 | 1.9 | 1.05 | 9.5 | 2.1 | 0.22 | 7.8 | 3.0 | 0.38 | 10.2 | 3.1 | 0.30 | 0.76 | 0.93 | 0.82 |
| 4 | 14.4 | 4.4 | 2.4 | 0.83 | 13.1 | 2.9 | 0.22 | 11.0 | 3.9 | 0.35 | 13.5 | 3.9 | 0.29 | 0.81 | 0.97 | 0.84 |
| 5 | 11.1 | 3.7 | 1.9 | 0.95 | 10.4 | 2.2 | 0.21 | 8.5 | 3.2 | 0.38 | 10.1 | 3.2 | 0.32 | 0.84 | 1.03 | 0.82 |
| 6 | 10.3 | 3.8 | 1.9 | 1.00 | 9.9 | 2.2 | 0.22 | 9.1 | 3.3 | 0.36 | 9.7 | 3.2 | 0.33 | 0.94 | 1.02 | 0.92 |
| 7 | 9.7 | 3.7 | 1.9 | 0.95 | 8.4 | 1.4 | 0.17 | 7.8 | 2.4 | 0.31 | 8.9 | 2.4 | 0.27 | 0.88 | 0.94 | 0.93 |
| 8 | 11.8 | 3.8 | 1.9 | 1.00 | 11.0 | 2.5 | 0.23 | 9.3 | 3.2 | 0.34 | 11.3 | 3.4 | 0.30 | 0.82 | 0.97 | 0.85 |
| 9 | 12.0 | 3.9 | 1.9 | 1.05 | 11.0 | 2.5 | 0.23 | 9.1 | 3.1 | 0.34 | 10.8 | 3.3 | 0.31 | 0.84 | 1.02 | 0.83 |
| 10 | 11.7 | 4.3 | 1.8 | 1.39 | 11.2 | 2.5 | 0.22 | 9.8 | 3.5 | 0.36 | 10.8 | 3.4 | 0.31 | 0.91 | 1.04 | 0.88 |
| Min. | 9.7 | 3.6 | 1.8 | 0.83 | 8.4 | 1.4 | 0.17 | 7.8 | 2.4 | 0.31 | 8.9 | 2.4 | 0.27 | 0.76 | 0.93 | 0.82 |
| Max. | 14.4 | 4.4 | 2.4 | 1.39 | 13.1 | 2.9 | 0.23 | 11.0 | 3.9 | 0.38 | 13.5 | 3.9 | 0.33 | 0.94 | 1.04 | 0.93 |
| Average | 11.5 | 3.9 | 2.0 | 1.00 | 10.6 | 2.3 | 0.21 | 9.0 | 3.2 | 0.36 | 10.8 | 3.3 | 0.30 | 0.84 | 0.99 | 0.85 |
| S.D. | 1.29 | 0.26 | 0.17 | 0.15 | 1.30 | 0.40 | 0.02 | 1.03 | 0.39 | 0.02 | 1.26 | 0.39 | 0.02 | 0.05 | 0.04 | 0.04 |

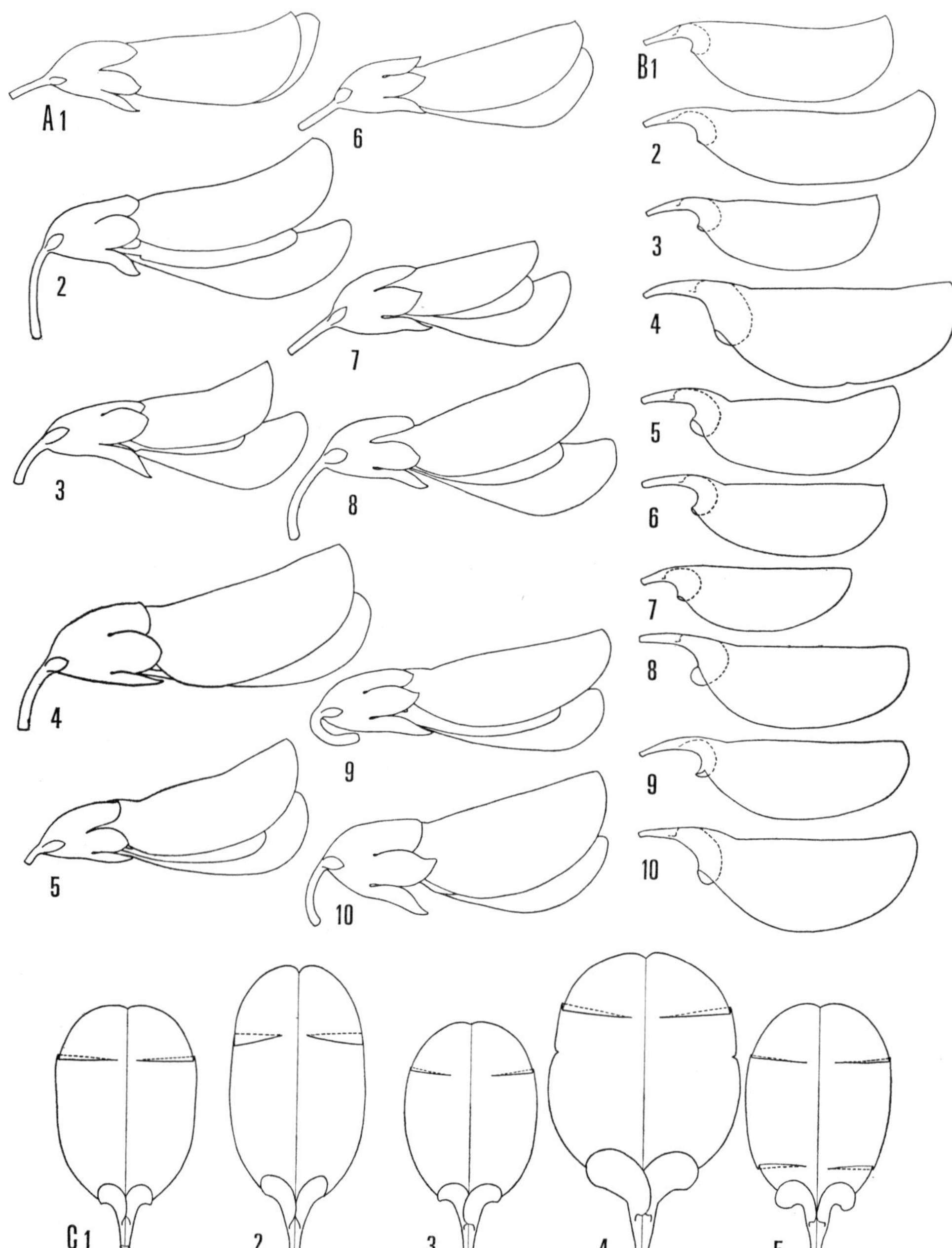

Fig. 15–1.  Variation of *L. homoloba*.

A, flowers, lateral view;  B, standards, lateral view;  C, standards, opend.  All×3.3.

1. Iwate (Toba, 5 Sept. 1932, TI).  2. Miyagi (Ohba & Akiyama 578, TI).  3. Ibaraki (Honda s.n., TI).  4. Tochigi (Ohba & Akiyama 1428, TI).  5. Ishikawa (Satomi s.n., KYO).  6. Nagano (Asano 21635, TI).  7. Kyoto (Togashi 10189, TI).  8. Shimane (Ohba & Akiyama 2108, TI).→

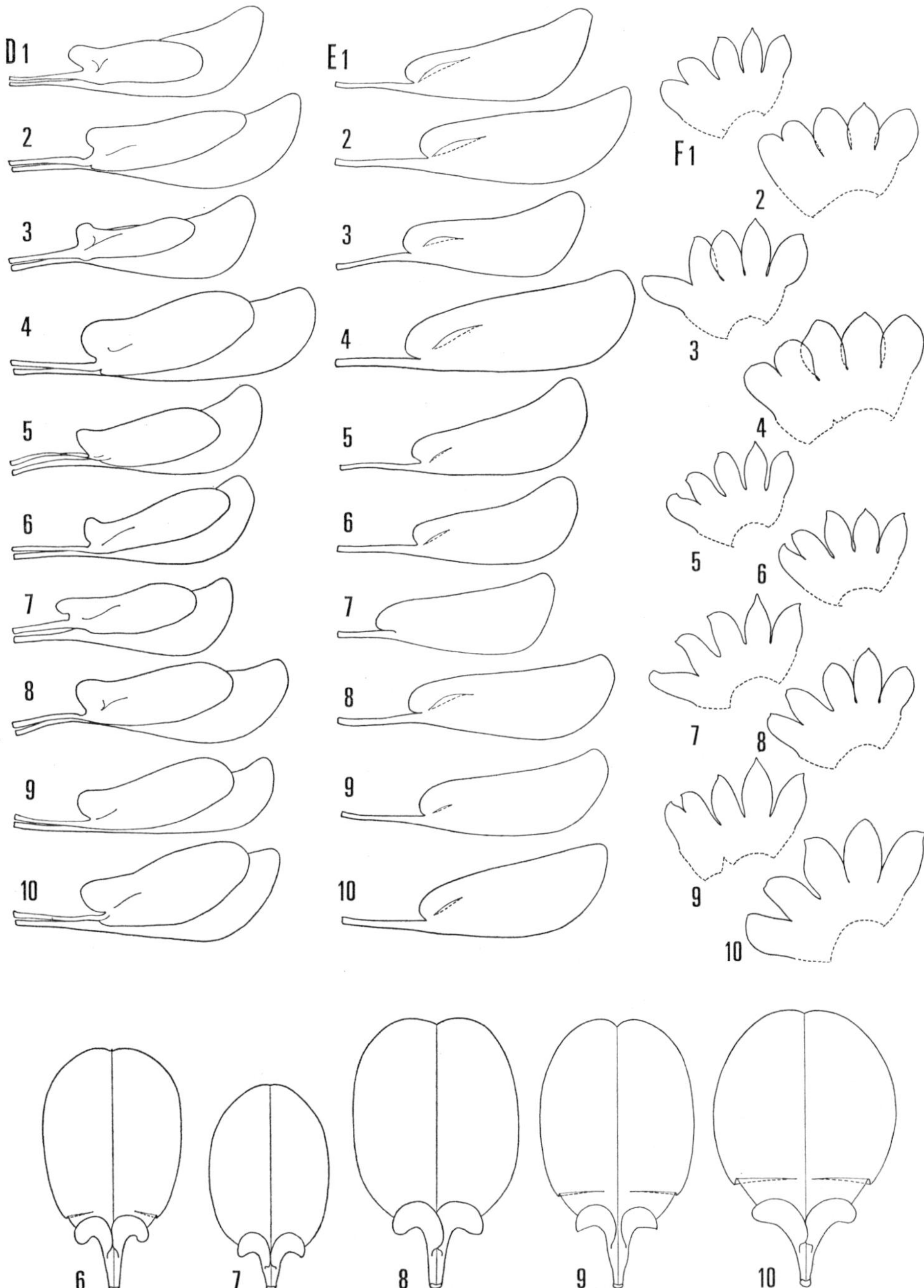

Fig. 15–2.  Variation of *L. homoloba*.
D, wings;  E, keel-petals;  F, calyces, dissected.  All×3.3.

9. Tokushima (Nikai 2574, TI—Lectotype of *L. homoloba* Nakai).  10. Kouchi (Ohashi, 20 June 1973, TI).

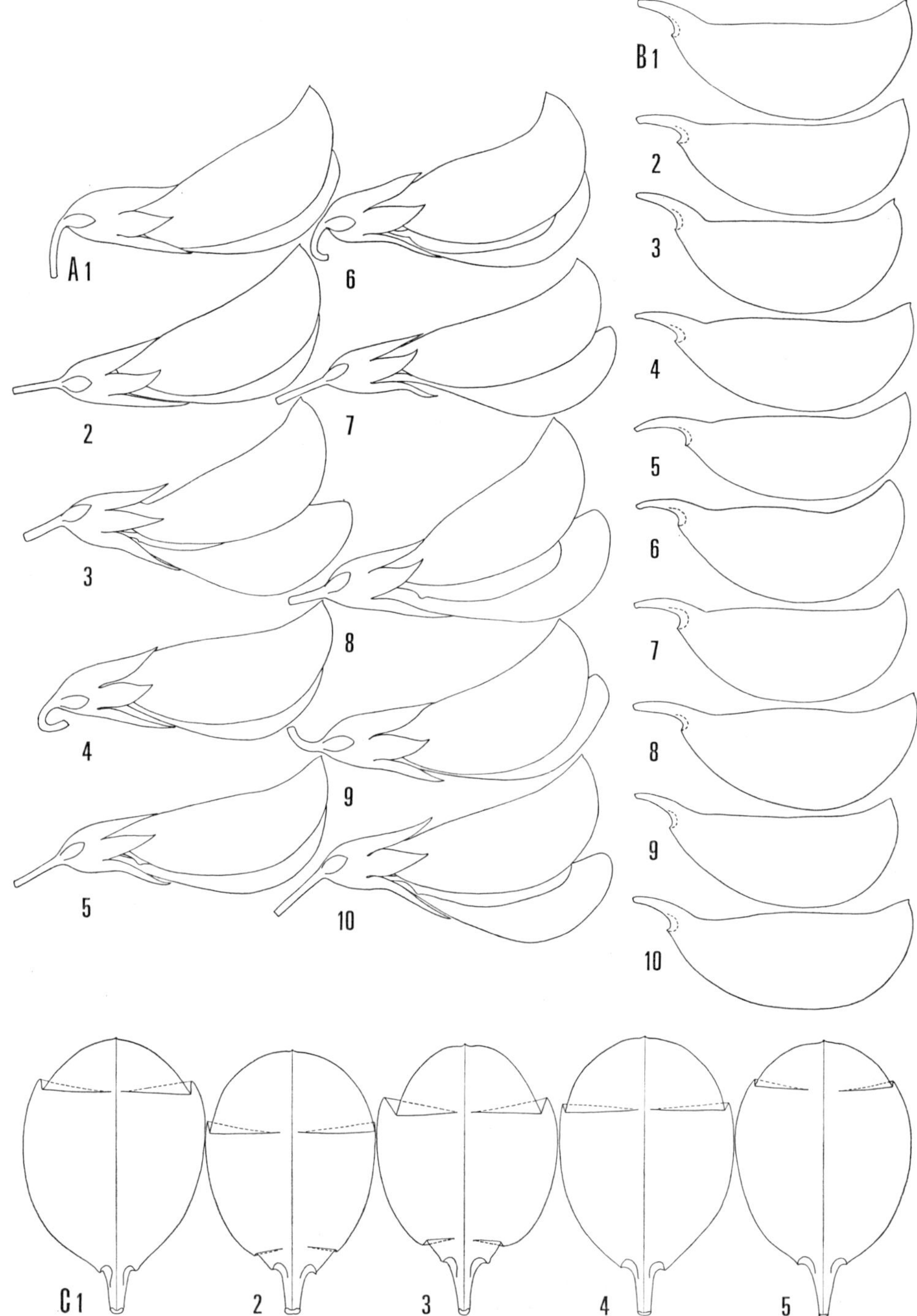

Fig. 16–1.  Phenotypic plasticity of *L. formosa* subsp. *velutina*.
A, flowers, lateral view;  B, standards, lateral view;  C, standards, opened.  All×3.3.

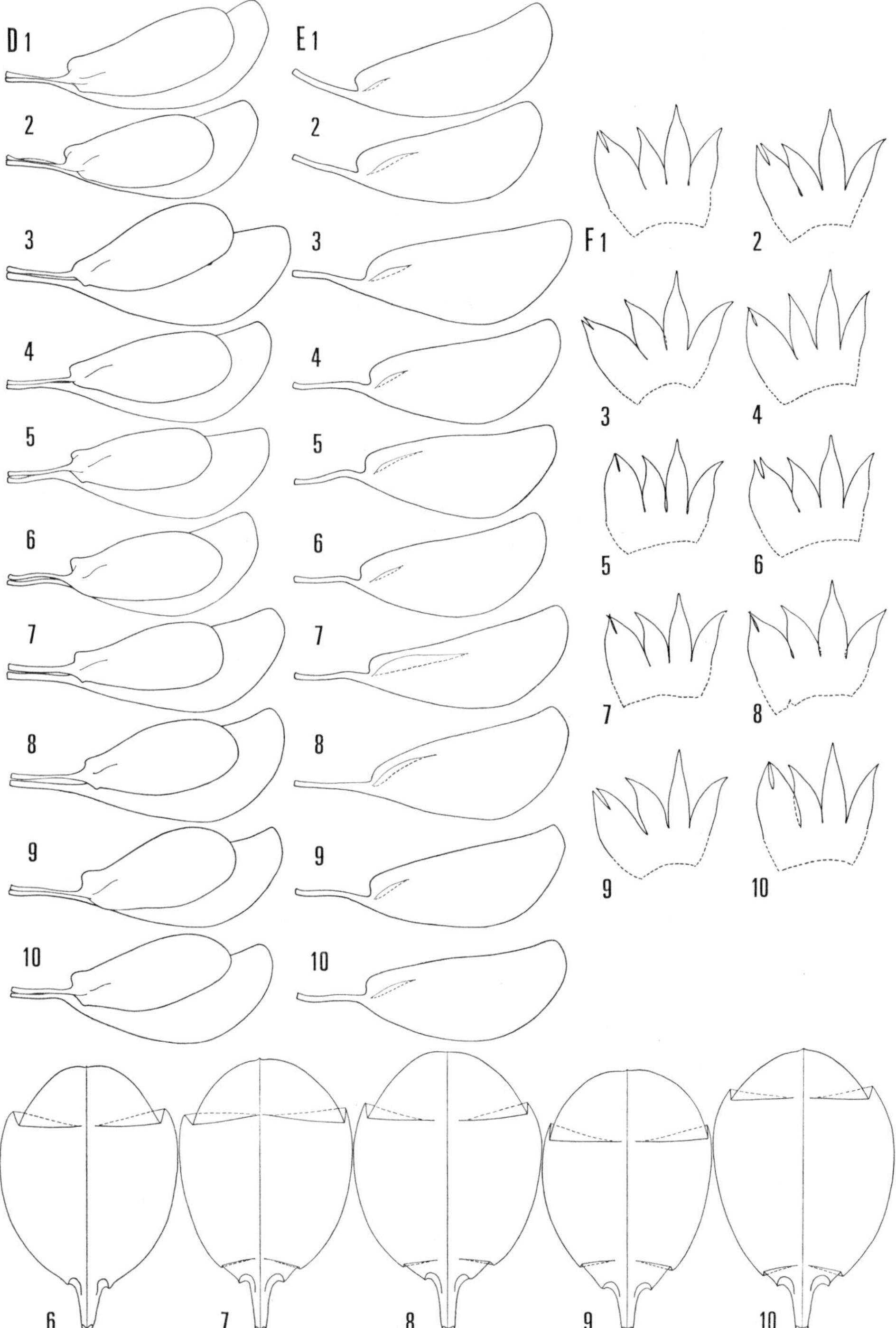

Fig. 16–2. Phenotypic plasticity of *L. formosa* subsp. *velutina*.
D, wings;   E, keel-petals;   F, calyces, dissected.   All × 3.3.

Table 15.   Phenotypic plasticity of *L. formosa* subsp. *velutina*.

| | F.1 | C.1 | CT.1 | CL.1/CT.1 | S.1 | SC.1 | SC.1/S.1 | W.1 | WC.1 | WC.1/W.1 | K.1 | KC.1 | KC.1/K.1 | W.1/K.1 | S.1/K.1 | W.1/S.1 |
|---|---|---|---|---|---|---|---|---|---|---|---|---|---|---|---|---|
| 1 | 12.0 | 4.7 | 2.1 | 1.24 | 11.5 | 1.8 | 0.16 | 9.4 | 2.7 | 0.29 | 10.8 | 2.8 | 0.26 | 0.87 | 1.06 | 0.82 |
| 2 | 11.4 | 4.0 | 1.5 | 1.67 | 11.1 | 1.7 | 0.15 | 9.1 | 2.8 | 0.31 | 10.4 | 2.8 | 0.27 | 0.88 | 1.07 | 0.82 |
| 3 | 12.2 | 4.3 | 1.6 | 1.69 | 11.3 | 1.8 | 0.16 | 9.7 | 2.9 | 0.30 | 11.4 | 3.1 | 0.27 | 0.85 | 0.99 | 0.86 |
| 4 | 12.0 | 4.4 | 1.9 | 1.32 | 11.5 | 1.9 | 0.17 | 9.0 | 2.8 | 0.31 | 11.5 | 3.2 | 0.28 | 0.78 | 1.00 | 0.78 |
| 5 | 11.8 | 4.3 | 1.8 | 1.39 | 11.0 | 1.9 | 0.17 | 8.6 | 2.5 | 0.29 | 10.1 | 2.8 | 0.28 | 0.85 | 1.09 | 0.78 |
| 6 | 11.8 | 4.1 | 1.7 | 1.41 | 11.5 | 2.1 | 0.18 | 8.8 | 2.9 | 0.33 | 11.1 | 3.0 | 0.27 | 0.79 | 1.04 | 0.77 |
| 7 | 12.2 | 4.5 | 1.9 | 1.37 | 11.7 | 2.1 | 0.18 | 9.7 | 3.0 | 0.31 | 11.5 | 3.2 | 0.28 | 0.84 | 1.02 | 0.83 |
| 8 | 12.1 | 4.4 | 1.6 | 1.75 | 11.7 | 1.8 | 0.15 | 9.5 | 2.9 | 0.31 | 11.1 | 3.2 | 0.29 | 0.86 | 1.05 | 0.81 |
| 9 | 12.1 | 4.6 | 2.0 | 1.30 | 11.0 | 1.7 | 0.15 | 9.4 | 2.8 | 0.30 | 11.0 | 3.1 | 0.28 | 0.85 | 1.00 | 0.85 |
| 10 | 12.0 | 4.4 | 1.7 | 1.59 | 11.9 | 2.0 | 0.17 | 9.4 | 2.7 | 0.29 | 11.3 | 2.9 | 0.26 | 0.83 | 1.05 | 0.79 |
| Min. | 11.4 | 4.0 | 1.5 | 1.24 | 11.0 | 1.7 | 0.15 | 8.6 | 2.5 | 0.29 | 10.1 | 2.8 | 0.26 | 0.78 | 0.99 | 0.77 |
| Max. | 12.2 | 4.7 | 2.1 | 1.75 | 11.9 | 2.1 | 0.18 | 9.7 | 3.0 | 0.33 | 11.5 | 3.2 | 0.29 | 0.88 | 1.09 | 0.86 |
| Average | 12.0 | 4.4 | 1.8 | 1.47 | 11.4 | 1.9 | 0.16 | 9.3 | 2.8 | 0.30 | 11.0 | 3.0 | 0.27 | 0.84 | 1.04 | 0.81 |
| S.D. | 0.24 | 0.21 | 0.19 | 0.18 | 0.31 | 0.15 | 0.01 | 0.37 | 0.14 | 0.01 | 0.47 | 0.17 | 0.01 | 0.03 | 0.03 | 0.03 |

15). The standard is nearly equal to or slightly longer than the keel-petal (Fig. 16A). The keel-petal is longer than the wing (Fig. 16D). The ratio of the length of the standard to the keel-petal is $1.04 \pm 0.03$, wing to standard $0.81 \pm 0.03$, and wing to keel-petal $0.84 \pm 0.03$ (Table 15).

The standard has a distinct claw; the shape of the lamina of the standard is obovate, the apex is retuse with a point, and the shape of the auricle at the base of the lamina is narrowly lunate (Fig. 16B & C). The ratio of the length of the standard-claw to the standard itself is $0.16 \pm 0.01$. The shape of the lamina of the wing is narrowly obovate with a slightly auriculate or cordate upper base and a minutely auriculate or tapering lower base (Fig. 16D). The ratio of the length of the wing-claw to the wing itself is $0.30 \pm 0.01$. The shape of the lamina of the keel-petal is narrowly obovate with a cordate upper and tapering lower base (Fig. 16E). The ratio of the length of the keel-petal-claw to the length of the keel-petal is $0.27 \pm 0.01$.

The shape of the lateral calyx-lobe is triangular-lanceolate with an acute apex (Fig. 16F). Two upper calyx-lobes are connated 2/3–3/4 of the way along its length. The calyx-lobes are apparently longer than the tube, and the ratio of the length of the lateral calyx-lobe to the tube is $1.47 \pm 0.18$.

Individual variation: A population consisting of eight blooming individuals was investigated (Fig. 17 and Table 16). As in the former cases almost all of the characters with constant expressions in phenotypic plasticity are stable within this population, but some characters become slightly more variable (Table 17).

The standard is slightly longer or, rarely, slightly shorter than the keel-petal (Fig. 17A). The ratio of the length of the standard to the keel-petal is $1.05 \pm 0.04$, wing to standard $0.80 \pm 0.05$, and wing to keel-petal $0.84 \pm 0.06$.

The shape of the auricle of the standard varies from narrowly lunate to lunate (Fig. 17C). The ratio of the length of the standard-claw to the standard itself is $0.18 \pm 0.02$. The shape of the upper base of the wing lamina is auriculate or truncate (Fig. 17D). The ratio of the length of the claw to the wing itself is $0.32 \pm 0.03$. The shape of the upper base of the lamina of the keel-petal is cordate or truncate (Fig. 17E). The ratio of the length of the claw to the keel-petal itself is $0.28 \pm 0.02$.

The shape of the lateral calyx-lobe is triangular to triangular-lanceolate (Fig. 17F). Two upper calyx-lobes are connated 1/3 to the entire way along its length. The ratio of the length of the lateral calyx-lobe to the tube is $1.57 \pm 0.11$.

Collective variation: Flowers from various localities were illustrated (Fig. 18) and measured (Table 18). The expressions of the characters investigated do not exceed the range of variation of those in the single population mentioned above. The relative length among various characters is slightly different from that in the single population (Table 16); in particular, the relative length between the standard and the keel-petal becomes variable: the former is slightly longer than, nearly equal to, or slightly shorter than the latter.

The lamina of standard varies from obovate to elliptic (Fig. 18C). The shape of the lamina base of wing is rather variable. The shape of the lamina of the keel-petal is narrowly obovate with a tapering lower base (Fig. 18E). The shape of the upper base of the keel-petal-lamina becomes variable.

50

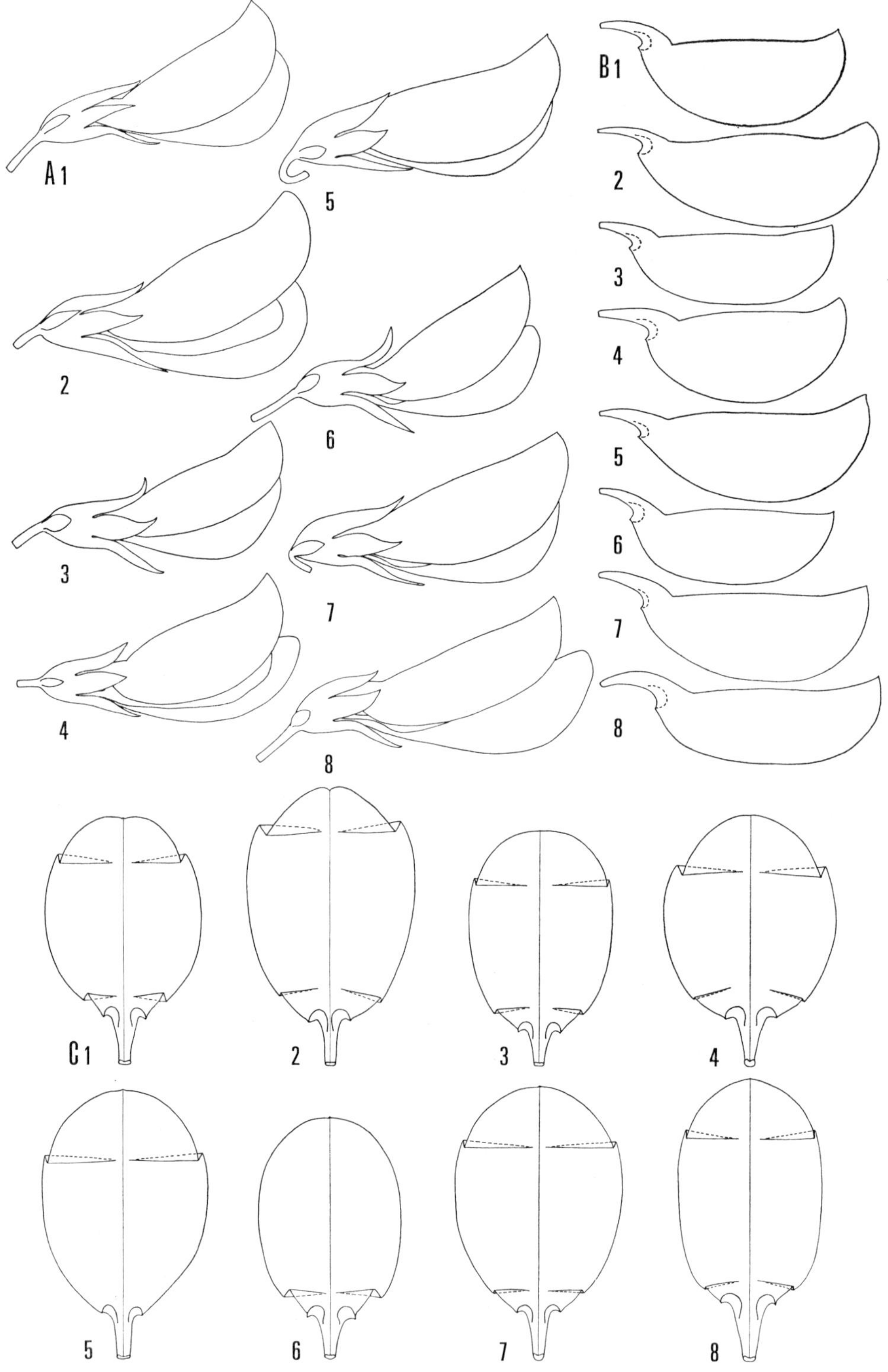
A 1
2
3
4
5
6
7
8
B 1
2
3
4
5
6
7
8
C 1
2
3
4
5
6
7
8

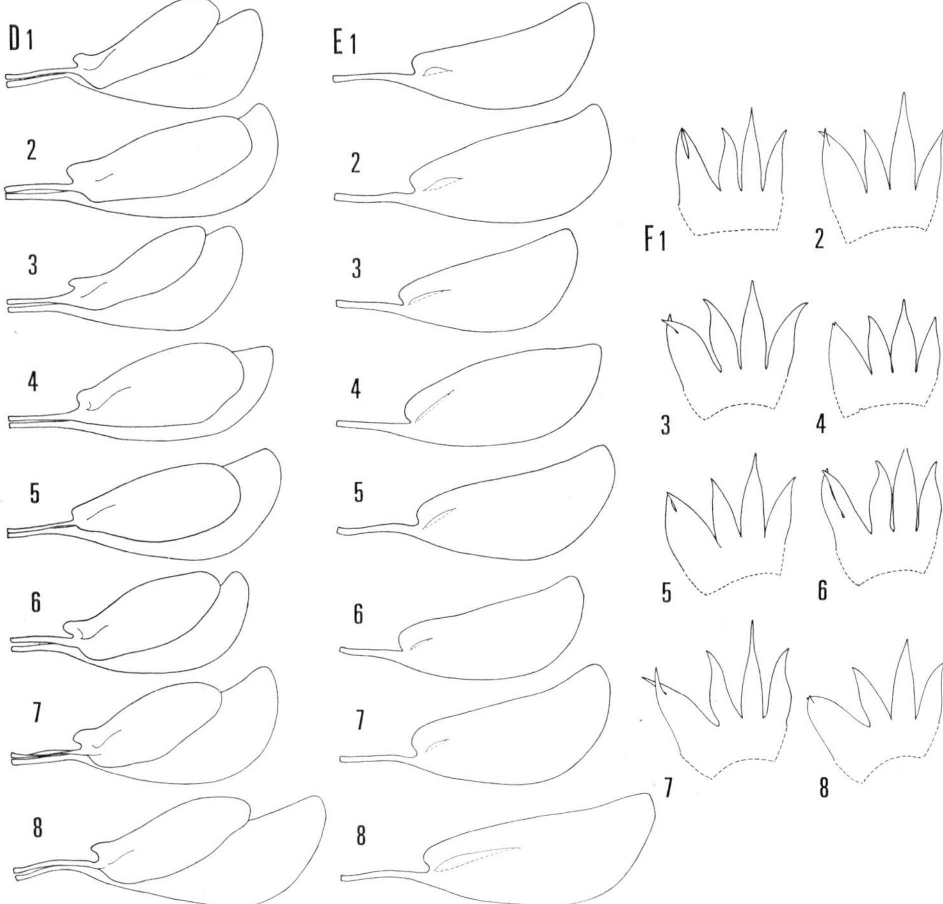

Fig. 17–2.   Individual variation in a population of *L. formosa* subsp. *velutina*.
D, wings;   E, keel-petals;   F, calyces, dissected.   All×3.3.

←Fig. 17–1.   Individual variation in a population of *L. formosa* subsp. *velutina*.
A, flowers, lateral view;   B, standards, lateral view;   C, standards,
opened.   All×3.3.

Table 16.  Individual variation of *L. formosa* subsp. *velutina*.

| | F.1 | C.1 | CT.1 | CL.1/CT.1 | S.1 | SC.1 | SC.1/S.1 | W.1 | WC.1 | WC.1/W.1 | K.1 | KC.1 | KC.1/K.1 | W.1/K.1 | S.1/K.1 | W.1/S.1 |
|---|---|---|---|---|---|---|---|---|---|---|---|---|---|---|---|---|
| 1 | 11.2 | 4.3 | 1.7 | 1.53 | 10.7 | 2.0 | 0.19 | 8.6 | 3.0 | 0.35 | 10.5 | 3.3 | 0.31 | 0.82 | 1.02 | 0.80 |
| 2 | 12.6 | 4.3 | 1.7 | 1.53 | 12.0 | 2.0 | 0.17 | 9.9 | 2.7 | 0.27 | 11.0 | 3.2 | 0.29 | 0.90 | 1.09 | 0.83 |
| 3 | 11.1 | 5.1 | 2.0 | 1.55 | 10.1 | 1.5 | 0.15 | 8.1 | 2.8 | 0.35 | 9.4 | 2.7 | 0.29 | 0.86 | 1.07 | 0.80 |
| 4 | 11.4 | 4.0 | 1.6 | 1.50 | 10.9 | 2.1 | 0.19 | 9.7 | 2.9 | 0.30 | 10.5 | 2.9 | 0.28 | 0.92 | 1.04 | 0.89 |
| 5 | 12.4 | 4.1 | 1.7 | 1.41 | 11.5 | 1.8 | 0.16 | 8.8 | 2.9 | 0.33 | 11.1 | 3.0 | 0.27 | 0.79 | 1.04 | 0.77 |
| 6 | 11.1 | 4.6 | 1.8 | 1.56 | 10.4 | 1.8 | 0.17 | 8.6 | 2.5 | 0.29 | 9.8 | 2.5 | 0.26 | 0.88 | 1.06 | 0.83 |
| 7 | 12.5 | 4.8 | 1.8 | 1.67 | 11.8 | 2.1 | 0.18 | 8.8 | 2.8 | 0.32 | 10.9 | 3.1 | 0.28 | 0.81 | 1.08 | 0.75 |
| 8 | 12.8 | 3.9 | 1.4 | 1.79 | 12.2 | 2.5 | 0.21 | 9.2 | 3.1 | 0.34 | 12.6 | 3.5 | 0.28 | 0.73 | 0.97 | 0.75 |
| Min. | 11.1 | 3.9 | 1.4 | 1.41 | 10.1 | 1.5 | 0.15 | 8.1 | 2.5 | 0.27 | 9.4 | 2.5 | 0.26 | 0.73 | 0.97 | 0.75 |
| Max. | 12.8 | 5.1 | 2.0 | 1.79 | 12.2 | 2.5 | 0.20 | 9.9 | 3.1 | 0.35 | 12.6 | 3.5 | 0.31 | 0.92 | 1.09 | 0.89 |
| Average | 11.9 | 4.4 | 1.7 | 1.57 | 11.2 | 2.0 | 0.18 | 9.0 | 2.8 | 0.32 | 10.7 | 3.0 | 0.28 | 0.84 | 1.05 | 0.80 |
| S.D. | 0.75 | 0.42 | 0.17 | 0.11 | 0.78 | 0.29 | 0.02 | 0.60 | 0.18 | 0.03 | 0.96 | 0.32 | 0.02 | 0.06 | 0.04 | 0.05 |

Table 17. Phenotypic plasticity and individual variation of *L. formosa* subsp. *velutina*.

| | Character | Phenotypic plasticity | Individual variation |
|---|---|---|---|
| Calyx | shape of lobe | triangular | triangular |
| | shape of apex | acute | acute |
| | length ratio of lateral lobe to tube | $1.47\pm0.18$ | $1.57\pm0.11$ |
| | ratio of connated part of upper lobes | 2/3–3/4 | 1/3–1 |
| Corolla | color | red-purple | red-purple |
| Length ratio | wing to keel-petal | $0.84\pm0.03$ | $0.84\pm0.06$ |
| | standard to keel-petal | $1.04\pm0.03$ | $1.05\pm0.04$ |
| | wing to standard | $0.81\pm0.03$ | $0.80\pm0.05$ |
| Standard | shape | obovate with distinct claw | obovate with distinct claw |
| | shape of apex | retuse with a point | retuse with a point or truncate |
| | shape of auricle | narrowly lunate | narrowly lunate–lunate |
| | length ratio of claw to standard | $0.16\pm0.01$ | $0.18\pm0.02$ |
| Wing | shape of lamina | narrowly obovate | narrowly obovate |
| | shape of upper base | slightly auriculate-cordate | auriculate-truncate |
| | shape of lower base | minutely auriculate-tapering | slightly auriculate-tapering |
| | length ratio of claw to wing | $0.30\pm0.01$ | $0.32\pm0.03$ |
| Keel-petal | shape of lamina | narrowly obovate | narrowly obovate |
| | shape of upper base | cordate-truncate | cordate-truncate |
| | shape of lower base | tapering | tapering |
| | length ratio of claw to keel-petal | $0.27\pm0.01$ | $0.28\pm0.02$ |

## *Lespedeza patens* Nakai

Phenotypic plasticity: Ten flowers were investigated (Fig. 19 and Table 19). The standard is shorter than the keel-petal (Fig. 19A), and the wing is shorter than the keel-petal and the standard (Fig. 19D). The ratio of the length of the standard to the keel-petal is $0.91\pm0.03$, wing to standard $0.72\pm0.01$, and wing to keel-petal $0.66\pm0.02$.

The standard has a distinct claw; the shape of the lamina is obovate and the apex is truncate with a point (Fig. 19C & D). The shape of the auricle is lunate. The ratio of the length of the claw to the standard itself is $0.13\pm0.01$. The shape of the lamina of the wing is obovate with an auriculate or cordate upper base and a minutely auriculate or cordate lower base (Fig. 19E). The ratio of the length of the claw to the wing itself is $0.34\pm0.01$. The shape of the lamina of the keel-petal is narrowly obovate with a truncate or cordate upper apex and tapering lower base (Fig. 19F). The ratio of the length of the claw to the keel-petal itself is $0.23\pm0.01$.

The shape of the lateral calyx-lobe is narrowly triangular with an acute apex in all flowers (Fig. 19G). Two upper calyx-lobes are connated 2/3–7/8 of the way along its length. The calyx-lobes are apparently longer than the tube; the ratio of the length of the lateral calyx-lobe to the tube is $1.56\pm0.21$.

Individual variation: The population studied consisted of nine blooming individuals (Fig. 20 and Table 20). As in previous cases almost all characters with constant ex-

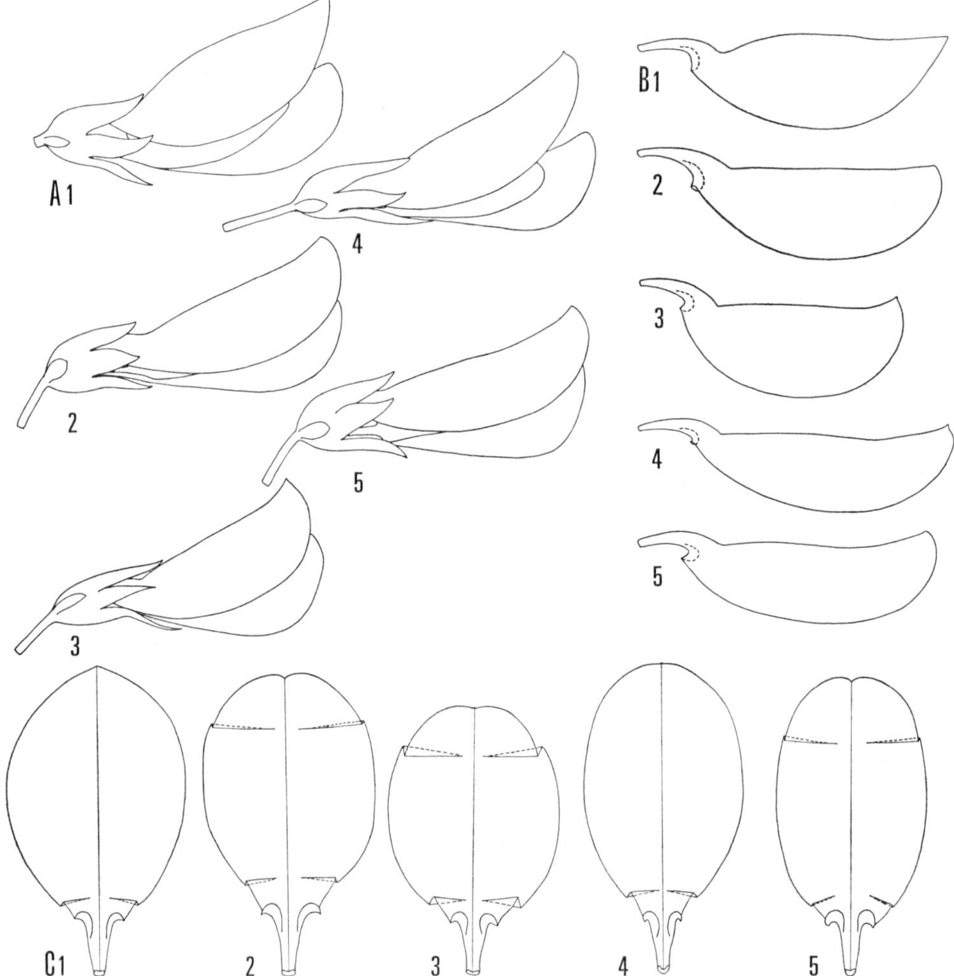

Fig. 18-1.   Variation of *L. formosa* subsp. *velutina*.
A, flowers, lateral view;   B, standards, lateral view;   C, standards, opened.   All×3.3.
1. Aichi (Torii, 3 Oct. 1965, TI).   2. Shiga (Murata 19030, TI). Okayama (Ohba & Akiyama 650, TI).   4. Okayama (Nikai 1072, TNS—Isotype of *L. kiusiana* Nakai).   5. Fukuoka (Ohba & Akiyama 2242, TI).

pressions in phenotypic plasticity are also stable within this population, although some become slightly variable (Table 21). The ratio of the length of the standard to the keel-petal is 0.93±0.03, wing to standard 0.74±0.04, and wing to keel-petal 0.69±0.02 (Table 21).

In the standard, the shape of the lamina varies from obovate to elliptic, the apex varies from truncate with a point to obtuse, and the auricle varies from lunate to narrowly lunate (Fig. 20B & C). The ratio of the length of the claw to the standard itself is 0.12±0.01. The shape of the base of the wing lamina is also variable (Fig. 20E).

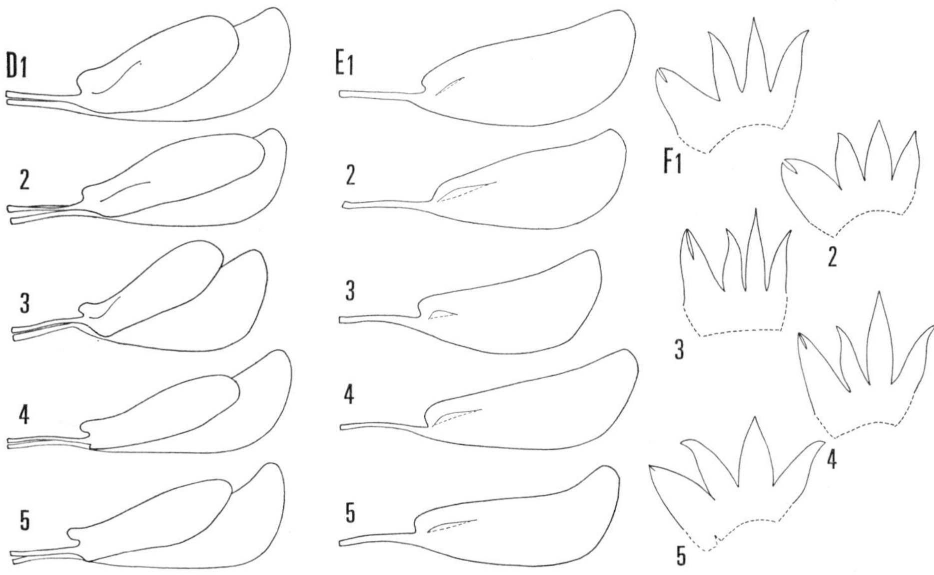



Fig. 18–2.   Variation of *L. formosa* subsp. *velutina*.
D, wings;   E, keel-petals;   F, calyces, dissected.   All×3.3.

The ratio of the length of the claw to the wing itself is 0.32±0.06. The shape of the upper base of the lamina of the keel-petal varies from cordate to truncate (Fig. 20F), and the ratio of the length of the claw to the keel-petal itself is 0.21±0.02. Two upper calyx-lobes are connated 2/3–7/8 of the way along its length (Fig. 20G). The ratio of the length of the lateral calyx-lobe to the tube is 1.95±0.43.

Collective variation: Flowers from various localities were illustrated (Fig. 21) and measured (Table 22). The expressions of the characters investigated never more than slightly exceed the range of variation of those in the single population mentioned above. The relative length among various characters is slightly different from that in the population. It is remarkable that the ratio of the length of the claw to the petal itself is 0.12±0.01 in the standard, 0.32±0.06 in the wing, and 0.21±0.02 in the keel-petal. The ratio of the length of the lateral calyx-lobe to the tube is 1.86±0.12.

### *Lespedeza Buergeri* Miquel

Phenotypic plasticity: Ten flowers were illustrated (Fig. 22) and measured (Table 23). The standard is shorter than the keel-petal (Fig. 22A). The wing is shorter than the keel-petal and nearly equal to or slightly longer than the standard (Fig. 22A & D). The ratio of the length of the standard to the keel-petal is 0.87±0.03, wing to standard 1.01±0.02, and wing to keel-petal 0.87±0.03.

The standard has a distinct claw; the shape of the lamina is round; the apex is retuse with a point (Fig. 22B & C). The shape of the auricle is broadly lunate. The ratio of the length of the claw to the standard itself is 0.16±0.01. The shape of the lamina of the wing is narrowly oblong with an auriculate upper base and a minutely auriculate or tapering lower base in all flowers (Fig. 22D). The ratio of the length of

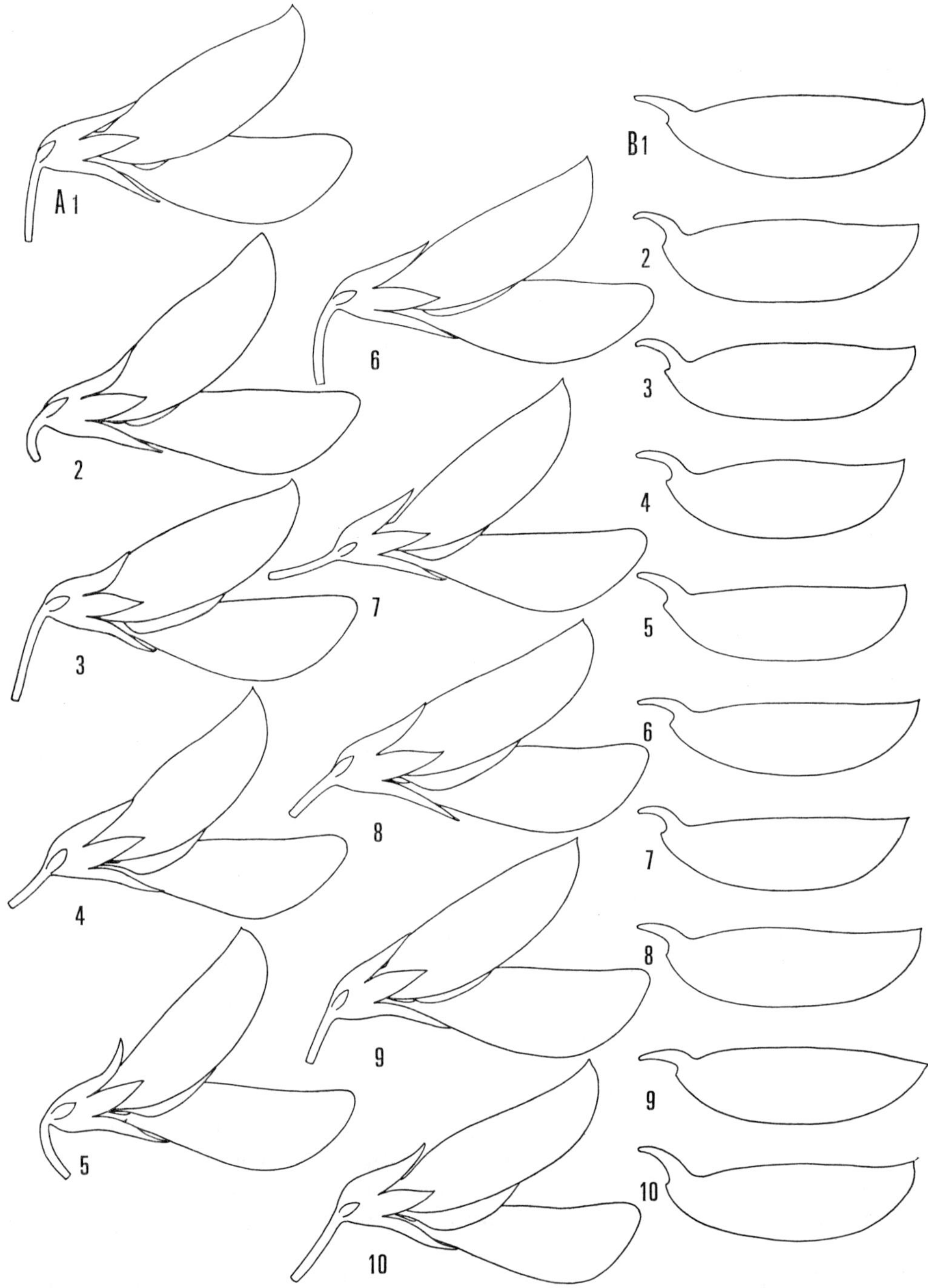

Fig. 19–1.　Phenotypic plasticity of *L. patens*.
A, flowers, lateral view;　B, standards, lateral view.　All ×3.3.

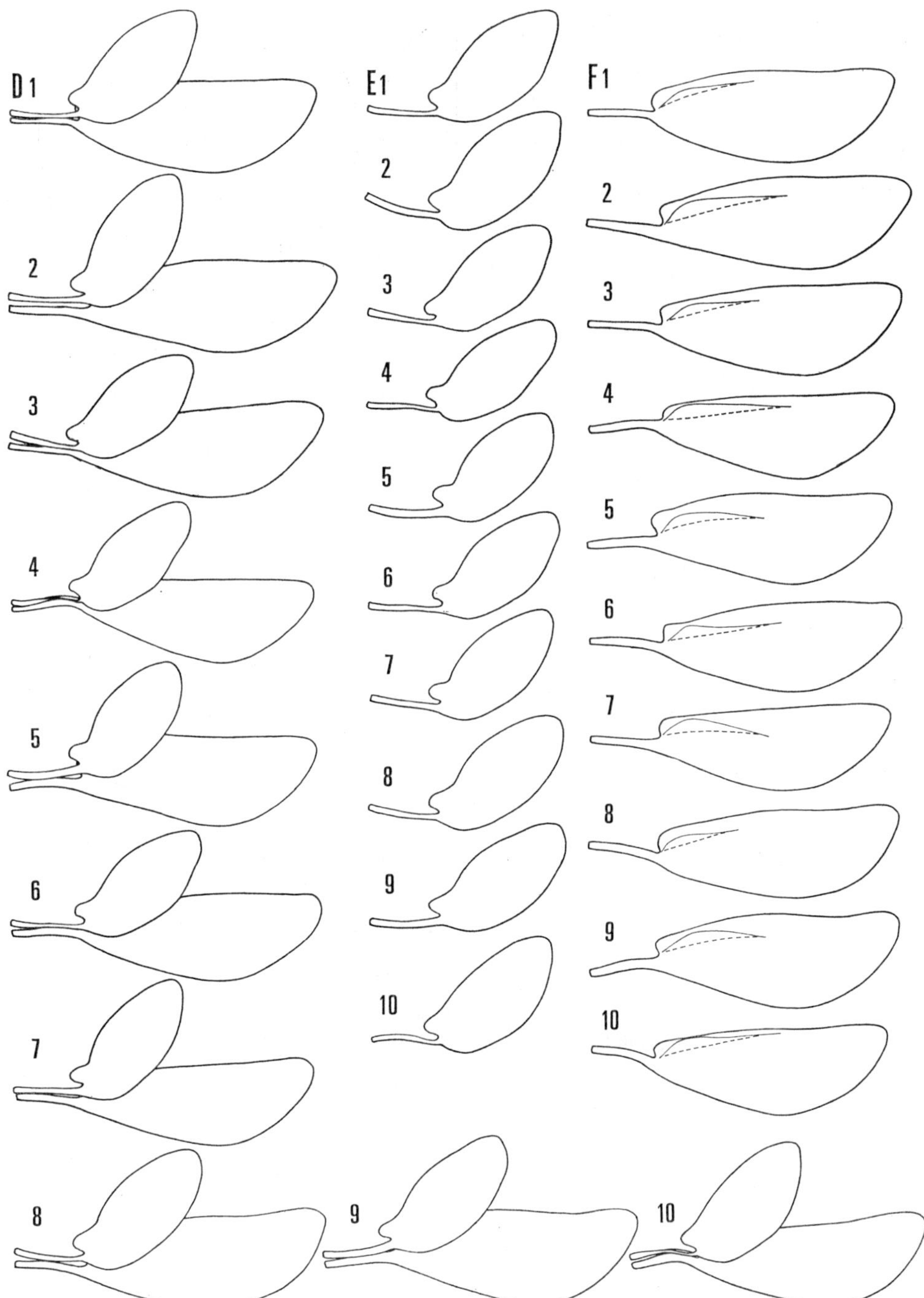

Fig. 19–2.   Phenotypic plasticity of *L. patens*.
D, wings and keel-petals;   E, wings;   F, keel-petals.   All × 3.3.

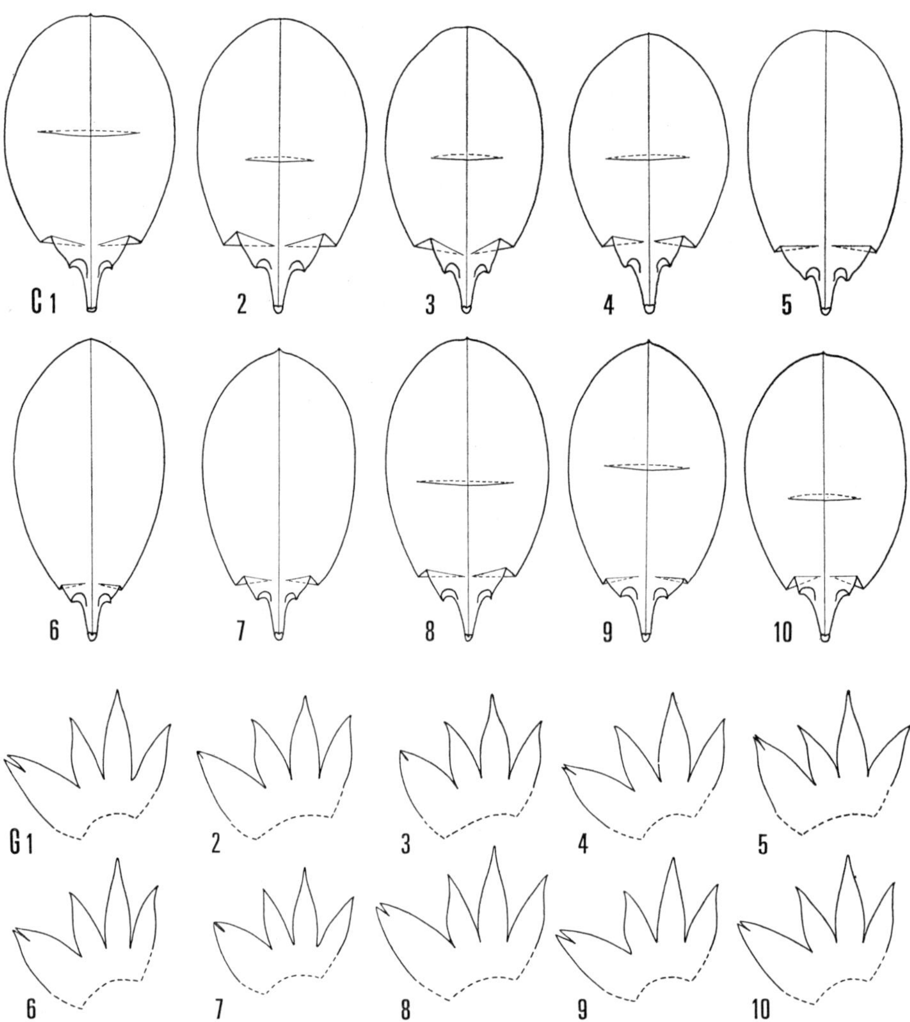

Fig. 19–3.   Phenotypic plasticity of *L. patens*.
C, standards, opened;   G, calyces, dissected.   All×3.3.

Table 18. Collective variation of *L. formosa* subsp. *velutina*.

| | F.1 | C.1 | CT.1 | CL.1/CT.1 | S.1 | SC.1 | SC.1/S.1 | W.1 | WC.1 | WC.1/W.1 | K.1 | KC.1 | KC.1/K.1 | W.1/K.1 | S.1/K.1 | W.1/S.1 |
|---|---|---|---|---|---|---|---|---|---|---|---|---|---|---|---|---|
| 1 | 12.8 | 4.3 | 1.6 | 1.69 | 12.4 | 2.2 | 0.18 | 9.6 | 3.2 | 0.33 | 11.7 | 3.4 | 0.29 | 0.82 | 1.06 | 0.77 |
| 2 | 12.7 | 3.9 | 1.7 | 1.29 | 12.1 | 2.4 | 0.20 | 10.1 | 3.1 | 0.31 | 11.3 | 3.6 | 0.32 | 0.89 | 1.07 | 0.83 |
| 3 | 11.7 | 4.3 | 1.7 | 1.53 | 10.7 | 2.0 | 0.19 | 8.6 | 3.0 | 0.35 | 10.5 | 3.3 | 0.31 | 0.82 | 1.02 | 0.80 |
| 4 | 12.5 | 4.4 | 1.7 | 1.59 | 12.4 | 2.2 | 0.18 | 9.6 | 3.4 | 0.35 | 12.0 | 3.5 | 0.29 | 0.80 | 1.03 | 0.77 |
| 5 | 12.3 | 4.3 | 1.6 | 1.69 | 11.9 | 2.1 | 0.18 | 9.2 | 2.9 | 0.32 | 11.3 | 3.1 | 0.27 | 0.81 | 1.05 | 0.77 |
| Min. | 11.7 | 3.9 | 1.6 | 1.29 | 10.7 | 2.0 | 0.18 | 8.6 | 2.9 | 0.31 | 10.5 | 3.1 | 0.27 | 0.80 | 1.02 | 0.77 |
| Max. | 12.8 | 4.4 | 1.7 | 1.69 | 12.4 | 2.4 | 0.20 | 10.1 | 3.4 | 0.35 | 12.0 | 3.6 | 0.32 | 0.89 | 1.07 | 0.83 |
| Average | 12.4 | 4.2 | 1.7 | 1.56 | 11.9 | 2.2 | 0.18 | 9.4 | 3.1 | 0.33 | 11.4 | 3.4 | 0.30 | 0.83 | 1.05 | 0.79 |
| S.D. | 0.44 | 0.19 | 0.05 | 0.16 | 0.70 | 0.15 | 0.01 | 0.56 | 0.19 | 0.02 | 0.56 | 0.19 | 0.02 | 0.04 | 0.02 | 0.03 |

Table 19. Phenotypic plasticity of *L. patens*.

| | F.1 | C.1 | CT.1 | CL.1/CT.1 | S.1 | SC.1 | SC.1/S.1 | W.1 | WC.1 | WC.1/W.1 | K.1 | KC.1 | KC.1/K.1 | W.1/K.1 | S.1/K.1 | W.1/S.1 |
|---|---|---|---|---|---|---|---|---|---|---|---|---|---|---|---|---|
| 1 | 13.2 | 4.3 | 1.7 | 1.53 | 12.2 | 1.7 | 0.14 | 8.6 | 2.9 | 0.34 | 12.7 | 2.8 | 0.22 | 0.68 | 0.96 | 0.70 |
| 2 | 13.1 | 4.4 | 1.6 | 1.75 | 12.0 | 1.6 | 0.13 | 8.7 | 3.0 | 0.34 | 13.5 | 3.1 | 0.23 | 0.64 | 0.89 | 0.73 |
| 3 | 13.3 | 4.1 | 1.5 | 1.73 | 11.7 | 1.6 | 0.14 | 8.1 | 2.8 | 0.35 | 13.0 | 3.0 | 0.23 | 0.62 | 0.90 | 0.69 |
| 4 | 12.8 | 4.3 | 1.7 | 1.53 | 11.2 | 1.6 | 0.14 | 8.3 | 2.9 | 0.35 | 12.7 | 3.1 | 0.24 | 0.65 | 0.88 | 0.74 |
| 5 | 13.1 | 4.3 | 1.8 | 1.39 | 11.3 | 1.5 | 0.13 | 8.3 | 3.0 | 0.36 | 12.7 | 2.9 | 0.23 | 0.65 | 0.89 | 0.73 |
| 6 | 13.2 | 4.2 | 1.4 | 2.00 | 11.8 | 1.7 | 0.14 | 8.6 | 3.1 | 0.36 | 13.0 | 3.1 | 0.24 | 0.66 | 0.91 | 0.73 |
| 7 | 12.8 | 3.7 | 1.5 | 1.47 | 11.4 | 1.3 | 0.11 | 8.2 | 2.8 | 0.34 | 12.5 | 2.9 | 0.23 | 0.66 | 0.91 | 0.72 |
| 8 | 13.3 | 4.7 | 2.0 | 1.35 | 11.9 | 1.5 | 0.13 | 8.6 | 2.9 | 0.34 | 12.9 | 3.0 | 0.23 | 0.67 | 0.92 | 0.72 |
| 9 | 13.3 | 4.3 | 1.7 | 1.53 | 12.1 | 1.6 | 0.13 | 8.7 | 2.9 | 0.33 | 13.0 | 2.9 | 0.22 | 0.67 | 0.93 | 0.72 |
| 10 | 12.5 | 4.0 | 1.7 | 1.35 | 11.6 | 1.6 | 0.14 | 8.4 | 2.7 | 0.32 | 12.3 | 2.7 | 0.22 | 0.68 | 0.94 | 0.72 |
| Min. | 12.5 | 3.7 | 1.4 | 1.35 | 11.2 | 1.3 | 0.11 | 8.1 | 2.7 | 0.32 | 12.3 | 2.7 | 0.22 | 0.62 | 0.88 | 0.69 |
| Max. | 13.3 | 4.7 | 2.0 | 2.00 | 12.2 | 1.7 | 0.14 | 8.7 | 3.1 | 0.36 | 13.5 | 3.1 | 0.24 | 0.68 | 0.96 | 0.74 |
| Average | 13.1 | 4.2 | 1.7 | 1.56 | 11.7 | 1.6 | 0.13 | 8.5 | 2.9 | 0.34 | 12.8 | 3.0 | 0.23 | 0.66 | 0.91 | 0.72 |
| S.D. | 0.27 | 0.26 | 0.17 | 0.21 | 0.34 | 0.12 | 0.01 | 0.22 | 0.12 | 0.01 | 0.33 | 0.14 | 0.01 | 0.02 | 0.03 | 0.01 |

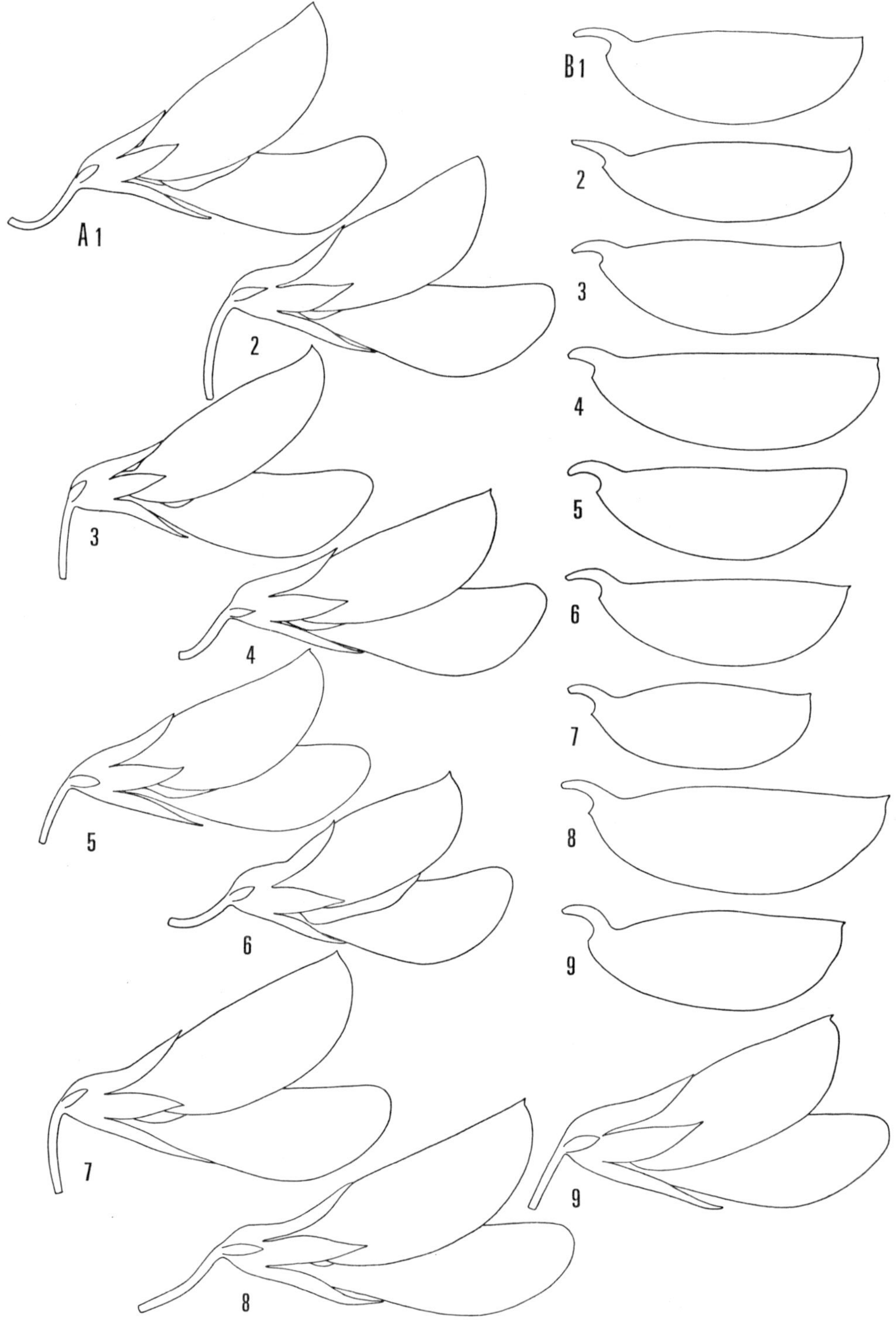

Fig. 20–1.   Individual variation in a population of *L. patens*.
A, flowers, lateral view;   B, standards, lateral view.   All × 3.3.

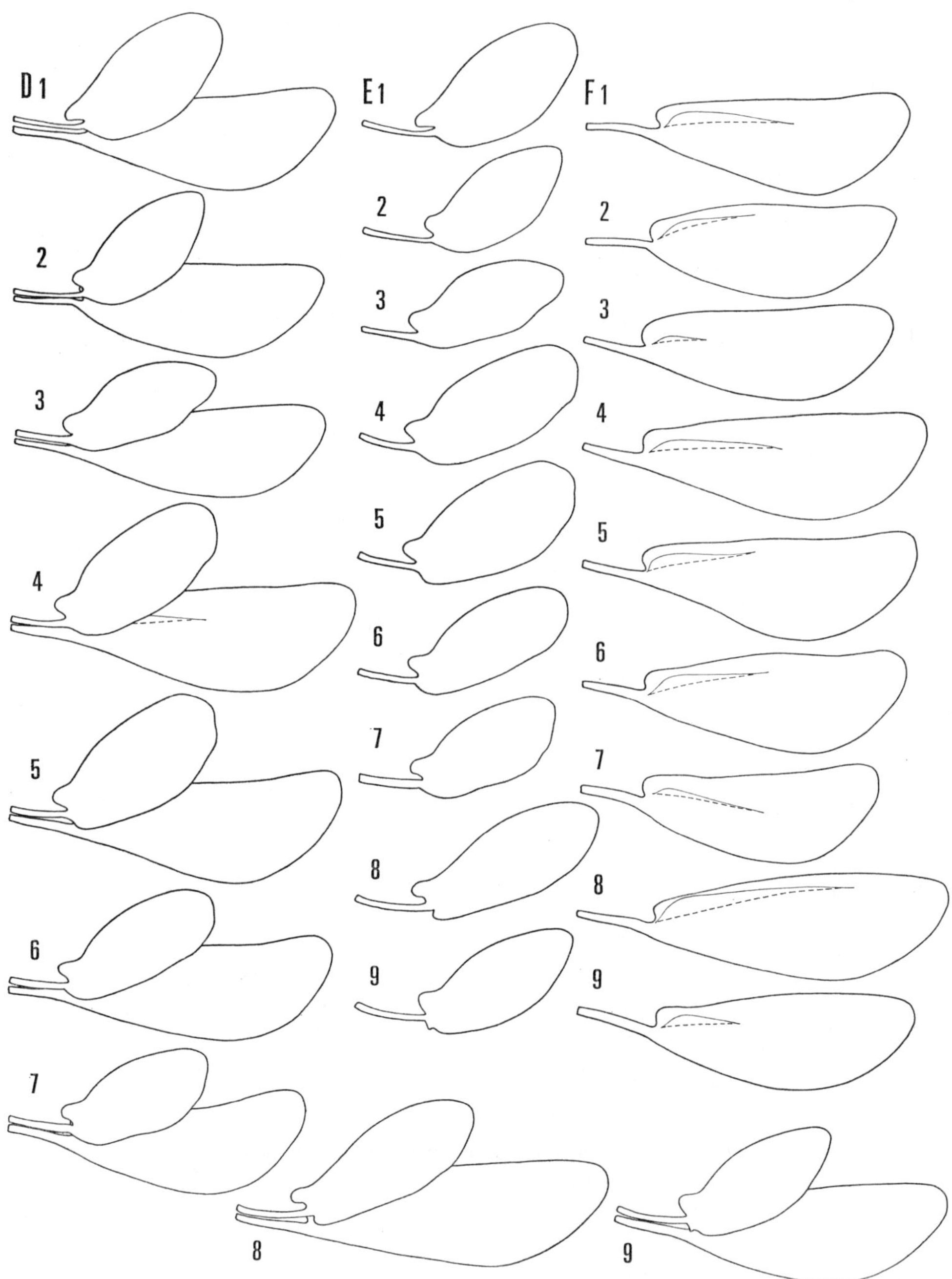

Fig. 20–2.  Individual variation in a population of *L. patens*.
D, wings and keel-petals;  E, wings;  F, keel-petals.  All×3.3.

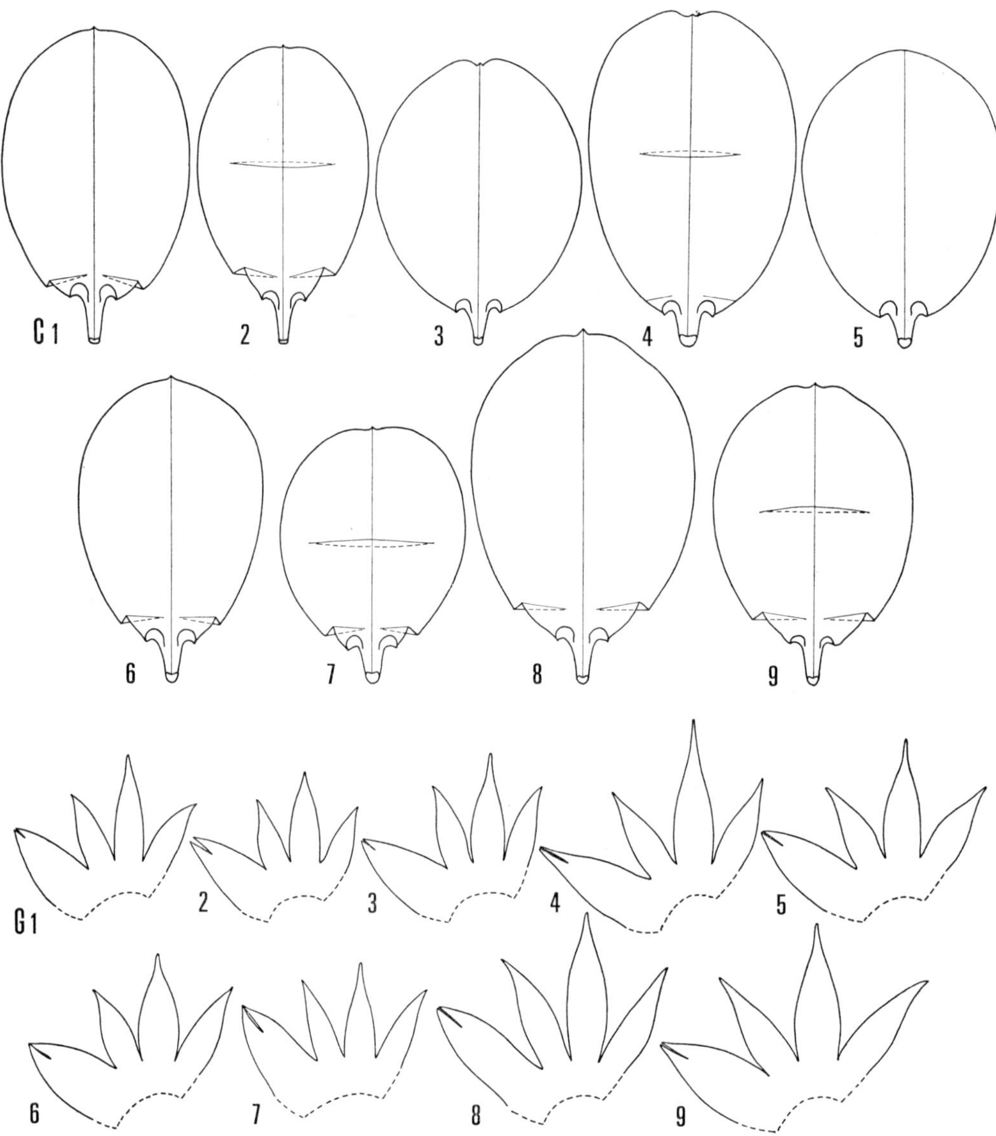

Fig. 20–3.  Individual variation in a population of *L. patens*.
C, standards, opened;   G, calyces, dissected.   All × 3.3.

Table 20.  Individual variation of *L. patens*.

|  | F.1 | C.1 | CT.1 | CL.1/CT.1 | S.1 | SC.1 | SC.1/S.1 | W.1 | WC.1 | WC.1/W.1 | K.1 | KC.1 | KC.1/K.1 | W.1/K.1 | S.1/K.1 | W.1/S.1 |
|---|---|---|---|---|---|---|---|---|---|---|---|---|---|---|---|---|
| 1 | 13.5 | 4.8 | 1.9 | 1.53 | 12.6 | 1.7 | 0.13 | 9.3 | 2.9 | 0.31 | 13.2 | 3.0 | 0.23 | 0.70 | 0.95 | 0.74 |
| 2 | 13.2 | 4.3 | 1.7 | 1.53 | 12.2 | 1.7 | 0.14 | 8.6 | 2.9 | 0.34 | 12.7 | 2.8 | 0.22 | 0.68 | 0.96 | 0.70 |
| 3 | 13.2 | 5.0 | 1.7 | 1.94 | 11.8 | 1.4 | 0.12 | 8.5 | 2.3 | 0.27 | 12.6 | 2.5 | 0.20 | 0.67 | 0.94 | 0.72 |
| 4 | 14.5 | 5.4 | 1.8 | 2.00 | 13.6 | 1.3 | 0.10 | 9.3 | 2.3 | 0.25 | 14.1 | 2.4 | 0.17 | 0.66 | 0.96 | 0.68 |
| 5 | 14.0 | 5.3 | 1.9 | 1.79 | 12.1 | 1.5 | 0.12 | 9.3 | 2.3 | 0.25 | 13.7 | 2.6 | 0.19 | 0.68 | 0.88 | 0.77 |
| 6 | 13.8 | 5.3 | 1.8 | 1.94 | 12.4 | 1.7 | 0.14 | 9.9 | 2.4 | 0.24 | 13.3 | 2.6 | 0.20 | 0.74 | 0.93 | 0.80 |
| 7 | 12.4 | 5.0 | 1.9 | 1.63 | 10.6 | 1.2 | 0.11 | 8.5 | 2.6 | 0.31 | 12.2 | 2.6 | 0.21 | 0.70 | 0.87 | 0.80 |
| 8 | 15.6 | 6.6 | 2.0 | 2.30 | 14.3 | 1.7 | 0.12 | 10.4 | 2.8 | 0.27 | 15.3 | 3.2 | 0.21 | 0.68 | 0.93 | 0.73 |
| 9 | 14.1 | 6.2 | 1.6 | 2.88 | 12.4 | 1.7 | 0.14 | 9.3 | 2.9 | 0.31 | 13.7 | 3.2 | 0.23 | 0.68 | 0.91 | 0.75 |
| Min. | 12.4 | 4.3 | 1.6 | 1.53 | 10.6 | 1.2 | 0.10 | 8.5 | 2.3 | 0.24 | 12.2 | 2.4 | 0.17 | 0.66 | 0.87 | 0.68 |
| Max. | 15.6 | 6.6 | 2.0 | 2.88 | 14.3 | 1.7 | 0.14 | 10.4 | 2.9 | 0.34 | 15.3 | 3.2 | 0.23 | 0.74 | 0.96 | 0.80 |
| Average | 13.8 | 5.3 | 1.8 | 1.95 | 12.4 | 1.5 | 0.12 | 9.2 | 2.6 | 0.28 | 13.4 | 2.8 | 0.21 | 0.69 | 0.93 | 0.74 |
| S.D. | 0.91 | 0.70 | 0.13 | 0.43 | 1.05 | 0.20 | 0.01 | 0.64 | 0.28 | 0.03 | 0.93 | 0.30 | 0.02 | 0.02 | 0.03 | 0.04 |

64

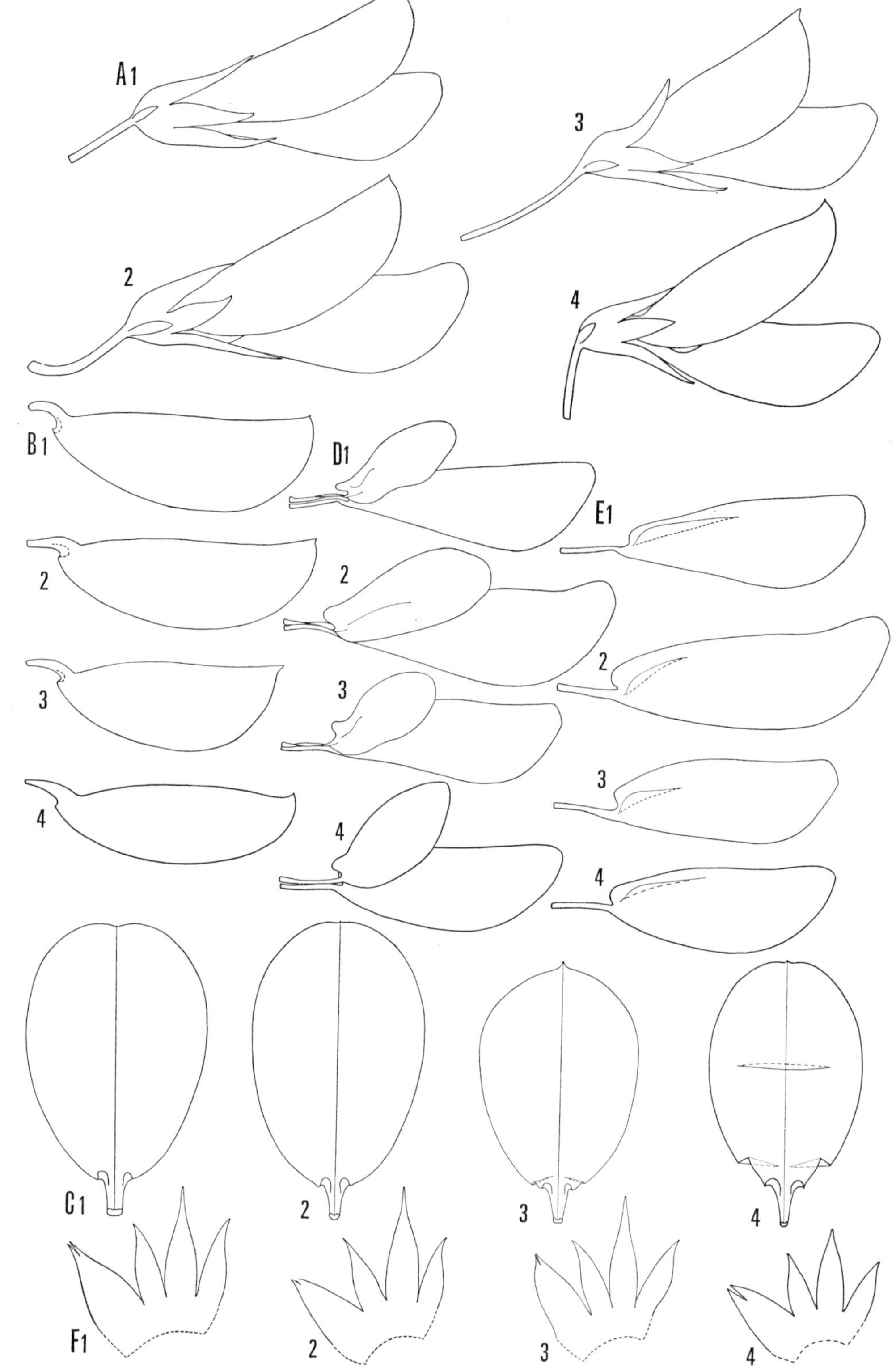

Table 21. Phenotypic plasticity and individual variation of *L. patens*.

| | Character | Phenotypic plasticity | Individual variation |
|---|---|---|---|
| Calyx | shape of lobe | narrowly triangular | narrowly triangular |
| | shape of apex | acute | acute |
| | length ratio of lateral lobe to tube | $1.56\pm0.21$ | $1.95\pm0.43$ |
| | ratio of connated part of upper lobes | 2/3–7/8 | 2/3–7/8 |
| Corolla | color | red-purple | red-purple |
| Length ratio | wing to keel-petal | $0.66\pm0.02$ | $0.69\pm0.02$ |
| | standard to keel-petal | $0.91\pm0.03$ | $0.93\pm0.03$ |
| | wing to standard | $0.72\pm0.01$ | $0.74\pm0.04$ |
| Standard | shape | obovate with distinct claw | obovate-elliptic with distinct claw |
| | shape of apex | truncate with a point | truncate with a point or obtuse |
| | shape of auricle | lunate | lunate-narrowly lunate |
| | length ratio of claw to standard | $0.13\pm0.01$ | $0.12\pm0.01$ |
| Wing | shape of lamina | obovate | obovate |
| | shape of upper base | auriculate-cordate | auriculate-cordate |
| | shape of lower base | minutely auriculate -cordate | auriculate-cordate |
| | length ratio of claw to wing | $0.34\pm0.01$ | $0.28\pm0.03$ |
| Keel-petal | shape of lamina | narrowly obovate | narrowly obovate |
| | shape of upper base | truncate-cordate | truncate-cordate |
| | shape of lower base | tapering | tapering |
| | length ratio of claw to keel-petal | $0.23\pm0.01$ | $0.21\pm0.02$ |

the claw of the wing itself is $0.29\pm0.01$. The shape of the lamina of the keel-petal is narrowly obovate with a cordate or truncate upper base and a tapering lower base (Fig. 22E). The ratio of the claw to the keel-petal itself is $0.29\pm0.01$.

The shape of the lateral calyx-lobe is triangular with an acute apex (Fig. 22F). Two upper calyx-lobes are connated 9/10 of the way along its length. The ratio of the length of the lateral lobes to the tube is $0.70\pm0.12$.

Individual variation: The population studied consisted of only two blooming individuals (Fig. 23 and Table 24). As in previous examples almost all characters with constant expressions phenotypically are stable within this population, but some characters are slightly more variable in expression and relative length. The ratio of the length of the standard to the keel-petal is about 0.91, wing to standard about 0.99, and wing to keel-petal about 0.89 (Table 25).

The standard varies from round to broadly obovate (Fig. 23A). The ratio of the length of the claw to the total length of each petal is about 0.14 in the standard, about 0.28 in the wing, and about 0.29 in the keel-petal. The ratio of the length of the

←Fig. 21. Variation of *L. patens*.

A, flowers, lateral view; B, standards, lateral view; C, standards, opened; D, wings; E, keel-petals; F, calyces, dissected. All×3.3.

1. Fukushima (Ohba & Akiyama 1850, TI). 2. Niigata (Ohba & Akiyama 1651, TI). 3. Nagano (Ohba & Akiyama 507, TI). 4. Ishikawa (Akiyama 1514, TI).

Table 22. Collective variation of *L. patens*.

| | F.l | C.l | CT.l | CL.l/CT.l | S.l | SC.l | SC.l/S.l | W.l | WC.l | WC.l/W.l | K.l | KC.l | KC.l/K.l | W.l/K.l | S.l/K.l | W.l/S.l |
|---|---|---|---|---|---|---|---|---|---|---|---|---|---|---|---|---|
| 1 | 13.8 | 5.4 | 1.8 | 2.00 | 13.0 | 1.5 | 0.12 | 7.9 | 2.8 | 0.35 | 13.6 | 3.0 | 0.22 | 0.58 | 0.96 | 0.61 |
| 2 | 15.4 | 5.3 | 1.9 | 1.79 | 13.3 | 1.5 | 0.11 | 9.4 | 2.4 | 0.26 | 14.6 | 2.7 | 0.18 | 0.64 | 0.91 | 0.71 |
| 3 | 13.2 | 5.3 | 1.9 | 1.79 | 12.6 | 1.7 | 0.13 | 6.9 | 2.4 | 0.35 | 12.4 | 2.8 | 0.23 | 0.56 | 1.02 | 0.55 |
| Min. | 13.2 | 5.3 | 1.8 | 1.79 | 12.6 | 1.5 | 0.11 | 6.9 | 2.4 | 0.26 | 12.4 | 2.7 | 0.18 | 0.56 | 0.91 | 0.55 |
| Max. | 15.4 | 5.4 | 1.9 | 2.00 | 13.3 | 1.7 | 0.13 | 9.4 | 2.8 | 0.35 | 14.6 | 3.0 | 0.23 | 0.64 | 1.02 | 0.71 |
| Average | 14.1 | 5.3 | 1.9 | 1.86 | 13.0 | 1.6 | 0.12 | 8.1 | 2.5 | 0.32 | 13.5 | 2.8 | 0.21 | 0.59 | 0.96 | 0.62 |
| S.D. | 1.14 | 0.06 | 0.06 | 0.12 | 0.35 | 0.12 | 0.01 | 1.26 | 0.23 | 0.06 | 1.10 | 0.15 | 0.02 | 0.05 | 0.05 | 0.08 |

Table 23. Phenotypic plasticity of *L. Buergeri*.

| | F.l | C.l | CT.l | CL.l/CT.l | S.l | SC.l | SC.l/S.l | W.l | WC.l | WC.l/W.l | K.l | KC.l | KC.l/K.l | W.l/K.l | S.l/K.l | W.l/S.l |
|---|---|---|---|---|---|---|---|---|---|---|---|---|---|---|---|---|
| 1 | 9.6 | 2.4 | 1.3 | 0.85 | 7.7 | 1.2 | 0.16 | 7.8 | 2.3 | 0.29 | 8.6 | 2.6 | 0.30 | 0.91 | 0.90 | 1.01 |
| 2 | 9.4 | 2.2 | 1.4 | 0.57 | 7.9 | 1.2 | 0.15 | 7.9 | 2.2 | 0.28 | 9.1 | 2.8 | 0.31 | 0.87 | 0.87 | 1.00 |
| 3 | 8.8 | 2.1 | 1.2 | 0.75 | 7.6 | 1.1 | 0.14 | 7.6 | 2.1 | 0.28 | 8.7 | 2.5 | 0.29 | 0.87 | 0.87 | 1.00 |
| 4 | 8.9 | 2.0 | 1.1 | 0.82 | 7.1 | 1.1 | 0.15 | 7.2 | 2.2 | 0.31 | 8.6 | 2.5 | 0.29 | 0.84 | 0.83 | 1.01 |
| 5 | 9.3 | 2.1 | 1.3 | 0.62 | 7.6 | 1.4 | 0.18 | 7.6 | 2.4 | 0.32 | 8.9 | 2.7 | 0.30 | 0.85 | 0.85 | 1.00 |
| 6 | 8.8 | 2.1 | 1.3 | 0.62 | 7.0 | 1.1 | 0.16 | 7.2 | 2.0 | 0.28 | 8.6 | 2.4 | 0.28 | 0.84 | 0.81 | 1.03 |
| 7 | 8.5 | 2.1 | 1.1 | 0.91 | 7.5 | 1.1 | 0.15 | 7.5 | 2.1 | 0.28 | 8.4 | 2.4 | 0.29 | 0.89 | 0.89 | 1.00 |
| 8 | 8.5 | 2.1 | 1.3 | 0.62 | 7.5 | 1.2 | 0.16 | 7.2 | 2.0 | 0.28 | 8.2 | 2.4 | 0.29 | 0.88 | 0.91 | 0.96 |
| 9 | 8.8 | 2.1 | 1.3 | 0.62 | 7.4 | 1.3 | 0.18 | 7.5 | 2.1 | 0.28 | 8.5 | 2.6 | 0.31 | 0.88 | 0.87 | 1.01 |
| 10 | 8.7 | 1.8 | 1.1 | 0.64 | 7.5 | 1.2 | 0.16 | 7.7 | 2.2 | 0.29 | 8.5 | 2.5 | 0.29 | 0.91 | 0.88 | 1.03 |
| Min. | 8.5 | 1.8 | 1.1 | 0.57 | 7.0 | 1.1 | 0.14 | 7.2 | 2.0 | 0.28 | 8.2 | 2.4 | 0.28 | 0.84 | 0.81 | 0.96 |
| Max. | 9.6 | 2.4 | 1.4 | 0.91 | 7.9 | 1.4 | 0.18 | 7.9 | 2.4 | 0.32 | 9.1 | 2.8 | 0.31 | 0.91 | 0.91 | 1.03 |
| Average | 8.9 | 2.1 | 1.2 | 0.70 | 7.5 | 1.2 | 0.16 | 7.5 | 2.2 | 0.29 | 8.6 | 2.5 | 0.29 | 0.87 | 0.87 | 1.01 |
| S.D. | 0.38 | 0.15 | 0.11 | 0.12 | 0.27 | 0.10 | 0.01 | 0.25 | 0.13 | 0.01 | 0.25 | 0.13 | 0.01 | 0.03 | 0.03 | 0.02 |

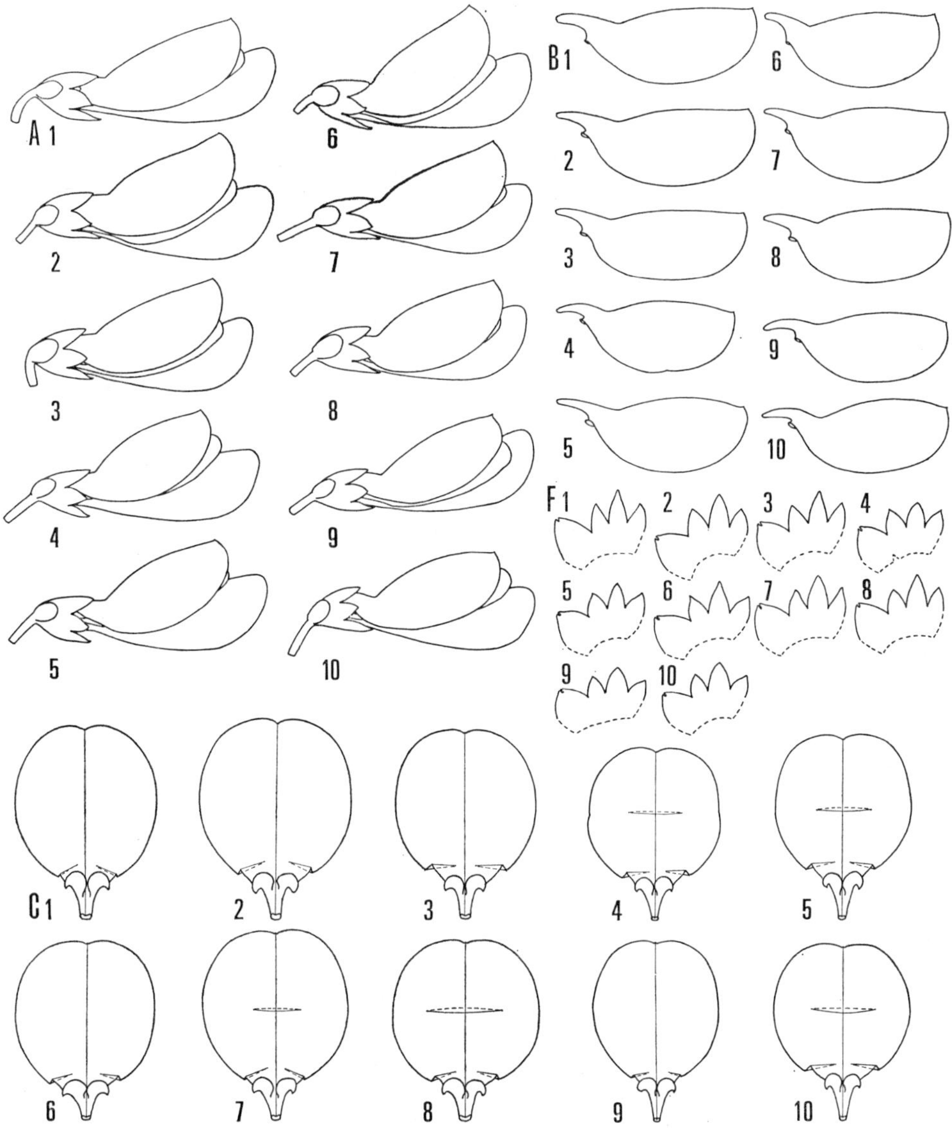

Fig. 22–1. Phenotypic plasticity of *L. Buergeri*.
A, flowers, lateral view;  B, standards, lateral view;  C, standards, opened;  F, calyces, dissected.  All×3.3.

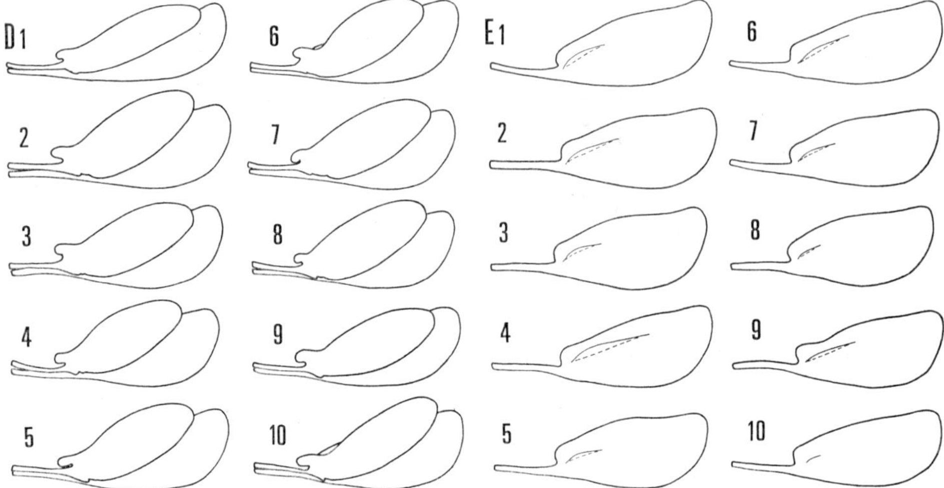

Fig. 22–2.  Phenotypic plasticity of *L. Buergeri*.
D, wings;  E, keel-petals.  All×3.3.

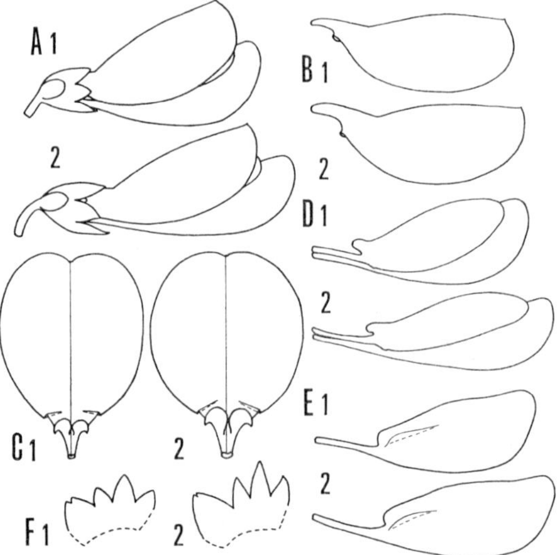

Fig. 23.  Individual variation in a population of *L. Buergeri*.
A, flowers, lateral view;  B, standards, lateral view;  C, standards, opened;
D, wings;  E, keel-petals;  F, calyces, dissected.  All×3.3.

Table 24. Individual variation of *L. Buergeri*.

| | F.l | C.l | CT.l | CL.l/CT.l | S.l | SC.l | SC.l/S.l | W.l |
|---|---|---|---|---|---|---|---|---|
| 1 | 9.6 | 2.4 | 1.3 | 0.85 | 7.7 | 1.2 | 0.16 | 7.8 |
| 2 | 8.4 | 1.9 | 1.3 | 0.46 | 7.6 | 1.0 | 0.13 | 7.3 |
| Min. | 8.4 | 1.9 | 1.3 | 0.46 | 7.6 | 1.0 | 0.13 | 7.3 |
| Max. | 9.6 | 2.4 | 1.3 | 0.85 | 7.7 | 1.2 | 0.16 | 7.8 |
| Average | 9.0 | 2.2 | 1.3 | 0.65 | 7.7 | 1.1 | 0.14 | 7.6 |

Table 24. Continued

| | WC.l | WC.l/W.l | K.l | KC.l | KC.l/K.l | W.l/K.l | S.l/K.l | W.l/S.l |
|---|---|---|---|---|---|---|---|---|
| 1 | 2.3 | 0.29 | 8.6 | 2.6 | 0.30 | 0.91 | 0.90 | 1.01 |
| 2 | 1.9 | 0.26 | 8.3 | 2.3 | 0.28 | 0.88 | 0.92 | 0.96 |
| Min. | 1.9 | 0.26 | 8.3 | 2.3 | 0.28 | 0.88 | 0.90 | 0.96 |
| Max. | 2.3 | 0.29 | 8.6 | 2.6 | 0.30 | 0.91 | 0.92 | 1.01 |
| Average | 2.1 | 0.28 | 8.5 | 2.5 | 0.29 | 0.89 | 0.91 | 0 99 |

Table 25. Phenotypic plasticity and individual variation of *L. Buergeri*.

| | Character | Phenotypic plasticity | Individual variation |
|---|---|---|---|
| Calyx | shape of lobe | triangular | triangular |
| | shape of apex | acute | acute |
| | length ratio of lateral lobe to tube | 0.70±0.12 | about 0.65 |
| | ratio of connated part of upper lobes | 9/10 | 3/4–9/10 |
| Corolla | color | yellow & purple | yellow & purple |
| Length ratio | wing to keel-petal | 0.87±0.03 | about 0.89 |
| | standard to keel-petal | 0.87±0.03 | about 0.91 |
| | wing to standard | 1.01±0.02 | about 0.99 |
| Standard | shape | round with distinct claw | round-broadly obovate with distinct claw |
| | shape of apex | retuse with a point | retuse with a point |
| | shape of auricle | broadly lunate | broadly lunate |
| | length ratio of claw to standard | 0.16±0.01 | about 0.14 |
| Wing | shape of lamina | narrowly oblong | narrowly oblong |
| | shape of upper base | auriculate | auriculate |
| | shape of lower base | minutely auriculate–tapering | minutely auriculate–tapering |
| | length ratio of claw to wing | 0.29±0.01 | about 0.28 |
| Keel-petal | shape of lamina | narrowly obovate | narrowly obovate |
| | shape of upper base | cordate-truncate | cordate-trucate |
| | shape of lower base | tapering | tapering |
| | length ratio of claw to keel-petal | 0.29±0.01 | about 0.29 |

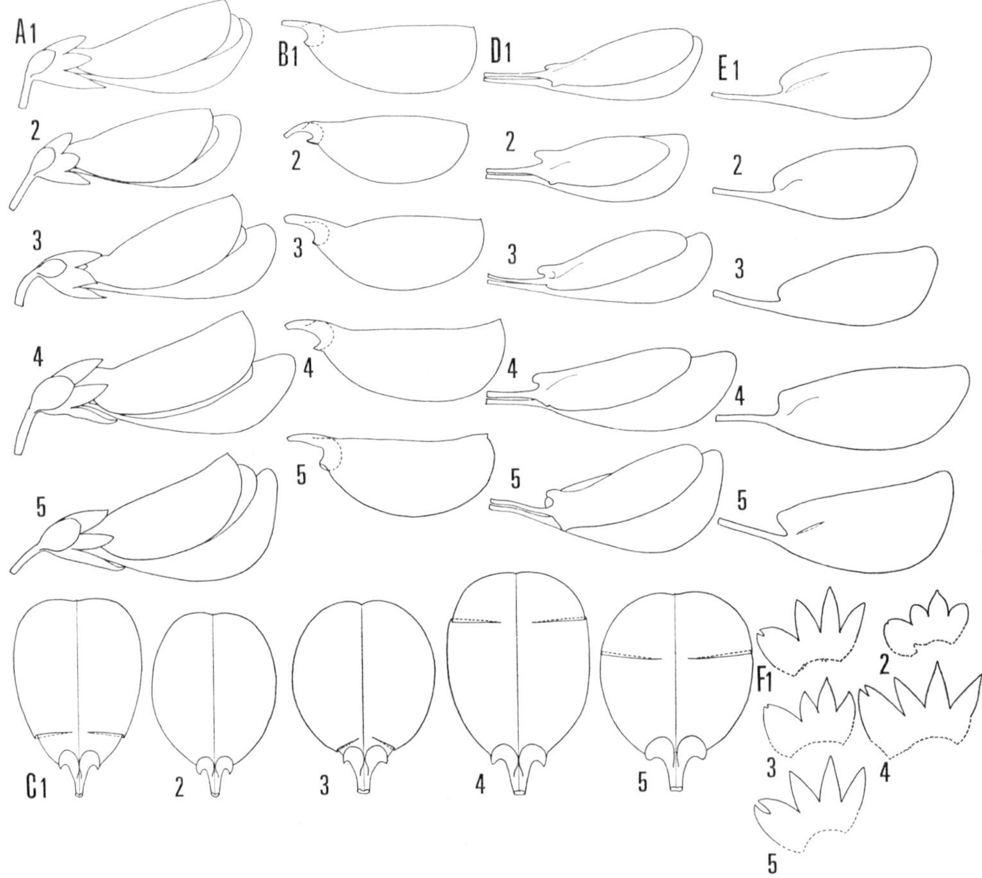

Fig. 24.   Variation of *L. Buergeri*.
A, flowers, lateral view;  B, standards, lateral view;  C, standards, opened;
D, wings;  E, keel-petals;  F, calyces, dissected.   All×3.3.
1. Iwate (Kanai, 2 Sept. 1959, TI).   2. Chiba (Ohba et al. 6808042, TI).   3.
Kanagawa (Ohba & Akiyama 615, TI).   4. Kouchi (Yamawaki, 14 Oct.
1944, TI).   5. Kagoshima (Ohba & Akiyama 875, TI).

lateral calyx-lobe to the tube is about 0.65.

Collective variation: Flowers from various localities were illustrated (Fig. 24) and measured (Table 26). The expression of the characters investigated does not exceed the range of variation of those in the single population mentioned above. The lamina of the standard varies from round to obovate. The relative length of various characters differs slightly from that in the population.

### Constancy of Floral Characters: Conclusions

Calyx: The shape of the lateral calyx-lobe and its apex has a narrow range of variation in a species. The shape of the lateral calyx-lobe is (broadly) elliptic in *L. homoloba*, while in the other species it varies from triangular to narrowly triangular. The shape of the apex is acuminate in *L. cyrtobotrya*, while in the other five species it is

Table 26. Collective variation of *L. Buergeri*.

| | F.1 | C.1 | CT.1 | CL.1/CT.1 | S.1 | SC.1 | SC.1/S.1 | W.1 |
|---|---|---|---|---|---|---|---|---|
| 1 | 8.9 | 2.7 | 1.2 | 1.25 | 7.8 | 1.2 | 0.15 | 8.6 |
| 2 | 8.5 | 1.8 | 1.1 | 0.64 | 7.3 | 1.1 | 0.15 | 7.3 |
| 3 | 9.6 | 2.4 | 1.3 | 0.85 | 7.7 | 1.3 | 0.17 | 7.8 |
| 4 | 10.3 | 2.8 | 1.2 | 1.33 | 8.8 | 1.4 | 0.16 | 8.4 |
| 5 | 9.8 | 2.9 | 1.2 | 1.42 | 7.9 | 1.5 | 0.19 | 8.6 |
| Min. | 8.5 | 1.8 | 1.1 | 0.64 | 7.3 | 1.1 | 0.15 | 7.3 |
| Max. | 10.3 | 2.9 | 1.3 | 1.42 | 8.8 | 1.5 | 0.19 | 8.6 |
| Average | 9.4 | 2.5 | 1.2 | 1.10 | 7.9 | 1.3 | 0.16 | 8.1 |
| S.D. | 0.72 | 0.44 | 0.07 | 0.34 | 0.55 | 0.16 | 0.02 | 0.57 |

Table 26. Continued

| | WC.1 | WC.1/W.1 | K.1 | KC.1 | KC.1/K.1 | W.1/K.1 | S.1/K.1 | W.1/S.1 |
|---|---|---|---|---|---|---|---|---|
| 1 | 2.7 | 0.31 | 8.7 | 2.8 | 0.32 | 0.99 | 0.90 | 1.10 |
| 2 | 2.2 | 0.30 | 8.0 | 2.4 | 0.30 | 0.91 | 0.91 | 1.00 |
| 3 | 2.3 | 0.29 | 8.6 | 2.6 | 0.30 | 0.91 | 0.90 | 1.01 |
| 4 | 2.2 | 0.26 | 10.1 | 2.5 | 0.25 | 0.83 | 0.87 | 0.95 |
| 5 | 2.7 | 0.31 | 8.9 | 2.8 | 0.31 | 0.97 | 0.89 | 1.09 |
| Min. | 2.2 | 0.26 | 8.0 | 2.4 | 0.25 | 0.83 | 0.87 | 0.95 |
| Max. | 2.7 | 0.31 | 10.1 | 2.8 | 0.32 | 0.99 | 0.91 | 1.10 |
| Average | 2.4 | 0.30 | 8.9 | 2.6 | 0.30 | 0.92 | 0.89 | 1.03 |
| S.D. | 0.26 | 0.02 | 0.77 | 0.18 | 0.03 | 0.06 | 0.01 | 0.06 |

acute or acute to obtuse. The relative length between the lobes and tube also shows a narrow range of variation. The relative length of the lobes is greater in *L. patens* than in *L. formosa* subsp. *velutina*. In *L. bicolor* and *L. homoloba* the lobes are nearly equal to the tube, shorter than the tube, or longer. But the degree of connation between the upper lobes is variable even in a single population as observed in *L. homoloba*, and this feature does not permit specific distinction.

Petals: The relative length among the petals (i.e., standard, wing, keel-petal) shows a narrower range of variation within species. Generally the wing is shorter, but in *L. cyrtobotrya* the wing is distinctly longer than the keel-petal. In *L. patens* the wing is characteristically shorter than the keel-petal compared with other species. In *L. bicolor* and *L. cyrtobotrya* the standard is longer than the keel-petal, while in *L. homoloba*, *L. patens*, and *L. Buergeri* it is usually shorter than the keel-petal and in *L. formosa* subsp. *velutina* it is nearly the same length as the keel-petal.

The shape of the standard and auricle is constant in each species. Two types of standard can be distinguished: one type is attenuate at the base, as in *L. bicolor* and *L. cyrtobotrya*; the other has a distinct claw, as in *L. homoloba*, *L. formosa* subsp. *velutina*, *L. patens*, and *L. Buergeri*. *L. patens* has a longer standard but the claw is much shorter than in *L. formosa* subsp. *velutina*. The shape of the auricle is (narrowly to broadly) lunate in general but in *L. homoloba* the auricle is well developed and reniform.

The shape of the lamina of the wing is rather constant and similar among species.

Table 27.  Specific characters and their expression in sect. *Macrolespedeza*.

| | Character | *L. bicolor* | *L. cyrtobotrya* | *L. homoloba* | *L. formosa* (subsp. *velutina*) | *L. patens* | *L. Buergeri* |
|---|---|---|---|---|---|---|---|
| Calyx | shape of lobe | triangular | triangular | (broadly) elliptic | triangular | narrowly tri-angular | triangular |
| | shape of apex | acute–obtuse | acuminate | obtuse with or without a point | acute | acute | acute–obtuse |
| | length ratio of lateral lobe to tube | 0.83±0.17 | 1.26±0.13 | 1.00±0.15 | 1.56±0.16 | 1.86±0.12 | 1.10±0.34 |
| Corolla | color | red-purple | red-purple | red-purple | red-purple | red-purple | yellow & purple |
| Length ratio | wing to keel-petal | 0.95±0.04 | 1.07±0.04 | 0.84±0.05 | 0.83±0.04 | 0.59±0.05 | 0.92±0.06 |
| | standard to keel-petal | 1.13±0.05 | 1.18±0.08 | 0.99±0.04 | 1.05±0.02 | 0.96±0.05 | 0.86±0.01 |
| | wing to standard | 0.84±0.04 | 0.91±0.06 | 0.85±0.04 | 0.79±0.03 | 0.62±0.08 | 1.03±0.06 |
| Standard | shape | obovate with attenuate base | obovate with attenuate base | elliptic-ovate-broadly ovate with distinct claw | obovate-elliptic with distinct claw | obovate-elliptic with distinct claw | elliptic-obovate with distinct claw |
| | shape of auricle | (narrowly) lunate | (narrowly) lunate | reniform | (narrowly) lunate | (narrowly) lunate | broadly lunate |
| | length ratio of claw to standard | — | — | 0.21±0.02 | 0.18±0.01 | 0.12±0.01 | 0.16±0.02 |
| Wing | shape of lamina | narrowly obovate | narrowly obovate | narrowly oblong | narrowly obovate | obovate | narrowly oblong |
| | length ratio of claw to wing | 0.41±0.02 | 0.42±0.03 | 0.36±0.02 | 0.33±0.02 | 0.32±0.06 | 0.30±0.02 |
| Keel-petal | shape of lamina | obovate | obovate | narrowly obovate | narrowly obovate | narrowly obovate | narrowly obovate |
| | length ratio of claw to keel-petal | 0.43±0.02 | 0.50±0.03 | 0.30±0.02 | 0.30±0.02 | 0.21±0.02 | 0.30±0.03 |

The relative length of the claw in both the wing and keel-petal shows a narrow range of variation within species. The claws of both wing and keel-petal are longer in *L. cyrtobotrya* and *L. bicolor* than in other species. The shape of the lamina of the keel-petal is constant but less characteristic, because its expression is similar among species.

In conclusion, characteristics by which species can be distinguished in sect. *Macro-lespedeza* are: the shape of the lateral calyx-lobe; the shape of the apex of the lateral calyx-lobe; the relative length of the lateral calyx-lobe to the calyx-tube; the relative length of the wing to the keel-petal; the relative length of the standard to the keel-petal; the shape of the standard; the shape of the auricle of the standard; the relative length of the standard-claw to the standard; the relative length of the wing-claw to the wing; and the relative length of the keel-petal-claw to the keel-petal (Table 27).

## Developmental Changes of Floral Characters

To study the developmental changes of floral characters, young flowers of various sizes from approximately 1 mm long to fully developed were taken from a single individual plant and examined (Figs. 25–31).

### Calyx

The shape of the lateral calyx-lobe when the flower length is about 1/3 that of a fully developed flower exhibits the characteristic shape of each species. The shape of the apex of the lateral calyx-lobe is acuminate in the very young stage in *L. cyrtobotrya* (Fig. 26F) while in the others it is acute or obtuse.

### Standard

The shape of the standard is nearly round without a claw or with an attenuate base until the length of the flower is about 5 mm. After that, in *L. bicolor* (Fig. 25C) and *L. cyrtobotrya* (Fig. 26C), the part near the auricles elongates, while in *L. homoloba* (Fig. 27C), *L. formosa* subsp. *velutina* (Fig. 28C), *L. patens* (Fig. 29C), and *L. Buergeri* (Fig. 30C) the part near the auricles elongates scarcely at all, although the basal part does elongate. As a result, fully developed *L. bicolor* and *L. cyrtobotrya* flowers have standards with an attenuate base and *L. homoloba* Nakai, *L. formosa* subsp. *velutina*, *L. patens*, and *L. Buergeri* have standards with a claw.

When flowers are less than 1–1.5 mm long, the auricle near the base of the standard is hardly recognizable under a binocular microscope, although it becomes distinct when flowers are more than 2 mm long. The development of the auricle in *L. bicolor*, *L. cyrtobotrya*, *L. formosa* subsp. *velutina*, and *L. patens* stops when the flower is about 5 mm long, but in *L. homoloba* and *L. Buergeri* it continues to develop almost to the anthesis (Fig. 31). The shape of the auricle of the former four species is lunate to narrowly lunate, while the auricle of the latter two is well developed and reniform or broadly lunate.

### Wing and Keel-petal

In *L. cyrtobotrya* (Fig. 26D) the wing is nearly equal to the keel-petal length in very early stages, but when the flower becomes about 5 mm long the wing is longer

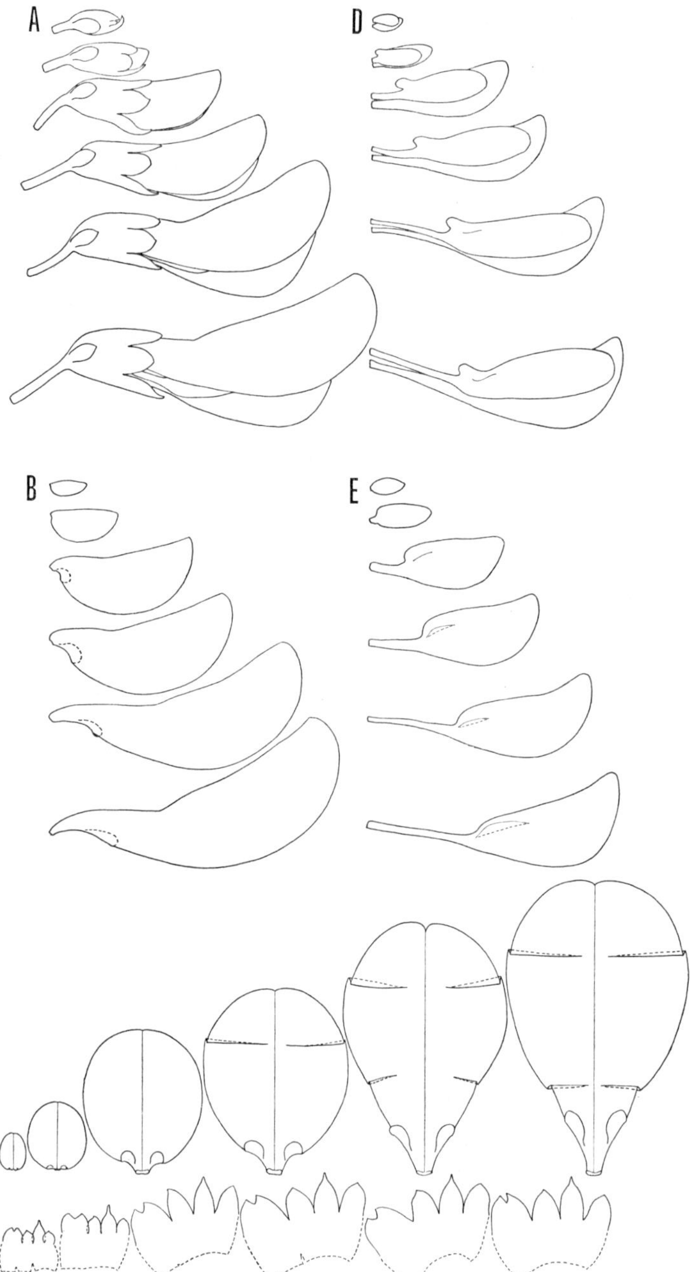

Fig. 25.   Developmental change in flowers of *L. bicolor* (Ohba & Akiyama 608, TI).
A, flowers, lateral view;   B, standards, lateral view;   C, standards, opened;
D, wings;   E, keel-petals;   F, calyces, dissected.   All×3.3.

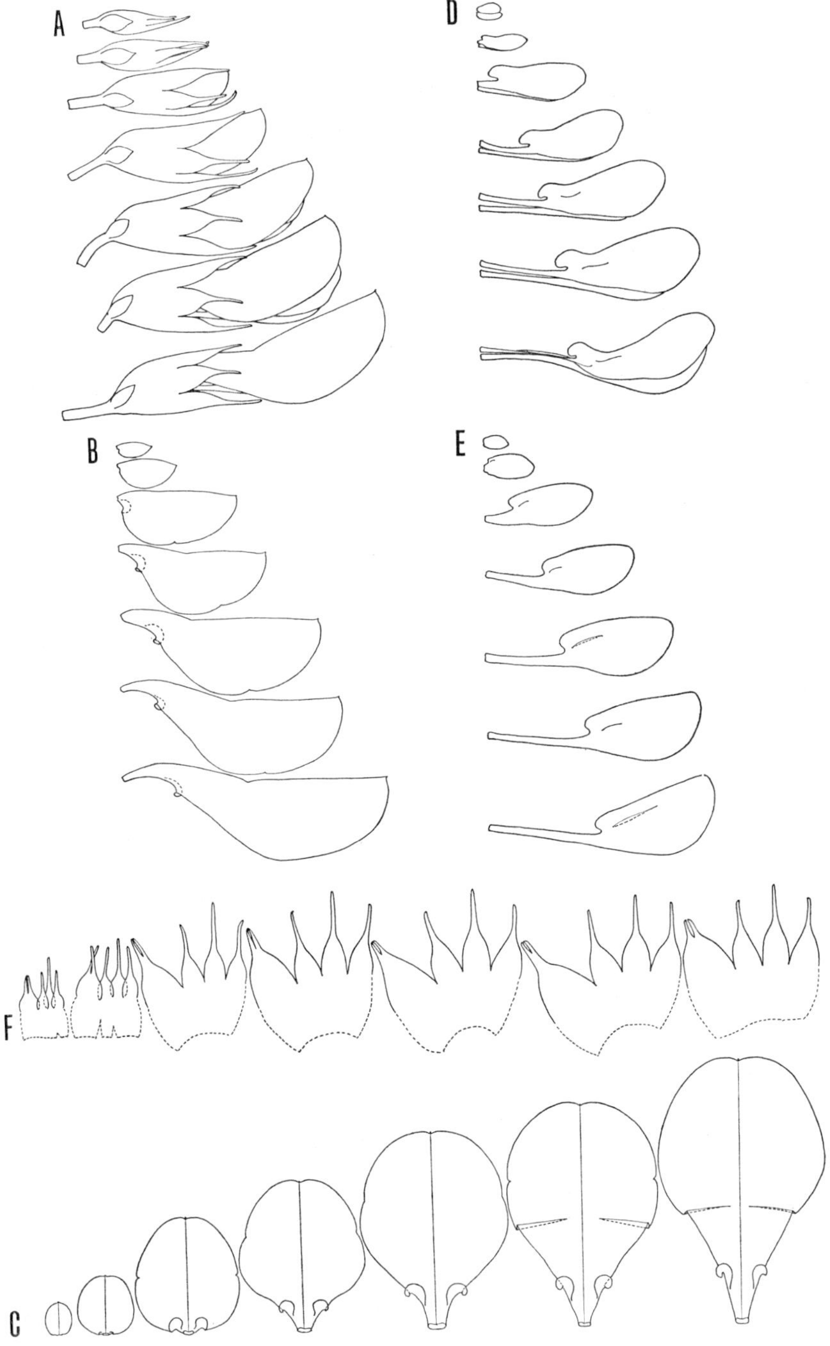

Fig. 26.  Developmental change in flowers of *L. cyrtobotrya* (Ohba & Akiyama 624, TI).
A, flowers, lateral view;  B, standards, lateral view;  C, standards, opened;
D, wings;  E, keel-petals;  F, calyces, dissected.  All ×3.3.

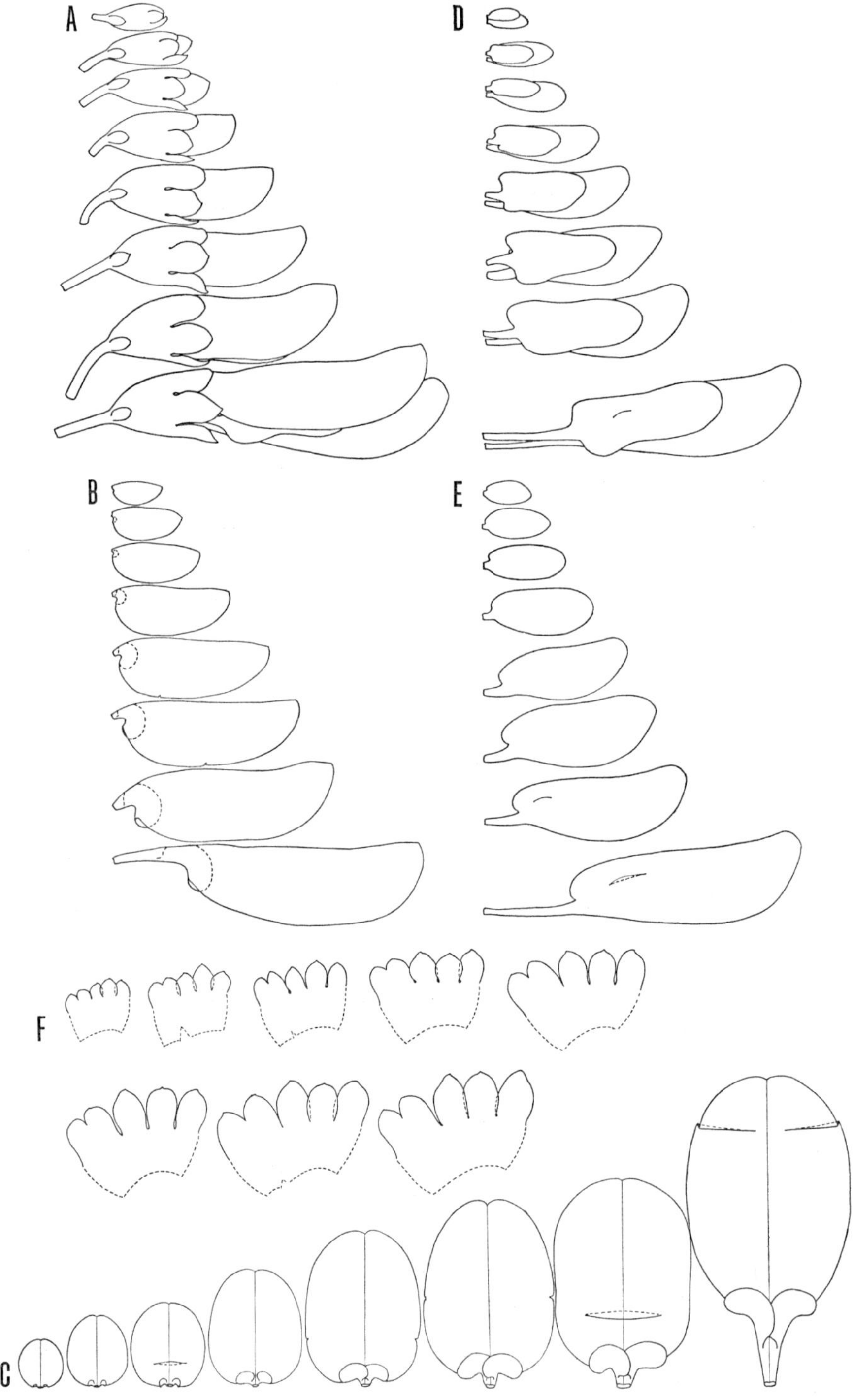

Fig. 28. Developmental change in flowers of *L. formosa* subsp. *velutina* (Ohba & Akiyama 655, TI).
A, flowers, lateral view; B, standards, lateral view; C, standards, opened; D, wings; E, keel-petals; F, calyces, dissected. All×3.3.

←Fig. 27. Developmental change in flowers of *L. homoloba* (Ohba & Akiyama 1423, TI).
A, flowers, lateral view; B, standards, lateral view; C, standards, opened; D, wings; E, keel-petals; F, calyces, dissected. All×3.3.

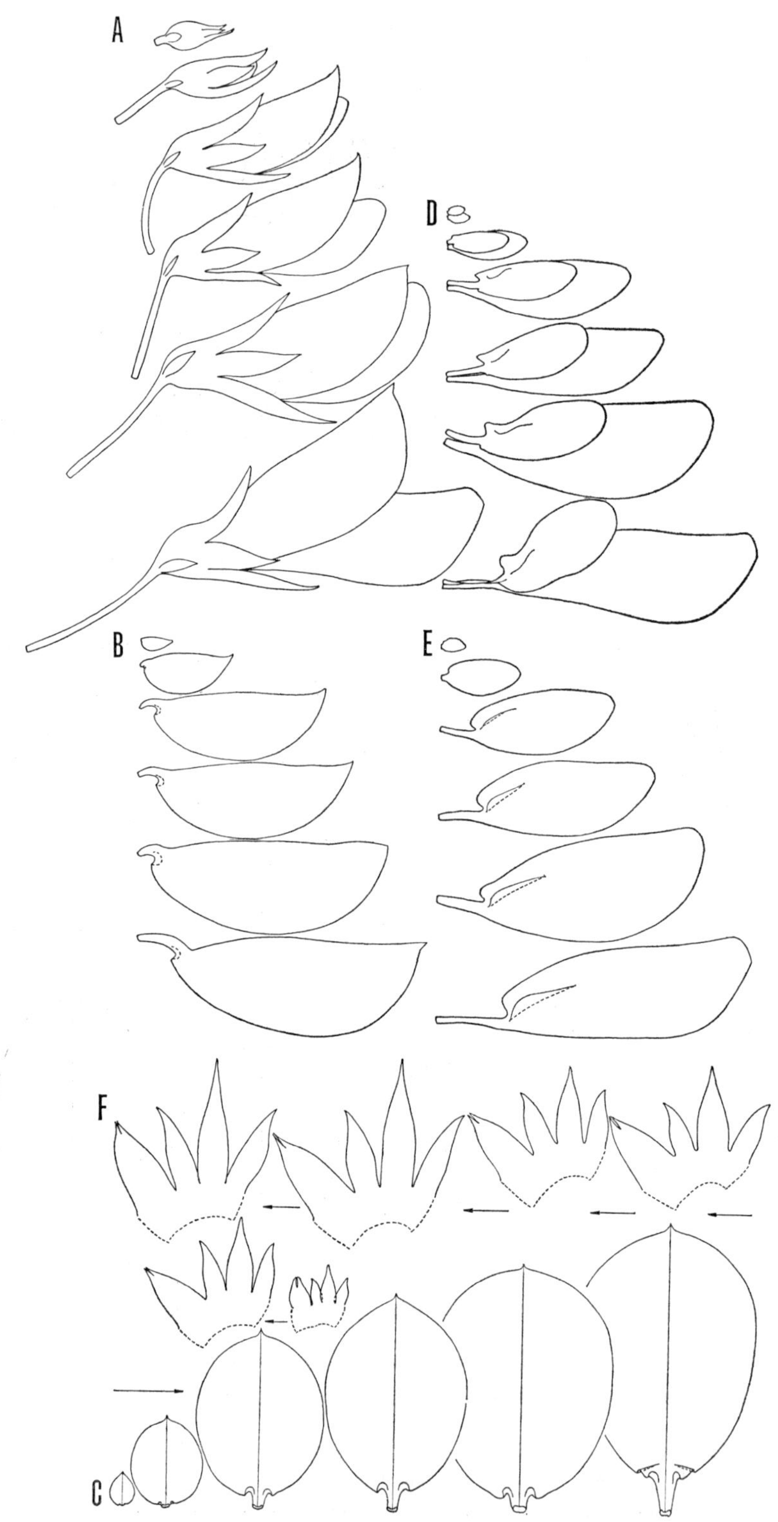

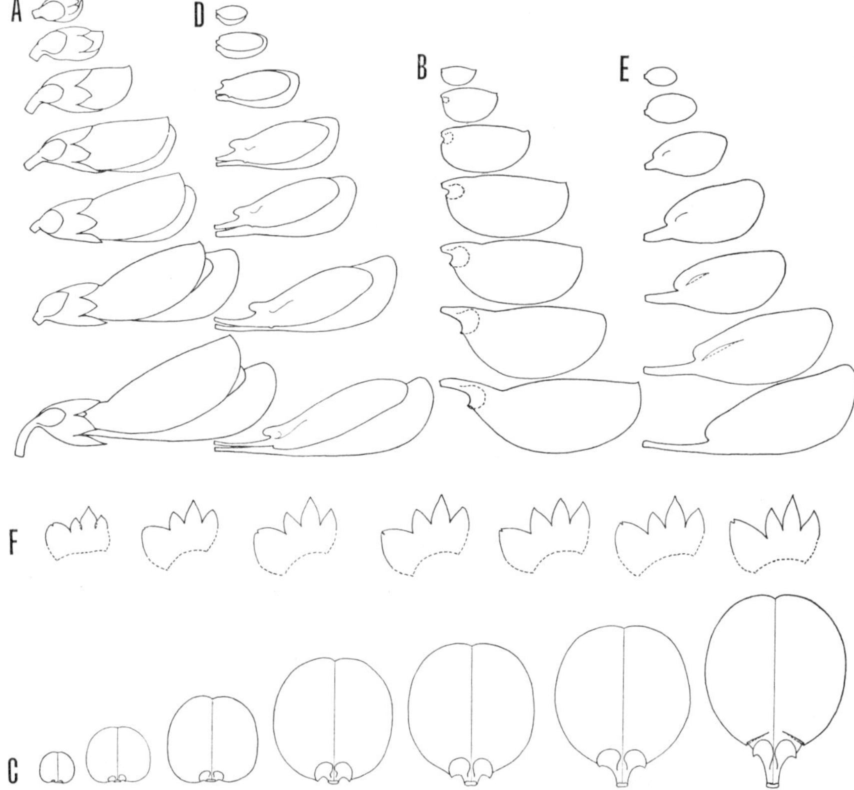

Fig. 30.  Developmental change in flowers of *L. Buergeri* (Ohba & Akiyama 615, TI).
A, flowers, lateral view;  B, standards, lateral view;  C, standards, opened;
D, wings;  E, keel-petals;  F, calyces, dissected.  All×3.3.

←Fig. 29.   Developmental change in flowers of *L. patens* (Ohba & Akiyama 507, TI).
A, flowers, lateral view;  B, standards, lateral view;  C, standards, opened;
D, wings;  E, keel-petals;  F, calyces, dissected.  All×3.3.

80

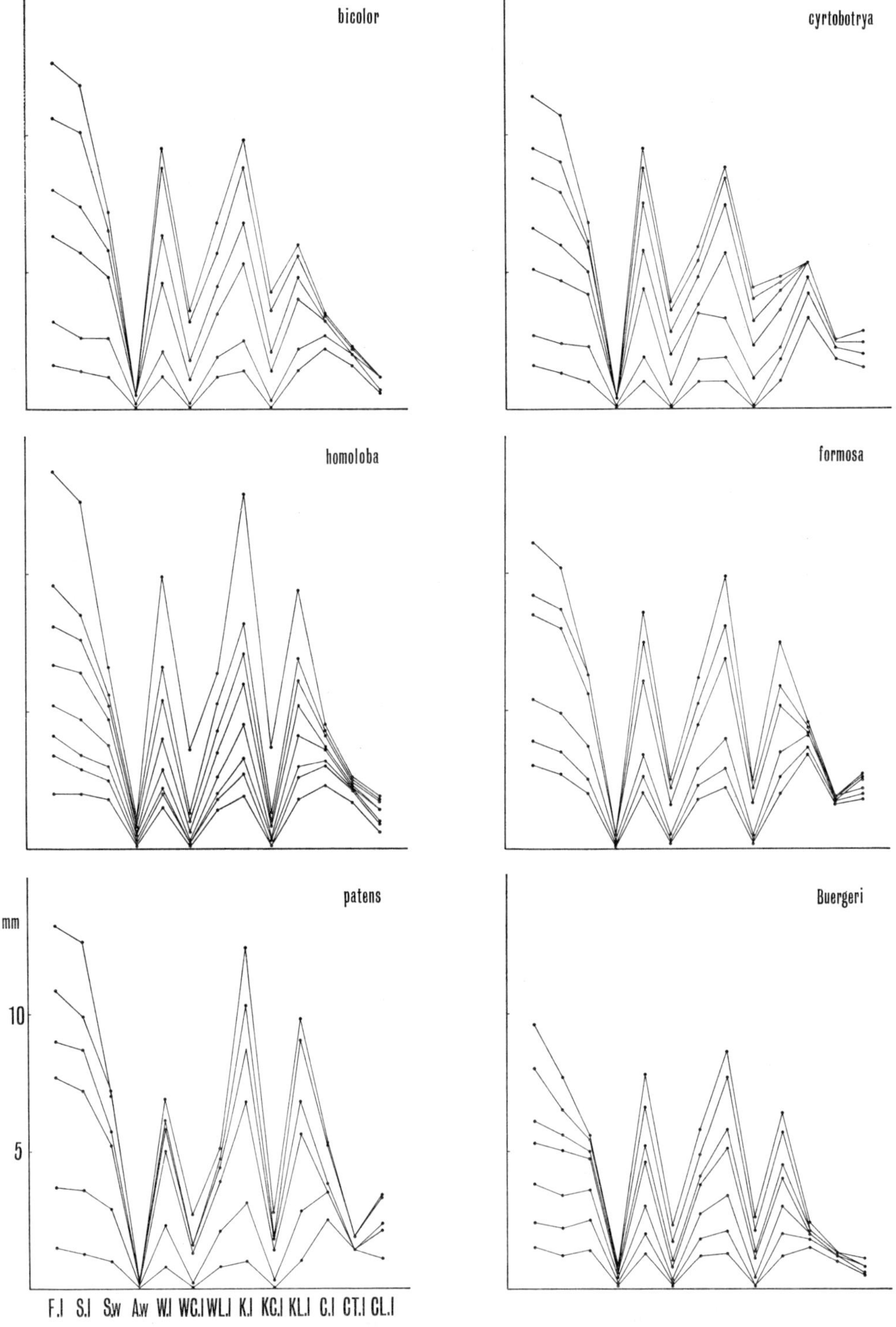


Fig. 31.   Developmental change in flowers of sect. *Macrolespedeza*.

than the keel-petal. In *L. bicolor* (Fig. 25D), *L. homoloba* (Fig. 27D), *L. formosa* subsp. *velutina* (Fig. 28D), *L. patens* (Fig. 29D), and *L. Buergeri* (Fig. 30D) the wing is shorter than the keel-petal throughout growth to anthesis.

The wing-claw and the keel-petal-claw are very short during the very young stage in all flowers. Later, in *L. bicolor* and *L. cyrtobotrya*, the claw becomes more well-developed than in *L. homoloba*, *L. formosa* subsp. *velutina*, *L. patens*, and *L. Buergeri*. When the flower is about 4 mm long, the ratio of the length of the wing-claw to the length of the wing and the ratio of the length of the keel-petal-claw to the length of the keel-petal are larger in *L. bicolor* and *L. cyrtobotrya* than in *L. homoloba*, *L. formosa* subsp. *velutina*, *L. patens*, and *L. Buergeri* (Fig. 30).

The characters described above are exhibited characteristically from the early stages. Thus, these characters are constant and useful in distinguishing species.


## Pollen Grains

The pollen grains of Japanese *Lespedeza* were studied by Ikuse (1956) in *L. bicolor*, *L. cyrtobotrya*, and *L. Buergeri*, and she reported no difference in pollen type or pattern and only slight differences in size. The pollen grains of these species are tricolporate and classified as type 6B[b] by her, while the form of the exine is subreticulate. The size of *L. bicolor* is 19.5–21 × 19.5 µm, somewhat smaller than those of *L. cyrtobotrya* and *L. Buergeri* which are 21–22 × 19.5–21 µm.

Ohashi (1971) examined the pollen grains of the tribe Coronilleae, including *L. bicolor* of sect. *Macrolespedeza*, some species of sect. *Lespedeza*, and some species of *Campylotropis*. In *Lespedeza* and *Campylotropis* the pollen grains are rather uniform and tricolporate. Ferguson & Skvarla (1981) surveyed the pollen grains of the subfamily Papilionoideae morphologically and pointed out that in the tribe Desmodieae (including *Lespedeza* and allied genera in the adapted system) the pollen is diverse in shape, of the tectum type in wall stratification. The tricolporate apertures generally lack an aperture membrane, and a thick endexine without the very thin foot layer which characterizes the pollen of the tribe. They did not examine pollen grains of *Lespedeza*, however.

In the present study we examined the pollen grains of *L. bicolor*, *L. Buergeri*, *L. cyrtobotrya*, *L. formosa* subsp. *velutina*, *L. formosa* subsp. *velutina* var. *satsumensis*, *L. homoloba*, *L. Maximowiczii*, *L. melanantha*, *L. patens*, and *L. Davidii*; four species of sect. *Lespedeza*: *L. pilosa*, *L. cuneata*, *L. tomentosa*, and *L. virgata*; and two species of the genus *Kummerowia* for comparison. Pollen grains were taken from herbarium specimens, coated with platinum, and examined with a scanning electron microscope (Hitachi S-700). The pollen grains of *Lespedeza* are tricolporate and subreticulate (Plates 5 & 6). Pollen morphology of *Lespedeza* shows no specific diversity. The size of the pollen grains of *Lespedeza* is also nearly the same among species (Plates 5 & 6A–E). The shape of the pollen grains of *Kummerowia* is also similar to that of *Lespedeza*, but the exine sculpture is somewhat different. The reticulation is not so clear as in *Lespedeza* (Plate 6F).

## Hairs on Leaf Surfaces

Species of sect. *Macrolespedeza* have hairs on the leaf surfaces. The pubescence of leaves and the length and density of hairs on leaf surfaces have been used as key characters in this section. To examine hairs on leaf surfaces, leaves were taken from herbarium specimens or living plants, coated with platinum, and examined with a scanning electron microscope (Hitachi S-700). The hairs on both leaf surfaces are appressed and covered with pusticulate (wart-like) protuberances. Except for the length and density there is no difference between the hairs on the upper and those on the lower surfaces of leaves (Plates 7–9). In *L. cyrtobotrya* many minute callose appendices are observed on both surfaces of leaves (Plate 7C–E & G). *L. patens* has been distinguished from *L. formosa* subsp. *velutina* by the glabrous upper surface of leaves, or in some plants by the pubescent upper surface of leaves with long, sparse hairs. According to our observations, however, *L. patens* usually has glabrous leaves on the upper surface (Plate 9G), and those individuals with pubescent leaves on the upper surface are scarcely distinguishable from *L. formosa* in the length and density of hairs on the upper surface of leaves (Plate 9H). Moreover, the hairiness of the upper surface of leaves of *L. formosa* is variable, and some plants have glabrous leaves on the upper surface (Akiyama & Ohba, 1988). The hairiness of the upper surface of leaves is also variable in other species, such as *L. bicolor*, *L. Buergeri*, and *L. Maximowiczii*. No taxonomical ranking, even as forma, can be based on this character. *L. homoloba* is characterized by its glabrous leaves on the upper surface except on and along the midrib, and by shorter hairs on the lower surface of leaves (Plate 8C & D).

## Seedlings

De Candolle (1825) made the first major survey of seedlings of the Leguminosae. He divided the greater part of the Papilionoideae into two, those with foliar cotyledons (phyllolobées) and those with storage cotyledons (sarcolobées). That classification was superseded by the system of Bentham (1865). In the Leguminosae the forms of seedlings of many species have been observed and described by various botanists (cf. Duke & Polhill, 1981). The seedlings of some species of *Lespedeza* were described briefly by de Candolle (1825). Vassilczenko (1937) reported the seedling morphology of *L. angustifolia* Ell. of sect. *Lespedeza*.

In the present study the seedlings of the following species were observed in the Botanical Gardens, the University of Tokyo: *L. bicolor*, *L. cyrtobotrya*, *L. homoloba*, *L. formosa* subsp. *velutina*, *L. patens*, and *L. Buergeri*, in sect. *Macrolespedeza*; and *L. cuneata*, *L. inschanica*, *L. juncea*, *L. pilosa*, *L. tomentosa*, and *L. daurica* in sect. *Lespedeza*. In all species of sect. *Macrolespedeza* so far examined, when the seed germinates, the primary root sprouts both seed coat and pericarp. After the primary root has elongated, the hypocotyle elongates orthotropously. The cotyledons are epigeous, exstipulate, asymmetrical, reniform with dense hairs on both surfaces and, have reticulated venation with a primary vein accompanied by a pair of lateral veinlets (Fig. 32). The shape of cotyledons is scarcely distinguishable among different species.

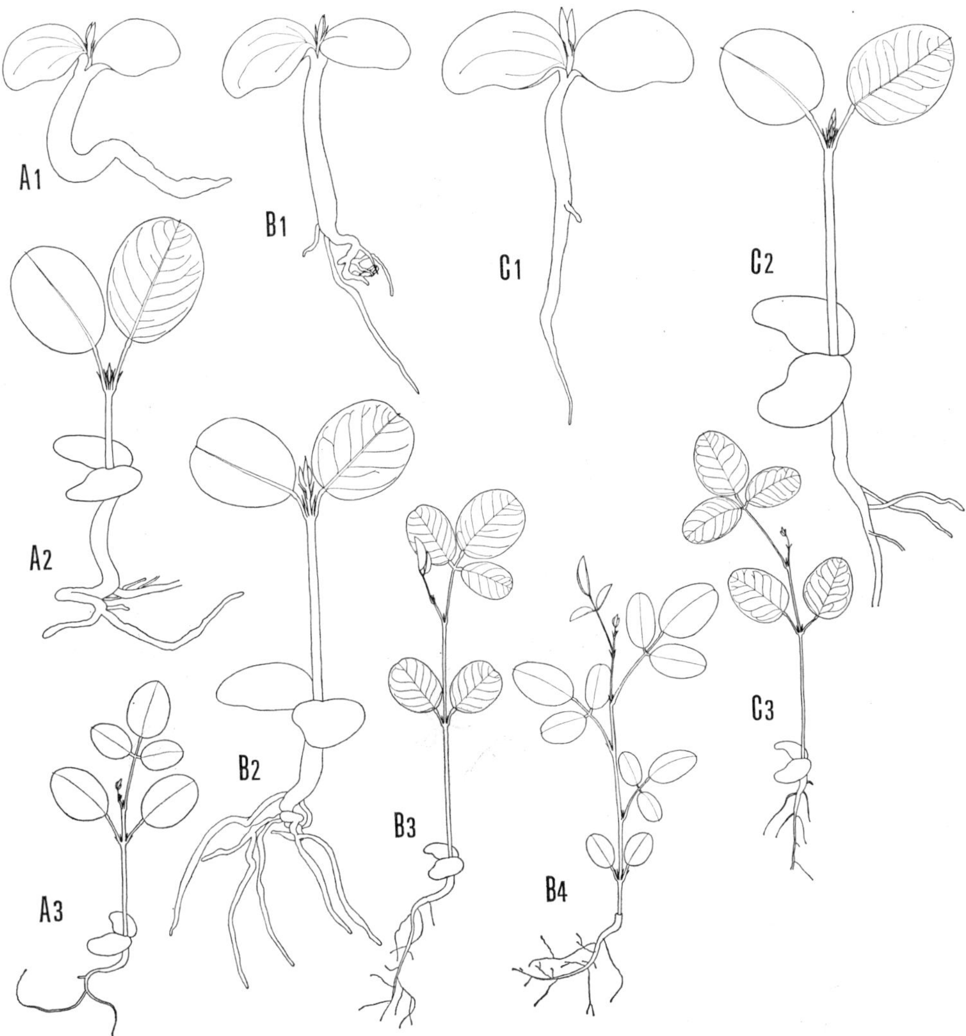

Fig. 32–1.  Seedling of sect. *Macrolespedeza*.
A, *L. bicolor*;  B, *L. cyrtobotrya*;  C, *L. homoloba*;  A1, B1, & C1 ×3.3;
A2, B2, & C2, ×1.7;  A3, B3, C3 ×1.3; others ×1.

The cotyledons, which are thick, colorless, and store nutrients before germination
commences, become green as photosynthetic organs. As soon as the cotyledons expand,
the initial stem appears and the first two opposite eophylls develop decussately (Fig.
32 A2, B2, C2, C2, E2). The first two eophylls are always unifoliolate, entire, with a
pair of stipules and have approximately six to ten lateral veinlets, which do not reach
the margins. The upper portion of the hypocotyl has dense to sparse hairs. The third
leaf is trifoliolate with a pair of stipules like the adult leaves (metaphylls) (Fig. 32 A3,
B3, C3, D3, E3). From the third leaf, the phyllotaxy changes from opposite to spiral

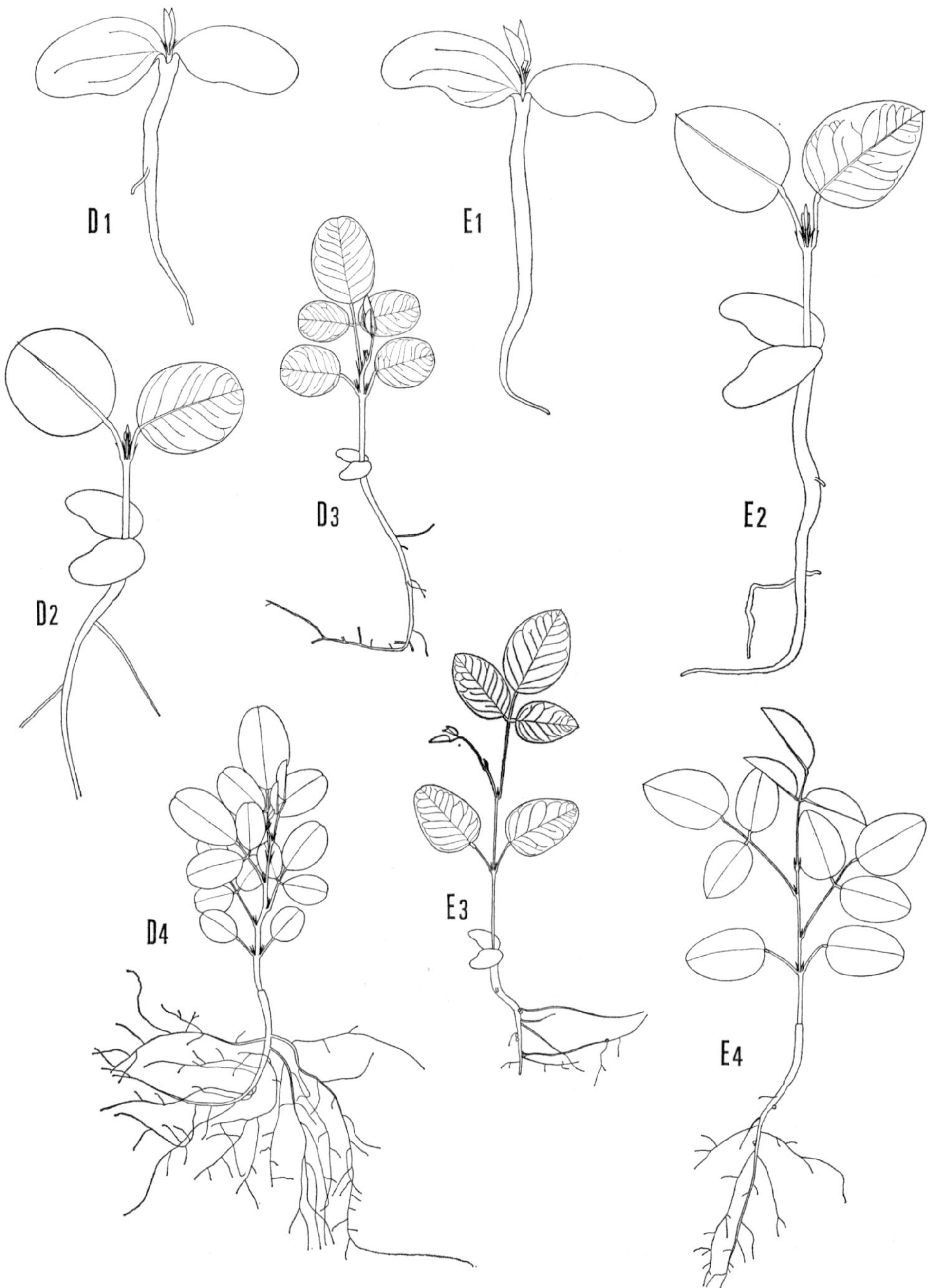

Fig. 32–2.  Seedling of sect. *Macrolespedeza*.
D, *L. patens*;  E, *L. Buergeri*.  D1 & E1 ×3.3;  D2 & E2 ×1.7; others
×1.

in divergence. There is no specific difference in seedling morphology and germination pattern in sect. *Macrolespedeza*. Even in *L. Buergeri* (Fig. 32E) the phyllotaxy in the seedling stage is spiral, although the phyllotaxy is distichous in the adult stage.

In the section *Lespedeza* more diversification in seedling morphology was observed. *L. cuneata*, its allies (*L. juncea* and *L. inschanica*), and *L. virgata* have seedlings like those of species of sect. *Macrolespedeza*. In *L. pilosa* and *L. tomentosa* the third leaf is unifoliolate. The seedling bears a few unifoliolate and sometimes bifoliolate leaves, and then trifoliolate leaves grow.

## Buds and Bud-scales

In the section *Macrolespedeza* winter buds are mainly axillary and consist of about ten to twenty bud-scales (Fig. 33). The bud-scales are arranged spirally (Fig. 33 B–E) in all species of the sect. *Macrolespedeza* except two species, *L. Buergeri* (Fig. 33 A) and *L. Maximowiczii* as pointed out by Momiyama (1933). The outer bud-scales are hard, ovate, and brownish. The inner ones are also ovate and only the exposed parts are brownish while the hidden parts are greenish and elongate the following spring. The inner bud-scales are sometimes two-lobed or three-lobed. Numerous juvenile normal leaves with three leaflets and two stipules are surround by the bud-scales. In some cases, the intermediate form (having two stipules and rudimentary leaflets) between the inner bud-scales and normal leaves is observed (Fig. 33 C). A small axillary bud is recognizable in the axil of the bud-scale (Fig. 33 B–E). But the axillary bud does not usually elongate.

Nakai (1939) distinguished sect. *Heterolespedza*, which consists of *L. Buergeri* Miq. and *L. Maximowiczii*, from sect. *Macrolespedeza* based on the difference of phyllotaxy of bud-scales: the phyllotaxy of *L. Buergeri* (Fig. 33 A) and *L. Maximowiczii* is distichous, while that of other species is spiral. But except for phyllotaxy the significant characters of bud-sclaes of these two species are not different from those of other species of sect. *Macrolespedeza*.

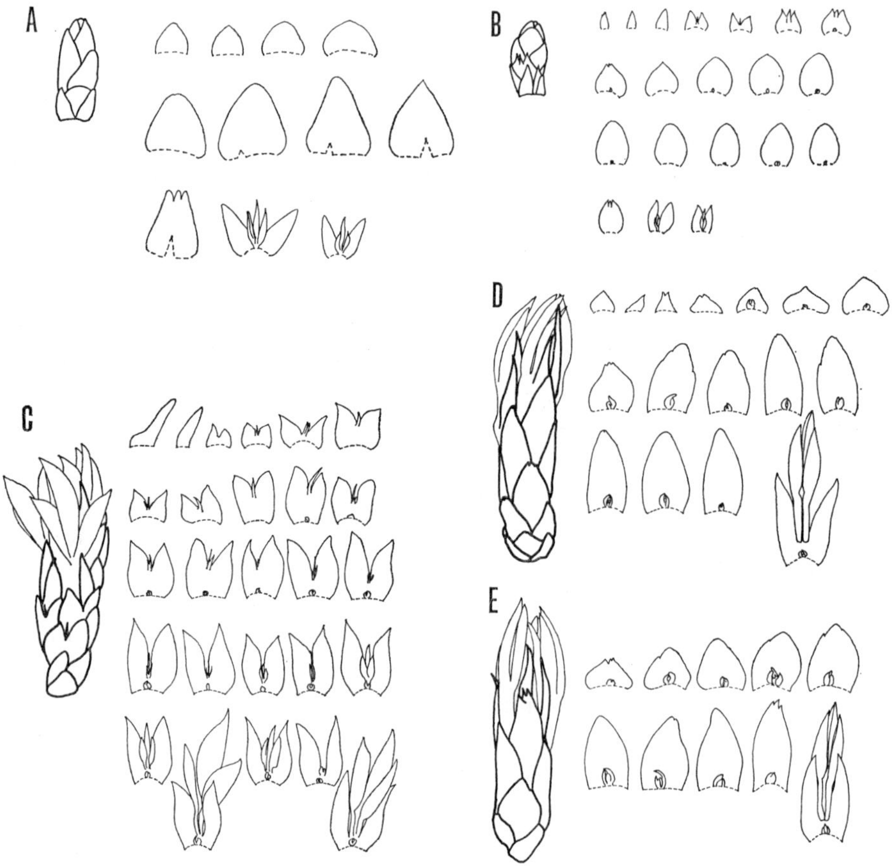



Fig. 33.  Winter buds and bud-scales of sect. *Macrolespedeza.*
A, *L. Buergeri* (December);  B, *L. bicolor* (December);  C, *L. cyrtobotrya* (April);  D & E, *L. homoloba* (April).  All × 3.3.

# Chromosome Numbers

Chromosome numbers of sect. *Macrolespedeza* have been reported by several workers. Kawakami (1930) reported the haploid number to be nine in *L. bicolor, L. cyrtobotrya, L. homoloba, L. Sieboldii* (=*L. Thunbergii*), and *L. Sieboldii* var. *albiflora* (=*L. japonica*). Cooper (1936) examined the chromosome numbers of the species of sect. *Lespedeza* and reported the diploid number to be 18 in *L. variegata* and *L. sericea*, 20 in *L. tomentosa*, and 36 in *L. daurica*. He suggested that nine is probably the basic number for the genus *Lespedeza*. Pierce (1939) reported the somatic chromosome number to be 22 in *L. bicolor, L. cyrtobotrya,* and *L. cyrtobotrya* var. *pedunculata, L. Maximowiczii* (non Schneid.) (=*L. bicolor*), *L. Thunbergii, L. japonica* var. *intermedia, L. robusta, L. floribunda,* and *L. virgata*, n=9 in *L. sericea*, 2n=20 in *L. violacea, L. Stuvei, L. repens, L. capitata, L. variegata, L. hirta, L. pilosa, L. procumbens, L. frutescens,* etc. He suggested that the basic chromosome number for the American species of *Lespedeza* is 10, and that for other species the basic number is 11 although n=9 has been reported for certain Japanese species. Young (1940) reported n=9, 10, and 2n=20 in American species, n=11 in *L. Thunbergii*, and 2n=22 in *L. Uyekii*. Lee (1969) reported 2n=20 in *L. bicolor* var. *typica, L. cuneata*, 2n=40 in *L. juncea* var. *inschanica*, and 2n=22 in *L. Uyekii*. Clewell (1971) reported n=11 in *L. bicolor, L. cyrtobotrya, L. homoloba,* and *L. Thunbergii*, and 2n=20, n=10, and 2n=22 in *L. cuneata*. Kondo *et al.* (1977) examined intraspecific variation of karyotypes in *Lespedeza* and *Desmodium* and reported 2n=22 in *L. cyrtobotrya* and 2n=20 in *L. cuneata* and *L. pilosa*. They reported intraspecific variation of karyotypes in *L. cyrtobotrya* with five karyotypes. They pointed out that *Lespedeza* and *Desmodium* had fewer differences than systematic treatments had previously suggested, although the chromosome number of the species of both genera seemed to be divided into two groups, one with 2n=20 and the other with 2n=22.

In the present study the chromosome numbers of the following species were examined: *L. bicolor, L. Buergeri, L. cyrtobotrya, L. formosa* subsp. *velutina, L. homoloba, L. patens,* and *L. Thunbergii*. Chromosome numbers were counted in root tips using the squashing method. Root tips were obtained from plants cultivated in the Botanical Gardens, the University of Tokyo, pretreated with 0.002M 8-hydroxyquinoline aqueous solution for 4 hours at room temperature, and fixed with Farmer's solution ethyl alcohol: glacial acetic acid=3:1) for more than 1 hour. Root tips were stained (absolute with 5% aceto-orcein solution in 1N hydrochloric acid and then squashed. Our results suggest that the basic number for sect. *Macrolespedeza* is not 9 but 11 (Table 28). Although polyploidy is not reported in sect. *Macrolespedeza*, Murata (1978) suggested the possibility for polyploidy of *L. patens*. According to our observations *L. patens* has the somatic number 2n=22 like other species of sect. *Macrolespedeza*; no polyploidy was observed. The lack of polyploidy and chromosome diversification is one reason why hybridization occurs frequently in sect. *Macrolespedeza*. The chro-

Table 28.  Chromosome numbers in sect. *Macrolespedeza*.

| Species | n | 2n | Investigator |
|---|---|---|---|
| *L. bicolor* Turcz. | 9 | | Kawakami 1930 |
| | | 22 | Pierce 1939 |
| (var. *typica* Maxim.) | 10 | | Lee 1969 |
| | 11 | | Clewell 1971 |
| | 11 | 22 | present study[1] |
| *L. Buergeri* Miq. | | 22 | present study[2] |
| *L. cyrtobotrya* Miq. | 9 | | Kawakami 1930 |
| | | 22 | Pierce 1930 |
| (var. *pedunculata* Nakai) | | 22 | Pierce 1930 |
| | | 22 | Kondo et al. 1977 |
| | | 22 | present study[3] |
| *L. formosa* (Vogel) Koehne subsp. *velutina* (Nakai) S. Akiyama et H. Ohba | | 22 | present study[4] |
| *L. homoloba* Nakai | 9 | | Kawakami 1930 |
| | 11 | | Clewell 1971 |
| | | 22 | present study[5] |
| "*L. japonica* L. H. Bailey var. *intermedia* Nakai" | | 22 | Pierce 1930 |
| | | 20 | Lee 1970 |
| "*L. Maximowiczii* Gandog." (=*L. bicolor*) | | 22 | Pierce 1930 |
| *L. patens* Nakai | | 22 | present study[6] |
| "*L. robusta* Nakai" | 9 | | Kawakami 1930 |
| "*L. Sieboldii* Miq." | 9 | | Kawakami 1930 |
| "var. *albiflora* Schneid." | 9 | | Kawakami 1930 |
| *L. Thunbergii* (DC.) Nakai | 11 | | Young 1940 |
| | 11 | | Clewell 1971 |
| | | 22 | present study[7] |
| "*L. Uyekii* Nakai" | | 22 | Young 1940 |
| | | 22 | Lee 1969 |

1) Aomori Pref., Shimokita-gun, Wakinosawa-mura (Akiyama CHR8405, TI).  2) Tokyo Pref., Nishitama-gun, Itsukaichi-machi (Akiyama CHR8408, TI).  3) Tokyo Pref., Nishitama-gun, Itsukaichi-machi (Akiyama CHR8414, TI).  4) Hyogo Pref., Shikama-gun, Yumesaki-mura (Akiyama CHR8412, TI).  5) Miyagi Pref., Kurokawa-gun, Ohira-mura (Akiyama CHR8423, TI).  6) Nagano Pref., Kamiminochi-gun, Otari-mura (Akiyama CHR8425, TI).  7) Cultivated in Tokyo (Akiyama CHR8428, TI).

mosome number 2n=22 occurs not only in sect. *Macrolespedeza* and sect. *Lespedeza*, but also in the genera *Campylotropis*, *Kummerowia*, and *Desmodium*.

# Relationships of Species of the Section *Macrolespedeza*

The section *Macrolespedeza* is a monophyletic group because its woody habit is regarded as an autapomorphic character. The secttion *Lespedeza* with its herbaceous habit is also a natural group characterized by another autapomorphy, its cleistogamous flower. The genus *Phylacium* is considered to be an outgroup of the genus *Lespedeza*.

In sect. *Macrolespedeza*, *L. Buergeri* and *L. Maximowiczii* have distichous phyllotaxy. The distichous character is considered to be derivative, because the phyllotaxy of the seedlings of *L. Buergeri* is spiral like that of the other species but later changes to distichous, and the spiral phyllotaxy is more common than the distichous one in *Lespedeza*. This character is considered to be derivative (apomorphic) and is found only in *L. Buergeri* and *L. Maximowiczii*. Thus these two species are taxonomically considered to form a subgroup, the series *Heterolespedeza*. *L. bicolor*, *L. cyrtobotrya*,

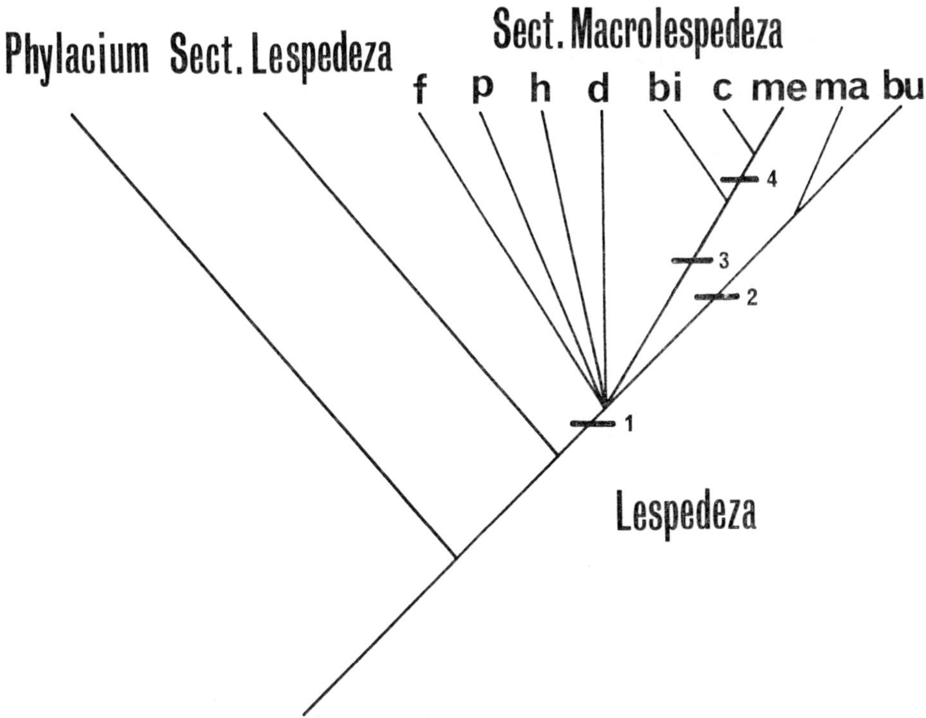



Fig. 34.　Relationship of the species of sect. *Macrolespedeza*.
bu, *L. Buergeri*; ma, *L. Maximowiczii*; me, *L. melanantha*; c, *L. cyrtobotrya*; bi, *L. bicolor*; d, *L. Davidii*; h, *L. homoloba*; p, *L. patens*; f, *L. formosa*.
1. Woody habit. 2. Distichous phyllotaxy. 3. Standard with attenuate base. 4. Wing longer than keel-petal.

and *L. melanantha* have standards with an attenuated base. This shape of standards is also considered to be derivative, because the standard with the clawed base is more common in *Lespedeza*. The standard with an attenuate base is more derivative than that with a clawed base and observed only in these three species. These three species are therefore put in a subgroup, series *Macrolespedeza*. In this series *L. cyrtobotrya* and *L. melanantha* have wings longer than keel-petals. This character state is also considered to be derivative, because *L. bicolor* and other species of sect. *Macrolespedeza* have wings shorter than keel-petals. The character state in *L. cyrtobotrya* and *L. melanantha* is regarded as autapomorphy, and thus *L. cyrtobotrya* is more closely related to *L. melanantha* than to *L. bicolor*. *L. patens* is very similar to *L. formosa* but no synapomorphic character has been found in the floral and vegetative characters. *L. patens* is different from the latter in its herbaceous nature. This may be due to adaptation to heavy snowfall in the winter season in its habitat. *L. homoloba* is characterized by well-developed auricles, but except for this character it is similar to *L. formosa*. The well-developed auricle of the standard may be a specialization. *L. Davidii* is quite different from other species in vegetative characters such as the shape and texture of leaves and the branches, but no nearest species can be detected in sect. *Macrolespedeza*. In *L. formosa*, *L. patens*, *L. homoloba*, and *L. Davidii*, no autapomorphic character has been found.

At present in sect. *Macrolespedeza*, ser. *Heterolespedeza* and ser. *Macrolespedeza* are revealed to be natural groups, but the relationship of the other four species has not been determined. We are tentatively proposing a new series, ser. *Formosae*, consisting of these four species.

# Systematic Treatments

**Lespedeza** Michx. [ex Richard], *Fl. Boreal.-Amer.* **2**: 70 (1803)—de Candolle, *Prodr.* **2**: 348 (1825)—G. Don, *Gen. Hist.* **2**: 307 (1838)—Endlicher, *Gen. Pl.* 1286, n. 6623 (1840)—Bentham in Benth. & Hook. f., *Gen. Pl.* **1**: 524 (1865)—Maxim. in *Acta Hort. Petrop.* **2**: 345 (1873)—Taubert in Engl. & Prantl, *Nat. Pfl.-fam.* III, **3**: 332 (1891)—Schindler in Engl., *Bot. Jahrb.* **49**: 570 (1913)—Nakai, *Lesp. Jap. Korea* (in *Bull. For. Exp. Stat. Chosen* **6**:) 1 (1927)—Hutchinson, *Gen. Fl. Pl.* **1**: 487 (1964)—Ohashi in *Journ. Fac. Sci. Univ. Tokyo*, III, **11**: 62 (1971)—Ohashi, Polhill & Schubert in Polhill & Raven (ed.), *Advan. Leg. Syst. Part* 1, 300 (1981).

Type species: *L. virginica* (L.) Britton (=*L. sessiliflora* Michx.).

Key to the sections of the genus *Lespedeza*
1. Perennial herb with both short inflorescence of cleistogamous flowers and longer inflorescence of chasmogamous flowers ....................................................Sect. *Lespedeza*[1]
1. Small shrub without cleistogamous flowers.............................Sect. *Macrolespedeza*

Section **Macrolespedeza** Maxim. in *Acta Hort. Petrop.* **2**: 346 (1873), ut "*Macro-Lespedeza*"—Schindler in Engl., *Bot. Jahrb.* **49**: 574 (1913)—Nakai, *Lesp. Jap. Korea*, 8 (1927)—Hatusima in *Mem. Fac. Agr. Kagoshima Univ.* **6**: 1 (1967), *in sensu emed., excl. L. floribunda* Bunge.

*Lespedeza* sect. *Archilespedeza* Taub. A. *Macrolespedeza* Maxim. apud Taubert in Engl. & Prantl, *Nat. Pfl.-fam.* III, **3**: 332 (1891).

*Lespedeza* sect. *Heterolespedeza* Nakai in *Journ. Jap. Bot.* **15**: 531 (1939)—Hatusima in *Mem. Fac. Agr. Kagoshima Univ.* **6**: 1 (1967).

*Lespedeza* subgen. *Macrolespedeza* (Maxim.) Ohashi in *Journ. Jap. Bot.* **57**: 29 (1982), *cum* sect. *Macrolespedeza* et sect. *Heterolespedeza*.

Type species: *L. bicolor* Turcz. (Nakai, 1939).

Key to the series of the section *Macrolespedeza*
1. Scales of winter buds and leaves spirally arranged ...........................................2
   2. Standard with attenuated base ....................................Series *Macrolespedeza*
   2. Standard with clawed base...................................................Series *Formosae*
1. Scales of winter buds and leaves distichously arranged ...........Series *Heterolespedeza*

## Series Macrolespedeza

Key to the species of the series *Macrolespedeza*
1. Wing shorter than keel-petal. Calyx-lobe ovate or triangular-lanceolate, obtuse or acute at the apex. Flowers arranged sparsely, 4–12 in an inflorescence ..............*L. bicolor*
1. Wing longer than keel-petal
   2. Calyx-lobe triangular, acuminate at the apex. Flowers arranged compactly, 4–10 in an inflorescence...................................................................... *L. cyrtobotrya*

---

[1] The sect. *Lespedeza* is not mentioned in this paper.

2. Calyx-lobe semicircular, round at the apex. Flowers 2–4(6) in an inflorescence
..................................................................................... *L. melanantha*

## Lespedeza bicolor Turcz.

In *Bull. Soc. Nat. Mosc.* **13**: 69 (1840)—Ledeb., *Fl. Ross.* **1**: 715 (1842)—Maxim., *Prim. Fl. Amur.*, 86 et 470 (1859); in *Acta Hort. Petrop.* **2**: 355 (1873), *incl. α. typica* (*excl. quoad specim. cit.* 'Colitur circa templa Nagasaki floribus candidis') et *β. intermedia* Maxim. (*incl. quoad specim. cit.* Tatarinow), *excl. γ. Sieboldii*—Regel in *Gartenflora* **60**: 270, t. 299 (1860); in *Mém. Acad. Imp. Sci. St.-Pétersb.* 7 sér. **4**: 48 (1861)—Miquel, *Ann. Bot. Ludg.-Bat.* **3**: 47 (1867) et *Prol. Fl. Jap.*, 235 (1867), *excl. quoad specim. cit.* Buerger, Siebold (*pro parte*) et Textor—Fr. Schmid tin *Mém. Acad. Imp. Sci. St.-Pétersb.* 7 sér. **12**: 38 (1868)—Koch, *Dendrol.*, **1**: 73 (1869) —Debeaux in *Act. Soc. Linn. Bordeaux* **31**: 140 (1876), *pro parte* (*incl. β. intermedia* Maxim., *excl. γ. Sieboldii* (Miq.) Maxim. et *syn. L. Sieboldii* Miq.)—Franchet & Savatier, *Enum. Pl. Jap.* **1**: 101 (1875)—Franchet in *Nouv. Arch. Mus. Paris*, sér. 2, **5**: 247 (1883); *Pl. David.* **1**: 95 (1884)—Forbes & Hemsley in *Journ. Linn. Soc. Bot.* **23**: 178 (1887), *pro parte*—Koehne, *Deutsch. Dendrol.*, 343 (1893), *pro parte* (cf. Nakai,

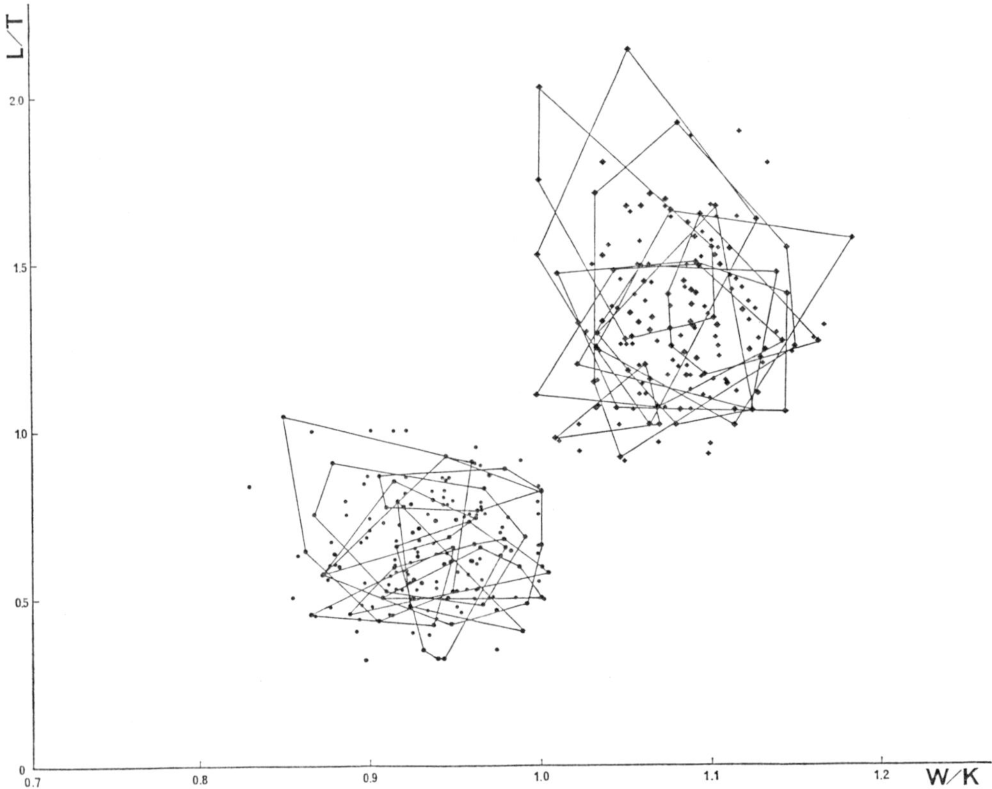

Fig. 35. Scatter diagram showing the variation of *L. bicolor* and *L. cyrtobotrya*.
W/K, Ratio of the length of wing to the length of keel-petal; L/T, ratio of the length of calyx-lobe to the length of calyx-tube.

1927)—Diels in Engl., *Bot. Jahrb.* **29:** 415 (1901)—Matsumura in *Bot. Mag. Tokyo* **14:** 51 (1902), *pro parte, incl.* var. *intermedia* Maxim. f. *acutifolia* Matsum., f. *grandifolia* Matsum. et f. *parvifolia* Matsum.; *Ind. Pl. Jap.* **2**(2): 267 (1912), *pro parte*—Komarov, *Fl. Manshuriae* **2:** 599 (1904)—Schneider, *Ill. Handb. Laubholz.* **2:** 112, fig. 70f, 71a–d (1907)—Nakai in *Journ. Coll. Sci. Univ. Tokyo* **26** (*Fl. Koreana* I): 155 (1909), *pro parte*; in *Journ. Coll. Sci. Univ. Tokyo* **32** (*Fl. Koreana* II): 467 (1911); in *Bot. Mag. Tokyo* **37:** 73 (1923); *Lesp. Jap. Korea,* 63 (1927)—Pampanini in *Nuov. Giorn. Bot. Ital.* n. ser. **17:** 398 (1910)—Makino in Iinuma, *Sōmoku-dzusetsu* (ed. Makino) **3:** 997, pl. 19 (1912), *pro parte*—Schindler in Engl., *Bot. Jahrb.* **49:** 583 (1913); in Sargent, *Pl. Wils.* **2:** 112, *pro parte*—Bean, *Trees & Shrubs Brit. Isl.* **2:** 16 (1914), *excl.* var. *alba* Bean—Bailey, *Stand. Cyclop. Hort.* **4:** 1845, fig. 2134 (1916), *excl.* "A white-fld. variety"—Makino & Nemoto, *Fl. Jap.,* 734 (1925)—Rehder in *Journ. Arn. Arb.* **7:** 173 (1926)—Vasil'ev in Komarov ed., *Fl. U.S.S.R.* **13:** 379, pl. 20-1 (1928)—Kitagawa, *Lineam. Fl. Mansh.,* 288 (1939); *Neo-Liueam. Fl. Mansh.,* 405 (1979)—Ohwi, *Fl. Jap.,* 678 (1953), *excl.* f. *alba* (Bean) Ohwi et *syn. L. melanantha* Nakai; *Fl. Jap.* rev. ed., 790 (1965), *excl.* f. *alba* (Bean) Ohwi et *syn. L. melanantha* Nakai; *Fl. Jap.* Eng. ed., 559 (1965), *excl.* f. *alba* (Bean) Ohwi; *Fl. Jap.* new ed., 790 (1975), *excl.* f. *alba* (Bean) Ohwi et *syn. L. melanantha* Nakai—Fu & Wang in Liou et al., *Ill. Fl. Lig. Pl. N.-E. China,* 342 (1955)—Fu in Acad. Sin. Bot., *Ill. Important Chin. Pl. Leguminosae,* 519 (1955)—Kitamura & Murata, *Col. Ill. Herb. Pl. Jap.* **2:** 100 (1961)—Kung in *Chinese Journ. Bot.* **1:** 22 (1963)—Hatusima in *Mem. Fac. Agr. Kagoshima Univ.* **6:** 14 (1967), *excl. syn. L. melanantha* Nakai, *L. melanantha* Nakai var. *longifolia* Uyeki, *L. melanantha* Nakai f. *rosea* Nakai, *L. homoloba* Nakai var. *higoensis* T. Shimizu—Acad. Sin. Bot., *Iconogr. Cormophyt. Sin.* **2:** 458, fig. 2646 (1972)—Lee, in *Bull. Seoul Nat. Univ. For.* No. 2, 5 (1965), excl. f. *alba* (Bean) Ohwi; *Ill. Fl. Korea,* 468 (1979)—Ohashi in Satake et al., *Wild Fl. Jap. Herb.* **2:** 205 (1982), *excl.* var. *higoensis* (T. Shimizu) Murata.

[Plates 10 & 11; Fig. 36]

Key to the varieties
1. Raceme longer than subtending leaf. Wing distinctly shorter than keel-petal. 1.5–2 m high ...............................................................................var. *bicolor*
1. Raceme shorter than subtending leaf. Wing nearly equal to or slightly Shorter than keel-petal. 15–30 cm high ...............................................................................var. *nana*

## Var. **bicolor**

*L. bicolor* Turcz. f. *microphylla* Miq. in *Ann. Bot. Lugd.-Bat.* **3:** 47 (1867); *Prol. Fl. Jap.,* 235 (1867), *pro maj. parte.*

*L. bicolor* Turcz. α. *typica* Maxim. in *Acta Hort. Petrop.* **2:** 356 (1873), *pro major parte.*

*L. bicolor* Turcz. β. *intermedia* Maxim. in *Acta Hort. Petrop.* **2:** 356 (1873), *pro parte, excl. quoad specim.* "Tschonoski s.n." et "Oldham 335"—Matsumurain *Bot. Mag. Tokyo* **16:** 69 (1902), *pro parte*; *Ind. Pl. Jap.* **2:** 267 (1912), *pro parte.*

*L. bicolor* Turcz. var. *intermedia* Maxim. f. *acutifolia* Matsum. in *Bot. Mag. Tokyo* **16:** 69 (1902)—Ohwi, *Fl. Jap.,* 678 (1953); *Fl. Jap.* rev. ed., 790 (1965); *Fl. Jap.* Eng.

ed., 559 (1965); *Fl. Jap.* new ed., 790 (1975), ut "*L. bicolor* Turcz. f. *acutifolia* Matsum."

*L. bicolor* Turcz. var. *intermedia* Maxim. f. *grandifolia* Matsum. in *Bot. Mag. Tokyo* **16**: 69 (1902).

*L. bicolor* Turcz. var. *intermedia* Maxim. f. *parvifolia* Matsum. in *Bot. Mag. Tokyo* **16**: 69 (1902).

*L. japonica* (non L. H. Bailey) Schindl. in Engl., *Bot. Jahrb.* **46**: Beibl. 106, 54 (1912), *nom. nud.*

*L. bicolor* Turcz. var. *japonica* Nakai in *Bot. Mag. Tokyo* **37**: 73 (1923).

*L. bicolor* Turcz. var. *sericea* Nakai, *Lesp. Jap. Korea*, 66 (1927).

*L. setiloba* Nakai, *Lesp. Jap. Korea*, 68 (1927).

*L. spicata* Nakai et F. Maekawa in *Bot. Mag. Tokyo* **48**: 52 (1934).

*L. spicata* Nakai et F. Maekawa f. *acutifolia* Nakai et F. Maekawa in *Bot. Mag. Tokyo* **48**: 53 (1934).

*L. Tobae* H. Koidzumi in *Journ. Pl. Iwateken* **2**: 75 (1937), *typum non vidi.*

*L. ionocalyx* Nakai in *Journ. Jap. Bot.* **15**: 532 (1939).

*L. bicolor* Turcz. f. *tomentella* Hatusima in *Mem. Fac. Agr. Kagoshima Univ.* **6**: 14 (1967).

*L. bicolor* Turcz. f. *sericea* (Nakai) Hatusima in *Mem. Fac. Agr. Kagoshima Univ.* **6**: 14 (1967), non Matsum. (1902)

*L. bicolor* Turcz. f. *Nakaiana* Murata in *Acta Phytotax. Geobot.* **34**: 129 (1983), *nom. nov.*

Japanese name. Yamahagi (Matsumura, 1902).

A perennial plant, 1.5–2 m high. Stems ascending, 1–3 cm in diameter; most of the terrestrial part lives a few years; branched in upper parts (a few branches coming from basal part of the main stem). Branches terete, ascending, sometimes dependent later; when young slightly angular with densely ascending appressed-sericeous (very rarely spreading) hairs (hairs 0.3 mm long, whitish), later glabrescent; distal parts die in winter, proximal parts (and the main stem itself) with axillary and adventitious winter buds; winter buds at lower parts mostly dormant.

Leaves trifoliolate, petiolate, stipulate, spirally arranged. Stipules free, linear-triangular–linear, 2–7 mm long, brownish, persistent. Petioles 0.5–4 cm long, hairy like the young branch. Rachides 5–15 mm long, similar to the petiole. Terminal leaflets (at middle parts of branches) petiolulate; petiolules 1–3 mm long, swollen; lamina 2–6 cm long, 1–3.5 cm wide, entire, (broadly) elliptic–(broadly) ovate or obovate, round or cuneate at the base, obtuse, retuse, or acute at the apex (the apex itself with or without a point), upper surface glabrescent (except along the midrib) or sericeous (hairs 0.3–0.5 mm long) at anthesis, lower surface appressed-sericeous (hairs 0.2–0.3(–0.5) mm long); lateral ones similar to terminal but somewhat smaller.

Inflorescence axillary racemous, solitary, 2–7(–10) cm long including the peduncle, somewhat loosely 4–12–flowered; peduncles 1–3 cm long, hairy like the young branch. Primary bracts ovate, about 1 mm long, pubescent, persistent; secondary bracts similar to primary ones.

Flowers 9.6–12.6 mm long at anthesis; pedicels 0.8–3 mm long, terete, curved, tomentose. Bracteoles at the base of the calyx broadly to narrowly ovate or obtuse,

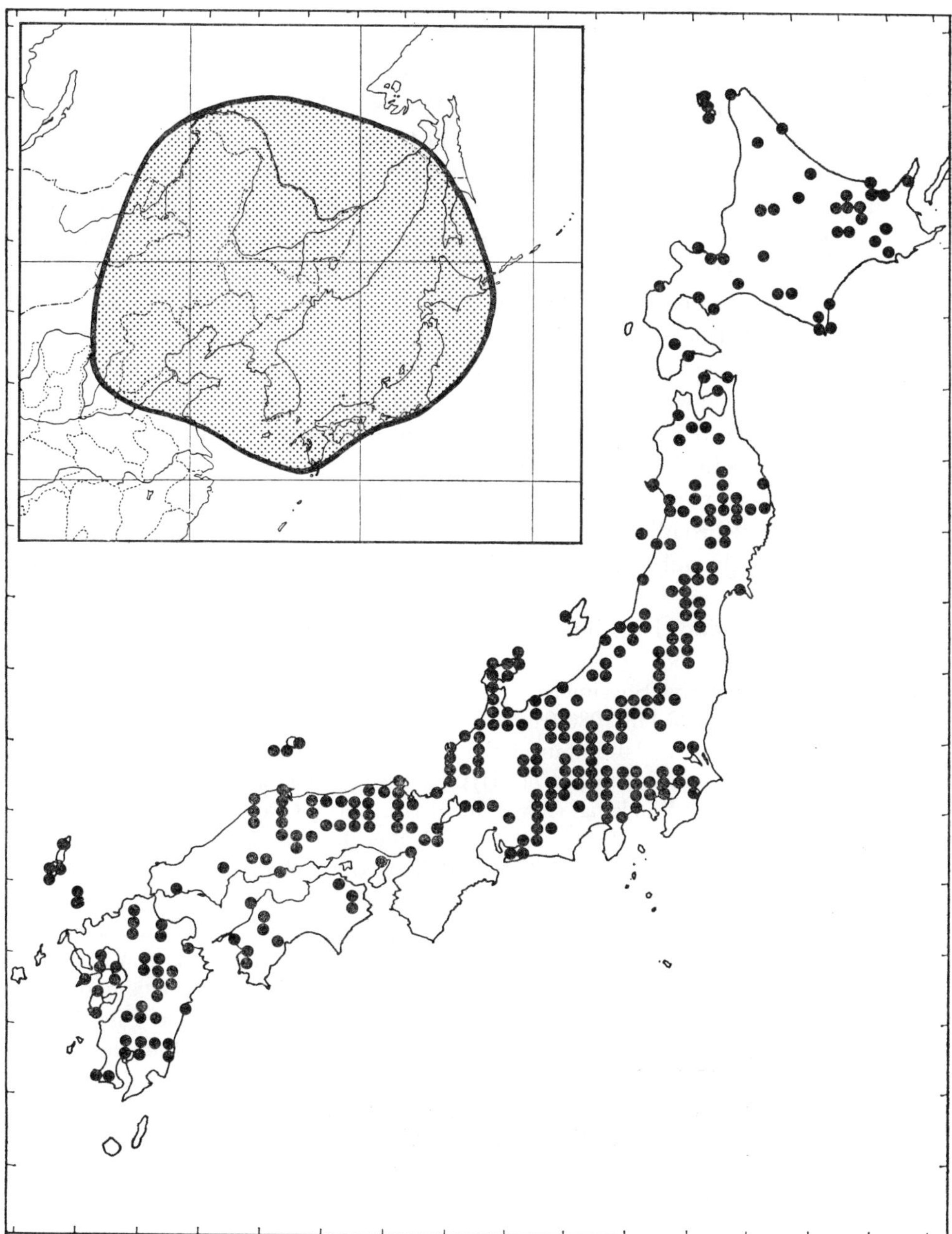

Fig. 36. Distribution of *L. bicolor*.

0.6–1.2 mm long, pubescent outside, glabrous inside, persistent. Calyx 3.1–4.7 mm long, tubulate, four-lobed near the middle part or lesser than or more than the middle part, densely appressed pubescent; tube 1.5–2.4 mm long; lobes subequal in length or the lower one longest; lateral ones 1.2–2.3 mm long, 0.9–1.3 mm wide, ovate or

96

triangular-lanceolate, obtuse or acute at the apex but not acuminate; upper one slightly two-cleft or not.

Standard longer than keel-petal, keel-petal longer than (rarely equal to or very rarely slightly shorter than) wings (S>K>W). Standard red-purple inside, paler outside, whitish near the base, obovate, 9.3–11.8 mm long, 5.4–7.2 mm wide, round or retuse at the apex with or without a point, attenuated at the base without distinct claw, inflexed-auriculate near the base, spreading (or reflexing) in anthesis from the part of 1/3 from the base. Wings deeper red-purple than standard, 7.3–9.5 mm long with distinct claw; lamina narrowly obovate to obovate, 4.4–6.8 mm long, 2.0–2.5 mm wide, auriculate at upper basal part; claw 3.2–3.9 mm long, whitish. Keel-petal 8.6–9.8 mm long with distinct claw; lamina obovate, 5.4–6.0 mm long, 2.3–2.9 mm wide, paler than wings, deepest in the apical part; claw 3.4–4.3 mm long, whitish. Stamens 10, nearly the same length, 8.4–9.6 mm long, diadelphous. Anthers elliptic, shallowly retuse at the apex, ca. 0.5 mm long; before anthesis yellow. Pistil 8.5–9.7 mm long; ovary elliptic, 1.5–2 mm long at anthesis, pubescent, subsessile; style 7–8 mm long, pubescent at the basal part, glabrescent at the apical part.

Fruits compressed broadly elliptic to round, 5–7 mm long, 4–6 mm wide, subsessile, slightly to densely pubescent or nearly glabrous. Seeds reniform, ca. 3 mm long, ca. 2 mm wide.

Distr. Japan (Hokkaido, Honshu, Shikoku, Kyushu), Korea, N. China, and Amur.

Voucher and representative Specimens.
**Japan** (Siebold s.n., L 908.118-1080 (right)); (Buerger s.n., GH).
HOKKAIDO.    SOUYA.    Isl. Rebun, Lake Kushu-ko (Midorikawa 123, TI); Mt. Rishiri-Fuji, Oshidomari (Tuyama, 7 Sept. 1974, TI); Is. Rishiri, Mt. Rishiri, Foot Oshidomari-guchi (Furuse 21579, TNS); Wakkanai (Hiroe 7656, TI); Esashi (Yamanoi, 8 Aug. 1971, TNS). ABASHIRI.    Kitami, Rubeshibe (Wilson 7382, A); Nokkeushi (Uno 21551 & 21480, A); Monbetsu-gun, Takinoue-machi (Okamoto, 20 July 1952, KYO); Tokoro-gun, Tokoro-cho (Okamoto, 25 Aug. 1963, KYO); Kitami-shi, Hokkousha (Mizushima, 4 Oct. 1945, TI); Tsubetsu-cho, Goban, alt. 250 m (Murata 21180, KYO); Abashiri-shi, Mokoto (Tanaka, 7 Sept. 1944, KYO); Shari-gun, Sharimachi, Iwaobetsu-onsen (Kobayashi, 19 July 1967, MAK). ABASHIRI–NEMURO. Shiretoko, Mt. Rausu-take (Okamoto, 24 Aug. 1958, KYO). KUSHIRO. Kushiro-shi, Otanoshige (Naruhashi 1225, KYO); Tomachise (Hara, Kurosawa & Tateishi, 1934, TI); Lake Akan-ko (Hara c.1907.b, TI); Kawakami-gun, Shibetcha-cho, Shibetcha Experimental Forest of Kyoto Univ. (Koyama 1733, TNS).    TOKACHI. Rikubetsu (Okamoto, 4 Sept. 1955, KYO); Kamiashoro (Okamoto, 7 Aug. 1959, KYO); Hiroo-gun, Mt. Rakkodake–Satsurakko, alt. 200–400 m (Koyama & Fukuoka 3484, TNS). HIDAKA. Horoman (Hara, Kurosawa & Tateishi, 1939b, 1940, TI); Shizunai-gun, Noya, Petegari-sanso, alt. 100–400 m (Koyama & Fukuoka 3106, TNS); Horoizumi, Syoya, Ruran (Hara, Kurosawa, & Tateishi, 1935, 1936, TI).    KAMIKAWA. Kamiotoineppu (Koidzumi, 8 Sept. 1930, TNS); Asahigawa (Asahikawa) (Koidzumi, Aug. 1916, TI; Aug. 1917, KYO); Kamikawa-gun, Kamuikotan (Uno s.n., A). SORACHI. Yubari-shi, Oyubari (Kuramoto 8108101–8108106, TI); Sorachi-gun, Toudai-enshuurin (Inokuma 2775, TOFO). ISHIKARI. Toyohirakawa (Matsumura, 1 Aug. 1899, TI—Syntype of *L. bicolor* Turcz. var. *japonica* Nakai); Sapporo (Yatabe, Aug. 1880, TI; Tanaka s.n., A; Miyabe, Aug. 1885, A); Mt. Moiwa-yama (Nakai, 25 Aug. 1920, TI); Teine (Takeda, 26 Sept. 1908, TNS); Zenibako (Matsuda, July 1887, KYO). SHIRIBESHI, Suttsu, Benkei-misaki (Hara & Kurosawa, 2 Sept. 1977, TI); Yoichi-machi (Okuda, 19 Aug.

1960, TNS); Otaru (Matsumura, 26 July 1878, TI—Syntype of *L. bicolor* Turcz. var. *intermedia* Maxim. f. *acutifolia* Matsum.). IBURI. Tomakomai-shi, Numanohata, Utonai-numa (Koyama 5644, KYO); Mukawa-machi, near Shiomi Station (Ohba & Akiyama 3501–3510, TI); Muroran (ut "Mororan") (Sargent, Sept. 1892, A); Lake Touya-ko (Nakano, Oct. 1899, TNS). HIYAMA. Ichinowatari (Miyabe & Tokubuchi, 15 July 1890, TI—Lectotype of *L. bicolor* Turcz. var. *intermedia* Maxim. f. *acutifolia* Matsum. and lectotype of *L. bicolor* Turcz. var. *japonica* Nakai; KYO). OSHIMA. Hakodate (Maximowicz in 1861, A; L 908.118-1089); Hakodate-shi, Hakodate-yama (Midorikawa 1890, TI); Ohnuma-koen (Wilson s.n., A).

HONSHU. AOMORI. Aomori (Kinashi, Oct. 1900, KYO); Hirosaki (Kinashi, Aug. 1904, KYO); Kamikita-gun, Shiwa-mura (Mizushima 31, TI); Higashitsugaru-gun, Takanozaki (Kurosawa, 28 July 1959, TI); Masukawa-eirinsho-bunai (Kusaka 1854, TOFO); Tappizaki (Hara, 27 July 1959, TI); Nishitsugaru-gun, Tateoka-machi, Hirataki-numa (Furuse, 19 Sept. 1955, KYO), Nishitsugaru-gun, Takayama-inari, seaside (Hara, Kurosawa, & Tateishi, 15 Sept. 1974, TI); Shimokita Penin. Ushitaki–Nodai (Naito, 12 Aug. 1963, TNS); Shimokita Penin. Shiriya–Kukidounosaki (Ohashi 4308, 4579, TI); Shimokita-gun, Sai-mura, Harada (Mori, 22 Aug. 1957, MAK); Mt. Iwaki (Yatabe, 24 July 1880, TI—Syntype of *L. bicolor* Turcz. var. *intermedia* Maxim. f. *grandifolia* Matsum.); Mutsu-shi (Midorikawa 1950–1953, TI); Shimokita-gun, Wakinosawa-mura (Midorikawa 1897, 1898, 1899, TI); Shimokita-gun, Higashitoori-mura, to Imooto-mura from Horobe (Furuse, 21 Sept. 1957, A). IWATE. Morioka (Toba 17, 647, TI; T. Saito, 2 Aug. 1932, KYO); Morioka-shi, Asagishi (Kikuchi, 24 Sept. 1967, TNS); Yuda (Toba 12, TI—Holotype of *L. spicata* Nakai et F. Maekawa, 18, TI—Holotype of *L. spicata* Nakai et F. Maekawa f. *acutifolia* Nakai et F. Maekawa, 11, 19, TI; 10 Sept. 1932, KYO); Mt. Kunimi (Iwabuchi, 28 July 1941, TNS); Araya Sinmachi (Sugawara, 1 Aug. 1963, TUSG); Hienuki-gun, Uchikawame-mura, Dake (Kanai, 1 Aug. 1959, TI); Kuji-shi, Ookawame (Murata & Tabata, 28 July 1967, KYO); Kitakami-shi, Kurosawajiri-machi (Oikawa, 7 Aug. 1965, MAK); Otobe-mura (Kitamura & Murata 526, KYO); Izawa-gun, Shitomae (Sugawara, 20 Aug. 1970, TUSG); Mt. Yakeishi-dake (Karizumi, 27 Aug. 1951, TNS); Kamihei-gun, Tsuchibuchi-mura (Hosoi 10988, TI); Iwate-gun, Tamayama-mura, Mt. Himegami-dake (Ogata 4145, TOFO); Shimohei-gun, Iwaizumi-cho, Ryuusendou (Shimizu, 14 Oct. 1957, KYO); Higashiiwai-gun, Mt. Tabashine-yama (Kikuchi, 17 Sept. 1967, TNS); Mt. Himekami (Muroi 4087, A); Kuzakai (Muroi 3971, A); Morioka, Mt. Iwayama (Muroi 4238, A). MIYAGI. Sendai (Iishiba, 23 Aug. 1928, TNS); Sendai-shi, Mt. Dainenji-yama (Suzuki, 10 Sept. 1952, TNS); Tamatsukuri-gun, Narugo-machi, Onikoube (Sugawara, Ohba et al. 708207, TI; TUSG); Katta-gun, Oogawara-machi, Mt. Zaou-san (Suzuki, 10 Sept., TNS); Shibata-gun, Kawasaki-machi, Kamafusa Dam (Ohba & Akiyama 573, TI); Shibata-gun, Kawasaki-machi, Maekawa, alt. 200–250 m (Ohba & Akiyama 582–586, TI); Katsuta-gun, Zaoumachi, Tougatta (Ohba & Akiyama 593, TI); Ishinomaki-shi, Tashiro-jima (Suzuki, 26 Aug. 1956, TNS); Natori-gun, near Okunikkawa Station (Kanai, 7 Sept. 1959, TI); Kami-gun, Kamiishi-mura (Suzuki, 18 Sept. 1936, TI); Kami-gun, Nakaniida-machi, Naruse River (Naito & Ishikawa, 1 Sept. 1978, TUSG). AKITA. Mt. Tegata-yama (Muramatsu, 30 Sept. 1930, TI); Kitaakita-gun, Yamase-mura (Furuie, 26 Aug. 1951, TNS); Oomagari-shi, Yotsuya-mura (Kobayashi, 29 Aug. 1936, TNS); Kawabe-gun, Yuuwa-mura, Tsubakigawa (Henmi, 31 July 1965, TNS); Yokote-shi, Mt. Mitake-san (Nakamura, Aug. 1959, MAK); Hiraka-gun, Masuda-machi (Ishida, 22 Aug. 1959, MAK); Kitaakita-gun, Nakamura–Mt. Moriyoshi-yama (Ohashi 4787, TI); Oga-shi, Kitaura, Nishimizuguchi (Fujii 560089, TI); Noshiro-shi, Otomo-numa (Henmi, 23 Aug. 1963, TNS); Yuri-gun, Kisakata-machi, Mt. Chokai-san, en route from Kisakata to Hokodate, alt. ca. 460 m, 520 m, 560 m, & 610 m (Kaji, 31 July 1982, TI); Oga, Mayama–Kamimayama (Inokuma 3107, TOFO). YAMAGATA. Yamagata-shi (Yuuki 3204, TI); Tobishima (Mori, 8 July 1962, MAK); Yunohama (Yuuki 3348, TI); Atsumi (Yuuki,

30 July 1936, TNS); Yonezawa-shi, Bansei (Katou, 18 Aug. 1935, KYO); Nishiokitama-gun, Oguni-machi, Taruguchi–Takikura (Ohba & Akiyama 2699–2703, TI); Nishiokitama-gun, Oguni-machi, Ichinosawa–Ashimizunakasato (Ohba & Akiyama 2686–2688, TI); Kitamura-yama-gun, Tateoka-machi, Komatsu kannon–Higashisawa (Kanai, 8 Sept. 1959, TI); Cho-kaizan (ut "Chokaitan") (Okubo, 29 July 1887, TI—Lectotype of *L. bicolor* Turcz. var. *intermedia* Maxim. f. *grandifolia* Matsum. and syntype of *L. bicolor* Turcz. var. *japonica* Nakai); Nishitagawa-gun, Mt. Maya-zan (Mori, 18 Aug. 1959, MAK); Yonezawa-shi, Onogawa, near Sasahara, alt. ca. 400 m (Ohba & Akiyama 2708–2711, 2712–2714, TI); Odate, Hadachi, Motaisan (Sugawara, Saito, & Sato, 3 Sept. 1969, TUSG); Ginzan-onsen (Okuyama 24143, TNS); Karuizawagoe (Okuyama 24142, TNS); Minamiokitama-gun, Mt. Azuma, Niitakayu (Uno, 4 Aug. 1952, A); Mt. Atsumi (Ikegami 17378, A). FUKUSHIMA. Yomogita-mura (Imai, 28 Aug. 1934, TNS); Mt. Bandai-san (Tashiro, 7 Aug. 1937, KYO); Kitaaizu-gun, Higashi-yama, Mt. Senakaaburi-yama (Hara & Kurosawa, 31 July 1957, TI); Aizuwakamatsu, Sea-buriyama, alt. 700 m (Yamazaki 6604, TI); Yama-gun, Inawashiro-machi, near Kawakami, alt. 700–750 m (Ohba & Akiyama 2724–2745, TI); Aidzu (Yatabe & Matsumura, Aug. 1879, TI—Syntype of *L. bicolor* Turcz. var. *intermedia* Maxim. f. *acutifolia* Matsum.); Aidzu, Yu-moto (Yatabe & Matsumura, 4 Aug. 1879, TI—Syntype of *L. bicolor* Turcz. var. *intermedia* Maxim. f. *grandifolia* Matsum.); Kawanuma-gun, Aidzubange-machi, Toudera–Ketanomiya (Ohba & Akiyama 1820–1828, TI); Date-gun, Yanagawa-machi, Mt. Togarimori-yama (Iga-rashi, 9 Aug. 1961, TNS); Fukushima-shi, Tsuchiyuonsen-cho, Tsuchiyu–Onuma, alt. 500–650 m (Kurosaki 13402, SHO); Mt. Shirabu-san (Okamoto, 4 Sept. 1938, KYO); Akaiya-chi, alt. 500 m (Yamazaki, 11 Aug. 1962, TI); Adachi-gun, Ohtama-mura, Adatara-onsen (Kanai & Tateishi 1141, TUSG); Kooriyama-shi, Nakada-machi, Shitaeda, alt. ca. 400 m (Ohba & Akiyama 2799–2801, TI); Kooriyama-shi, Nakada-machi, Shitaeda, alt. ca. 400 m (Ohba & Akiyama 2802–2804, 2805, TI); Minamiaidzu-gun, Shimogo-machi, Ono, alt. 400–500 m (Ohba & Akiyama 1929–1935, TI); Kooriyama-shi, near Nishita-machi, Sanchome, alt. 200–300 m (Ohba & Akiyama 2761–2773, TI). GUNMA. Minakami (Hisauchi, 5 Aug. 1933, TI; 20 Aug. 1933, TNS); Tanigawa-onsen (Maekawa 58-A34, TI); Mt. Shirane-san (Sakurai, Sept. 1885, TNS); Mt. Myogi (Ishii, 13 Sept. 1967, TNS); Mt. Haruna-san (Ohwi, 6 Sept. 1950, A; TI; TNS); Mt. Arafune-yama–Hitsuse (Kimura, 6 Oct. 1953, TI); Kouzu-bokujo (Okuyama, 22 July 1940, TNS); Ushi-gun, Iriyama-touge (unknown collector, 13 Sept. 1934, TI); Azuma-gun, Shinkazawa (Kanai, 19 July 1958, TI); Shimizu-touge (Sato, 23 July 1933, TNS); Tone-gun, Kawaba-mura, Mt. Joshu-Hotaka-yama, Sainokawara (Kanai, 5 July 1957, TI); Tone-gun, Katashina-mura, Tokura (Makino in 1929, KYO); Tone-gun, Katashina-mura, nr. Lake Maru-numa, alt. ca. 1500 m (Ohba & Akiyama 1496–1507, TI); Tone-gun, Tone-mura (Ohba & Akiyama 1516–1519, TI); Tone-gun, Shirasawa-mura, Namae alt. ca. 550 m (Ohba & Akiyama 1520–1523, TI); Tone-gun, Minakami-machi, Yubiso, alt. ca. 600 m (Ohba & Akiyama 1575–1576, TI); Matsuida (Wakana, 26 Sept. 1954, TNS); Maebashi-shi, Taguchi, Mt. Tachibana-yama (Shiobara, Aug. 1928, TNS); Tano-gun, Ueno-mura, Fujidoo, alt. 750–900 m (Murata, Ohba, & Midorikawa 6262, TI); Kamizawa, alt. 1000 m (Wilson in 1914, A). TOCHIGI. Shiobara (Momiyama 1099, 1200, 1201, 1202, TI); Oku-shiobara (Suzuki, 9 Aug. 1951, A); Shioya-gun, Kuriyama-mura, Yunishigawa (Kubota, 31 Aug. 1965, TNS); Nikko (Sargent, 2 Sept. 1892, A); Nikko-shi, Arasawa-machi, alt. 650–700 m (Ohba & Aki-yama 606 & 608, TI). IBARAKI. Tsukuba (Owatari, 21 July 1895, TI—Syntype of *L. bicolor* Turcz. var. *intermedia* Maxim. f. *grandifolia* Matsum. and syntype of *L. bicolor* Turcz. var. *japonica* Nakai); Shida-gun (Hitachi) (Matsumura s.n. TI—Holotype of *L. bicolor* var. *intermedia* f. *parvifolia* Matsum.); Ishioka-shi, Ishioka-machi (Tsurumachi, 11 Sept. 1911, MAK). SAITAMA. Hannou (Hisauchi, 24 Sept. 1919, TI); Mt. Nobi-yama (Miyazaki in 1933, TNS); Niiza-machi, Katayama (Ohashi 9094, 9095, A; TI); Mt. Arashiyama (Hisau-

chi, 17 Sept. 1933, TI; TNS). CHIBA. Kashiwa (Okuyama 21279A, B, TNS); Kounodai (Makino, Sept. 1894, MAK); Inba-gun, Shimoshidzu-mura (Makino, 10 Sept. 1895, MAK); Teganuma (Makino in 1924, MAK); Inbanuma, Usui (Hisauchi, 23 Sept. 1933, TI); Ooami (Yamazaki, 6 Sept. 1956, TI); Shisui (Okuyama 9945, TNS); Ichikawa-shi, Ohmachi (Ohba 626905, TI). TOKYO. Inokashira (Nakai in 1928, TI); Ooi-futou (Midorikawa 1295, TI); Shibuya-ku, Yoyoki (Makino, Sept. 1900, KYO); Setagaya-ku, Chitose (Makino, 13 Oct. 1935, KYO); Suginami-ku, Iogi (Makino, 15 Sept. 1932, KYO); near Sekido (Mizushima, 15 Sept. 1946, TI); Chofu–near Ikou-ji (Mizushima 2458, TI); Kitatama-gun, Shinkoganei (Kanai, 10 July 1949, TI); Hino-shi, Minamitaira (Ohba 1329, 1347, TI); Hikida–Ohme (Mizushima 422, A; TI); Nishitama-gun, Itsukaichi-machi, Yokosawa (Ohba 1785, 1832, TUSG). KANAGAWA. Mukogaoka, alt. 50 m (Suzuki 331, A); Mizonokuchi (Hisauchi, 6 Sept. 1931, TI); Noborito (Hisauchi, 12 Sept. 1932, TI); Kawasaki-shi, Ikuta (Makino, 16 Sept. 1948, KYO); Fujisawa-shi, Kameino, Mutsuai, Exp. Forests, Nippon Univ. (Takada, 6 July 1964, TNS); Ashigarashimo-gun, Hakone-machi, Sengokuhara (Ohba & Akiyama 648, TI); Hakone, Miyagino-mura (Mizushima 17, A). NIIGATA. Mt. Hachikoku-san (Ikegami, 23 Sept. 1944, TNS); Mt. Gozu-yama (Noda, 27 Aug. 1957, TNS); Ojiya, alt. 150 m (Yamazaki 8989, A; TI); Higashikanbara-gun, Tsugawa, Suwatoge, alt. 400 m (Yamazaki 255, TI); Kitakanbara-gun, Nakaura-mura (Kitakanbara-gun, Toyoura-mura) (Goto s.n., MAK); Minamikanbara-gun, Shitada-mura, Taya (Ohba & Akiyama 1747–1756, TI); Nakakanbara-gun, Muramatsu-machi, Maki (Ohba & Akiyama 1759, 1762, TI); Kunigami-mura, Mt. Kunigami-yama (Eguchi, Sept. 1951, TI); Nishikambara-gun, Mineoka (Ikegami 16443, A); Niigata-shi, Yamanoshita (Ikegami 9414, TNS); Nagaoka-shi, (Kusumi, 23 Sept. 1972, TNS); Kitauonuma-gun, Kawaguchi-machi, alt. ca. 100 m (Ohba & Akiyama 1586–1593, 1594–1612, TI); Koshi-gun, Yamakoshi-mura (Ohba & Akiyama 1666–1677, 1690, TI); Joetsu-shi, Kasugayama-joshi, alt. 50 m (Ohba & Akiyama 467–468, TI); Kamo-shi, Nishiyama (Ohba & Akiyama 1757–1758, TI); Shibata-shi, w. foot of Mt. Iide, Higashiakatani–Yunohira Spa (Fukuoka & Konta, 18 July 1964, KYO); Minamiuonuma-gun, Mikuni-mura, Hiuchi Pass, alt. 920 m (Yamazaki, Kanai, Hasegawa, & Ohkubo 9036, A; TI); Minamiuonuma-gun, Yuzawa-machi, near Asagai, alt. ca. 900 m (Ohba & Akiyama 1578–1585, TI); Nishikubiki-gun, Aomi-cho, Mt. Kurohime (Hiroe 14301, KYO); Isl. Sado, Aikawa (Ikegami, 24 Sept. 1939, TNS). NAGANO. Akashina, alt. 600–1000 m (Wilson 7470, A); Mt. Hachibuse-yama (Momose, 24 Aug. 1933, TI); Mt. Kiso-koma, Naidaijo, alt. ca. 1400 m (Okamoto, 21 July 1950, KYO); Mt. Norikura, alt. 1850 m (Furuya, 15 Aug. 1950, A; TI); Mt. Nyugasa-yama, alt. 1000 m (Yamazaki, 25 Sept. 1954, TI); Iiyama-shi, Sekiya, alt. ca. 500 m (Ohba & Akiyama 493–495, TI); Nagano-shi, Susobana-kyo, alt. ca. 500 m (Ohba & Akiyama 498, 499, 500, 503–505, 506, TI); Kamiminochi-gun, Togakushi-mura, Toyooka, alt. 600–800 m (Ohba & Akiyama 513, 515, TI); Kamiminochi-gun, Kinasa-mura, Kinasa, alt. 700–800 m (Ohba & Akiyama 519, 522, 524, TI); Kamiminochi-gun, Ogawa-mura, nr. Meotoiwa, alt. ca. 1000 m (Ohba & Akiyama 526–528, TI); Kamiminochi-gun, Shinano-machi, near Lake Nojiriko, alt. ca. 700 m (Midorikawa 1005, TI); Oomachi-shi, Taira, nr. Lake Aoki, alt. ca. 750 m (Ohba & Akiyama 453–456, TI); Iiyama-shi, Sekiya, alt. ca. 500 m (Ohba & Akiyama 493–495, TI); Kitaazumi-gun, Hakuba-mura, Shirasawa, alt. ca. 900–950 m (Ohba & Akiyama 529–530, TI); Kitaazumi-gun, Hakuba-mura, Minekata, alt. 800–900 m (Ohba & Akiyama 531, TI); Minamiazumi-gun, Ariake (Hurusawa, Sept. 1939, TI); Sarashina-gun, Oooka-mura, Hikageshiden (Mineura, 1 Aug. 1958, MAK); Komoro-shi (Makino in 1940, KYO); Nishichikuma-gun, Kiso-mura, Hosojima, alt. 1100 m (Kanai & Yamashita, 27 July 1964, TI); Wadatouge (Okamoto, 4 Aug. 1961, TNS); 2 km west of Kiyosato (Tuyama, 26 Sept. 1974, TI); Sugadaira, alt. ca. 1350 m (Teramoto, 27 Aug. 1947, TI); Chiisagata-gun, Shioda-cho, Bessho-onsen (Kitagawa 5362, KYO); Karuizawa (Okuyama, 17 April 1936, TNS); Mt. Ichinoji-

100

yama (Shiroishi, 25 Aug. 1935, TNS); Kutsukake–Kose, alt. 1000 m (Mizushima 10154, 10156, A; TI); Karuizawa, Mt. Yagasaki-yama (Kanai, 30 Aug. 1955, TI); Mt. Asama-yama (Ito, Aug. 1933, TI); Minamisaku-gun, Azusayama–Mikuniyama, alt. 1400 m (Murata, 7 Aug. 1958, KYO); Minamisaku-gun, Kitaaiki-mura, Shiroiwa, alt. 1000 m (Murata, Ohba, & Midorikawa 1216, TI); at the foot of Mt. Senjo-dake, Todai, alt. 1400 m (Uematsu, 24 Sept. 1949, TI); Ochikura (Hisauchi 1782, TI); Shimoina-gun, Iida, Mt. Hontakamori, alt. 800 m (Yamazaki & Asano 7524, TI); Shimoina-gun, Yamato-mura (Asano, Sept. 1963, TI); Shimoina-gun, Ooshika-mura, Nakao (Asano 9217, 4392, TI).  YAMANASHI. Karisaka-touge, alt. 1500 m (Uematsu, 28 Sept. 1949, TI); Mt. Daibosatsu-toge (Okuyama 14618, TNS); Kitakoma-gun, Masutomi-mura, Mt. Kinpu-san, alt. 1600 m (Kanai, 25 Sept. 1954, TI); Mitake, Shosen-kyo (Mizushima 189, TI); Higashiyamanashi-gun, Mt. Kentoku-san, alt. 1100 m (Fujita, 10 Oct. 1954, TI); Mt. Minobu-san (Oiwake–Akazawa) (Moriya, 9 Aug. 1966, TNS); Mt. Yatsu, Ōizumi (Okuyama 12840, 12545b, TNS); N. side of Mt. Fuji, Yoshida-guchi, Umagaeshi–Ohishi-chaya, alt. 1400 m (Tateishi 374, TI); Fujiyoshida-shi, NE slope of Mt. Fuji, Nakanochaya Lodge, along Yoshida-guchi mountaineering path, alt. 1100 m (Konta, 24 Aug. 1976, MAK); W. side of Mt. Fuji (Ohba 69875, TI); Minamitsuru-gun, Yamanaka Lake (Makino in 1936, KYO); Lake Kawaguchi-ko (Hisauchi 855, TI); Mt. Misaka (Kanai, 18 Sept. 1955, TI); at the foot of Mt. Houou-san (Hisauchi, 10 July 1938, TI); Mt. Houousan, Aoki-kousen, alt. 1200 m (Yamazaki, 7 Sept. 1954, TI).  SHIZUOKA. Inasa-gun, Uritouge (Momiyama, 19 July 1951, TI); Sakuma-machi, Ohashi-bokujo (Okuyama 23266A, B, TNS).  GIFU. Mt. Ibuki-yama, Sangou-me (Yamazaki, 3 Oct. 1959, TI); Ohgaki-shi, Nagamatsu (Yamazaki, 6 Aug. 1969, TI; TUSG); Ohno-gun, Asahi-mura, near Akigami Damsite (Onogi, 5 Sept. 1961, TNS); Mt. Norikura-dake, at the foot (Ito, 18 Aug. 1891, TNS); Tokishi, Tajimi-machi (Shiota 7033, A); Toki-shi, Shimoishi-machi (Shiota 5393, A).  AICHI. Tsukude-mura, Tashiro (Torii 6204, KYO; 6205, TUSG); Kamitsugu-mura, Orimoto-touge (Torii 6188, TUSG); Nishiura-machi (Torii 6179, TUSG); Hourai-machi, Ikeba (Torii, 29 July 1962, KYO); Minamishitara-gun, Mt. Tonbaku (Torii, 3 Nov. 1950, A); Kitashitara-gun, Inabu-machi, Kurakaitsu (Torii, 15 Sept. 1957, KYO); Toyone-mura, Itauba (Torii, 22 Sept. 1958, KYO); Kitashitara-gun, Miwa-mura, Uren-gawa (Yamazaki, 30 July 1957, TI); Shinshiro-shi, Uri-touge (Torii, 4 Aug. 1963, TI); Toyohashi-shi (Torii 6185, TUSG); Seto-machi (Shiota, 21 Sept. 1930, KYO).  TOYAMA. Shimoniikawa-gun, Unazuki-machi, Kurobe (Makino in 1935, KYO; MAK); Sakai-gawa (Hayashi, middle of Aug. 1930, KYO); Kaminiikawa-gun, Sasadzu (Shinno, 21 Sept. 1935, TNS); Nakaniikawa-gun, Kamiichi-machi, Asou (Kurosaki, Oct. 1966, KTO); Nei-gun, Kureha-mura, Itakura, Hime-jinja (Kotou, 29 Sept. 1940, TNS); Toyama-shi, west side of Mt. Kureha-yama (Shinno, 3 Oct. 1933, MAK).  ISHIKAWA. Konosu (Kitamura, 22 July 1929, KYO); Ogi-machi (? Kogi-machi), Ohsawa (Hori 564, TOFO—Holotype of *L. bicolor* Turcz. f. *tomentella* Hatus. (ut "Kogi-machi")); Sosogi (Okuyama 14697, TNS); Kashima-gun, Midukami–Sekidosan (Masamune 12315, MAK); Suzugun, Mt. Yamabushi-yama (Satomi, 8 Sept. 1960, KYO); Ishikawa-gun, Tsurugi-cho, west of Kuragadake (Fukuoka, 19 Oct. 1958, KYO); Hakui-gun, Mt. Houdatsu-san (Satomi, 29 Aug. 1961, KYO); Mt. Ioo, Miage–Nishiodaira (Fukuoka 4420, KYO; TNS); Mt. Kuraga-dake (Satomi, 13 Oct. 1953, TNS); Mt. Kigo (Satomi, 2 Sept. 1951, TI); Kanazawa-shi, Yuuhidera, alt. 50 m (Ohashi 9078, A; 9079, TI; 9088, A; TI).  FUKUI. Ohno-shi, Rokuroshi (Murata & Shimizu, 22 Aug. 1954, KYO); Fukui-shi, Daian-ji (Ohashi, 14 Oct. 1964, TI); Mikata-gun, Mikata-cho, Tsunegami-misaki (Nakai, 26 July 1969, KYO); Yoshida-gun, Shimoshigi-mura (Hori, 18 Sept. 1952, TNS); Itai-gun, Oshima-mura (Tashiro, 19 Sept. 1933, KYO; TNS); Imatate-gun, Mt. Hino-san (Hori, Aug. 1952, TNS).  SHIGA. Mt. Ibuki-yama (Tashiro, 2 Sept. 1928, KYO); Ohtsu-shi, Mt. Ohsaka-yama (Hashimoto 11346, TNS).  KYOTO. Mt. Kurama-yama (Hiroe 6445, TNS); Hukutiyama, Mutobe (Togasi, 21 Sept. 1949, TI); Mt.

Kimio-yama (Araki 15492, TI); Mt. Ooeyama, at the summit, alt. 830 m (Murata, 11 Oct. 1967, KYO); Maidzuru-shi, Mt. Akaiwa-yama, Shimomidani–the summit, alt. 200–600 m (Murata 22379, KYO; TNS; TUSG); Yosa-gun, Futyu-mura (Murata 4950, KYO); Takeno-gun, Yasaka-cho, Noma, near Sukawa, alt. 200–300 m (Murata, 29 Sept. 1964, KYO). HYOGO. Mt. Rokkou-san, near Okuike (Yoshina, 5 Aug. 1934, KYO); Mikata-gun, Mt. Hachibuse-yama (Makino, 9 Aug. 1937, KYO; Araki 14461, KYO; Nishihara, 19 Aug. 1937, TNS; Muroi 5432, A); Mikata-gun, Yohka (Muroi 5552, A); Yabe-gun, Sekinomiya-cho, Mt. Hachibuse, Okubo–Takamaru, alt. 700 m (Murata 21139, KYO; TI; TNS); Mikata-gun, Onsen-cho, Umigami–Mt. Ogino-sen (Murata 20749, KYO); Kinosaki-gun, Kasumi-cho, Yoroi, at the seaside (Murata, 23 Oct. 1966, KYO); Taki-gun, Hioki-mura, Kitajima–Kumobe-mura, Izumi (Araki 14561, KYO; TI); Nishihama-mura, Moroyose-kaigan (Hiratsuka 24, TI); Mt. Myoken-san (Tashiro, 26 Aug. 1932, TNS); Hikami-gun, Kasuga-machi, Kurotsuhou (Hosomi, 19 July 1936, KYO); Mt. Hyono-sen, at the foot (Makino, 8 Aug. 1937, KYO); Siso-gun, Tikusa-mura, Nabegatani (Tagawa, 21 Aug. 1954, KYO); Siso-gun, Haga-cho, Tokura, alt. 500–700 m (Murata, 17 Aug. 1968, KYO; Murata & Hotta 13, TI; TNS); Sayou-gun, Nanko-cho, Mt. Hunakoshi-yama, alt. 300–650 m (Murata 33767, KYO); Kanzaki-gun, Ohkawauchi-cho, Tonomine Hills (Mizushima & Koyama, 17 Oct. 1966, TNS); Asako-gun (Araki 13737, KYO); Isl. Awaji, Mihara-gun, along the road extending north of Fukiage-cho (Fukuoka 9873, KYO). OKAYAMA. Niimi (Makino in 1935, MAK); Jobou-gun, Kayou-machi, Suzuyama (Nanba 9417, KAG); Tsuyama-shi, Kouchida (Nanba 9425, KAG); Katsuta-gun, Kitayoshino-mura (Hiratsuka 17, TI); Maniwa-gun, Yatsuka-mura (Tashiro s.n., KYO); Maniwa-gun, Yubara-cho, near Yubarako (Murata 27145 & 28 July 1962, KYO); Maniwa-gun, Mt. Kamihiruzen, Nakahukuda–the summit, alt. 700 m (Murata, 29 July 1962, KYO); Atetsu-gun, Ohnobe (Osaka, 15 Aug. 1929, KYO); Kawakami-gun, Hirakawa-mura (Tashiro, 13 Aug. 1932, KYO); Tsukubo-gun, Asabara, Sugōson (Uno, 20 Oct. 1950, A). HIROSHIMA. Taishaku-kyo (Naohara, 4 Sept. 1931, KYO); Mt. Azuma-yama (Kishino, 18 Aug. 1932, KYO); Hiroshima-shi, Kabe-machi, Nabarakyo, alt. ca. 300 m (Shimoda, 2 Oct. 1981, TI); Takada-gun, Mukaihara, alt. ca. 300 m (Umebayashi, 13 Sept. 1977, KYO); Jinseki-gun, Toyomatsu-mura (Fukuoka & Kurosaki, 1968, KYO). TOTTORI. Tottori (Hiratsuka 12, TI); Iwami-gun, Kokufu-machi, Mt. Ohgino-sen (Tanaka, 27 Aug. 1978, KYO); Hino-gun, Nichinan-machi, nr. Abire, alt. ca. 600 m (Ohba & Akiyama 2153–2154, TI); Kedaka-gun, Shikano-machi, Mt. Washimine-sen (Tanaka, 5 Aug. 1975, KYO); Touhaku-gun, Misasa-mura, Ohse (Makino, 10 Aug., MAK); Touhaku-gun, Misasa-machi, Tawara, alt. 420 m (Ishizawa 17212, KYO); Mt. Daisen (Makino, Aug. 1906, MAK; Ikoma, 25 Aug. 1935, 9 Aug. 1936, KYO); Saihaku-gun, Daisen-cho, Mt. Daisen, near Daisen-ji Temple, alt. 1000 m (Terabayashi 378, KYO). SHIMANE. Nita-gun, Yokota-machi, near Nakamura, alt. ca. 400 m (Ohba & Akiyama 2131–2134, TI); Takobana (Okuyama & Utsumi 11626, TNS); Torikami-mura (Maruyama, Aug. 1937, TNS); Nishisusa (Maruyama, 3 Sept. 1944, TNS); Mt. Sanbe-san (Koidzumi, 26 July 1942, KYO); Hirata-tyo, Gakuenji, alt. 300 m (Murata, 27 Aug. 1958, KYO); Hirata-shi, Inome (Oka 39936, TNS); Matsue-shi, Tsuda-ku (Hayashi, 10 July 1936, KYO); Isl. Oki, Saigou-machi, Araki (Oka 26020, TNS); Isls. Oki, Mt. Daimanji-san (Hori, 17 Aug. 1889, MAK); Isls. Oki, Shimamae, Kuroki-mura (Maruyama, 25 July 1937, KYO). YAMAGUCHI. Kuga-gun, Nishiki-machi, Usa (Oka 20598, TNS; 20600, KYO); Yoshiki-gun, Adzisu-machi (Shiomi s.n., KAG).

SHIKOKU. TOKUSHIMA. Myozai-gun, Ano-mura, Akawa, Hiroishi (Inobe, 15 July 1948, TNS); Myozai-gun, Ano-mura, Akawa (Inobe 681, TNS); Tokushima-shi, Nyuda-machi, Tsukinomiya (Takatou, 31 Aug. 1969, TNS); Oe-gun, Misato-mura, near Miyakura (Ohba & Akiyama 3764–3769, TI). EHIME. Onsen-gun, Nakajima-mura, Uwama (Yamamoto, 9 Sept. 1945, TI); Nishiuwa-gun, Iikata-mura, Kameura (Nomura, 16 Sept. 1953, KYO). KOUCHI. Mt. Chokou-san (unknown collector, 18 Oct. 1943, TI); Mt. Sasayama (Okuyama 16019,

TNS); Takaoka-gun, Higashi-tunoyama-mura, Mt. Irazu-yama (Tagawa, 5 Sept. 1939, KYO); Nanokawa (Watanabe, Sept. 1888, A).

KYUSHU. *In regionibus littoralibus secundum fretum v.o. Capellen* (Pierot 872, L 908. 118-1056). FUKUOKA. Mt. Hiko-san (Yoshioka, 23 Sept. 1938, TOFO); Mt. Houman-zan (Nabeshima, 9 Oct. 1938, KYO); Buzen-shi, Mts. Inugatake, Osarebuchi-dani–Osao-touge (Hotta, 10 Aug. 1961, KYO); Asakura-gun, Kamiakidzuki-mura (Nakajima 26, TI). SAGA. Kashima-shi, Hiratani (Baba, 26 Aug. 1957, TNS); Oochi-machi, Yamase (Baba 19, KAG). NAGASAKI. (Oldham 330, L 908. 118-1055); Minamitakagi-gun, Hijikuro (Kunimi-machi, Hijikuro) (Makino, 10 Sept. 1906, MAK); Nagasaki, Hikoyama (Okada, 5 Oct. 1902, MAK); Mt. Tara-dake (Baba, 20 Aug. 1937, KYO); *In vallibus montis ignivomi Wunzen* (Pierot 312, L 908. 118–1077); Kitatakagi-gun, Oguri-mura (Yokoo, 7 Oct. 1938, TNS); Iki-gun, Gonoura-cho, Kamakifure-higashi, seaside (Mimoro 1710, KYO); Katsumoto-machi (Shinagawa, 24 Sept. 1955, TNS). Isl. Tsushima (Tashiro, 8 Aug. 1935, KYO); Kuroha-touge (Tashiro, 11 Aug. 1935, KYO); Shimoagata-gun, Uchiyama (Nakajima, 27 July 1933, KYO); Mt. Tara-san (Tashiro, 8 Aug. 1935, TNS); Shimoagata-gun, Izuhara-cho, on higher elevation of Mt. Aria-ke, alt. 550 m (Koyama 2658, TNS); Shimoagata-gun, Mt. Shiratake (Nakajima, 1 Aug. 1933, TI); Kuta-mura (Tashiro, 8 Aug. 1935, KYO); Shimoagata-gun, Himi, alt. 50 m (Ohashi, Tateishi & Ohba 67, 68, TI); Shimoagata-gun, Mizushima-cho, NW foot of Mt. Matsu-yama, alt. 100 m (Koyama 2753, TNS); Kamiagata-gun, WSW middle elevation of Mt. Mitake, alt. 200–400 m (Koyama 2837, TNS). KUMAMOTO. Mt. Sobo-san (Ohba, 15 Aug. 1934, KYO); Go-kanosho, alt. 1000 m (Hatusima & Sako 27122, KAG); Aso-gun, Yunotani (Sugino, 23 Oct. 1930, KYO); Aso-gun, Mt. Fukaba-yama (Yamashiro, 16 Aug. 1958, TNS); Namino (Momiyama, 7 Oct. 1958, TI); Aso-gun, Nojiriichi (Shimada 7848B, TNS); Aso-gun, Kugino-mura (Shimada 7857B & 16 July 1954, TNS); Aso, Mt. Neko-dake (Kôzuma, 25 Aug. 1926, TNS); Ashi-kita-gun, Sashiki (Kaneta 108, TI); Tenjin-touge (Yamashiro 4294, TNS); Aida (Mayebara 2111 & 31 July 1927, TI; 18 Sept. 1927, KYO); Watari (Mayebara, 28 Sept. 1924, KYO); Mt. Ichifusa-yama (Mayebara, 8 Oct. 1919, TNS); Kuma-gun, Kume-mura (unknown collector, 25 Aug. 1930, TI); Kamimura (Mayebara, 23 Aug. 1935, KYO); Amakusa-gun, Reihoku-cho, Tomioka (Tomiyama, Sept. 1968, KYO); Kikuchi-gun, Nishiharu-mura, alt. 400 m (Shimada, 15 Oct. 1961, KYO). OITA. Beppu-shi, Hotokezaki (Yamazaki, 28 Oct. 1928, KYO); Shimoge-gun, Tsutami-mura, Masaki (Yamazaki, 16 Sept. 1928, KYO); Mt. Kujû (Makino in 1911, KYO). MIYAZAKI. Mt. Sobo-san (Tashiro, 31 Aug. 1933, KYO); Higashiusuki-gun, Toogoo-machi, Yamage–Oouchihara Dam (Murata 8585, TI); Mt. Souseki-san (Tashiro, 28 Aug. 1934, KYO). KAGOSHIMA. Aira-gun, Saishuyama-mura (Tashiro, 23 Sept. 1917, TNS); Mt. Kirishima, alt. 100–1000 m (Tashiro in 1917, A); Hayato-cho, alt. 200 m (Hatusima & Sako 25441, KAG); Aira-gun, Yoshimatsu, alt. 400 m (Hatusima & Sako 30273, KAG); Hioki-gun, Mt. Kinpou-san, at the foot (Hatusima 40548, KAG); Nishitara-mura (Doi, 26 Aug. 1932, KYO); Kaseda-shi, Toujinbaru (Ohba & Akiyama 2620–2626, TI).

**Korea.** CHOLLANAM. Isl. Heucksan Do (Daikokuzantō) (Ishidoya & Chung 3552, TI—Holotype of *L. bicolor* Turcz. *sericea* Nakai).

**Amur** (Komarov in 1895, LE).

**Ussuri** (Poretzky 122a, LE).

**China.** Mt. Po-hua-shan (Bretschneider 1877, LE). Manshuria Rossica. Bakuczan. In cozyletis montis (Komarov, 7 July 1895, TI); Manchuria Chinensis, prov. Murdenensy (Komarov, 16 July 1897, TI); Mandsuria austro-orientalis (Maximowicz in 1860, CAL); Coast of Manchuria (Wilford in 1859, CAL); Komoru-Ho, Hei-Ho (Takahashi 2344, TI); Hopeh in monte Hsiano-wu-tai-shan (Yabe, 28 July 1906, TI—Lectotype of *L. ionocalyx* Nakai); Liao-yang (Yabe, 15 Aug. 1909, TI); Lushan (Sato, 14 Aug. 1928, TI).

INNER MONGOLIA. Pailingmiao–Wuchuan (Togashi, 1 Sept. 1943, TI).

BEIJING.　Pekin (Tatarinow in 1856, LE—Lectotype of *L. bicolor* Turcz. *β. intermedia* Maxim.); Peking, Nankou Pass (Wawura 986, LE).　SHANDONG. (Maingay 35, CAL).

This species is distributed widely in Japan, but in the western and southern parts it grows at higher elevations. This species is distinguished from *L. cyrtoborya* by its wing being shorter than its keel-petal (Fig. 35). At Mt. Apoi, Hokkaido, however, a variety with a longer wing is found. It is thought that the relative length of the wing and the keel-petal is directly related to pollination by bees. This variety, therefore, is more or less isolated from typical ones reproductively. In Shikoku, especially the interior of Tokushima Prefecture and Ehime Prefecture, flowers with wings longer than the keel-petal as in *L. cyrtobotrya* are found. This *Lespedeza* with these extraordinal flowers (see p. 18; Fig. 8) differs from var. *nana* by the taller habit and might be better treat as a local variety.

*L. bicolor* is reported from Taiwan (Huang & Ohashi, 1977), but we were unable to examine any specimens of *L. bicolor* from Taiwan. It is doubtful whether this species is distributed in Taiwan.

Forma **niveoflora** S. Akiyama et H. Ohba, nom. nov.
*L. bicolor* Turcz. f. *albiflora* Tatewaki in *Trans. Sapporo Nat. Hist Soc.* **16**: 3 (1939), non Matsum. (1902)
Japanese name. Shirobana-yezo-yamahagi (Tatewaki, 1939).
Flores albi.
Distr. Japan (Hokkaido).
Specimens examined.
HOKKAIDO.　TOKACHI. Ikeda-machi, Samamai (Okuda 4464, 25 Aug. 1937, SAPT—Holotype of *L. bicolor* f. *albiflora* Tatewaki); Ikeda (cult.) (Yokoyama, 25 Aug. 1937, SAPT).

*L. bicolor* f. *albiflora* Tatewaki is a later homonym of *L. bicolor* f. *albiflora* Matsumura (1902). Matsumura's *albiflora* is the white-flowered form of *L. japonica*, cv. Japonica, not of *L. bicolor*. This was previously known only from Hokkaido but it is expected to distribute sporadically in the range of this species.

Var. **nana** Nakai in *Tennenkinenbutsu-oyobi-meisho-chosahokoku* **12** (*Veg. Apoi*): 29 (1930)—Hara in *Bot. Mag. Tokyo* **49**: 795 (1935).
Japanese name. Chabo-yamahagi (Nakai, 1930).
A perennial plant, 15–30 cm high. Terminal leaflet broadly elliptic to elliptic, 2–2.5 cm long, 1–2 cm wide. Raceme 1.5–2 cm long, shorter than subtending leaf. Calyx 3.5–4.5 mm long; lateral lobes triangular, acute, or more or less acuminate at the apex. The wing is nearly equal to or slightly longer than the keel-petal.
Distr. Japan (Hokkaido, Mt. Apoi).
Specimens examined.
HOKKAIDO.　HIDAKA. On the upper part of Mt. Apoi (Nakai, Aug. 1928, TI—Holotype of *L. bicolor* Turcz. var. *nana* Nakai; Hara, 24 Aug. 1933, TI; Ohba & Akiyama, 22 Aug. 1984, no. 1–3, TI).

This variety grows only on the upper slopes of Mt. Apoi. At the foot of this mountain and other localities in Hokkaido var. *bicolor* is abundant. Some var. *nana* have a calyx with an acuminate apex and short inflorescence, and are very similar to *L. cyrtobotrya*. But var. *nana* differs from that in the texture of leaf, the shape of fruit, and some other vegetative features.

## Lespedeza cyrtobotrya Miquel

In *Ann. Mus. Bot. Lugd.-Bat.* **3**: 48 (1867) et *Prol. Fl. Jap.* 236 (1867), *pro parte*, *excl. syn. Desmodium racemosum* var. *albiflorum* Sieb., *p.p.* et *quoad specim. cit.* Oldham *L. virgata*? (Schindler, 1913)—Maximowicz in *Acta Hort. Petrop.* **2**: 357 (1873)—Franchet & Savatier, *Enum. Pl. Jap.* **1**: 102 (1875)—Forbes & Hemsley in *Journ. Linn. Soc. Bot.* **23**: 180 (1887), ut "*L. cyclobotrya* Miq."—Matsumura in *Bot. Mag. Tokyo* **16**: 70 (1902); *Ind. Pl. Jap.* **2**(2): 268 (1912)—Schneider, *Ill. Handb. Laubholzk.* **2**: 113, fig. 70h (1907)—Nakai in *Journ. Coll. Sci. Tokyo* **26** (*Fl. Koreana* I): 155 (1909); *Lesp. Jap. Korea*, 42 (1927)—Yabe, *Enum. Pl. S. Manch.*, 77 (1912)—Schindler in Engl., *Bot. Jahrb.* **49**: 582 (1913); in Sargent, *Pl. Wils.* **2**: 112 (1914)—Bean, *Trees & Shrubs Brit. Isl.* **2**: 17 (1914)—Makino & Nemoto, *Fl. Jap.*, 735 (1925)—Makino, *Shokubutsu Dzukwan*, 393, fig. 753 (1925)—Rehder in *Journ. Arn. Arb.* **7**: 172 (1926)—Vasil'ev in Komarov ed., *Fl. U.S.S.R.* **13**: 379, pl. 20-2 (1928)—Kitagawa, *Lineam. Fl. Mansh.*, 288 (1939); *Neo-Lineam. Fl. Mansh.*, 405 (1979)—Ohwi, *Fl. Jap.*, 678 (1953); *Fl. Jap.* Eng. ed., 559 (1965); *Fl. Jap.* rev. ed., 789 (1965); *Fl. Jap.* new ed., 790 (1975)—Fu & Wang in Liou et al., *Ill. Fl. Lig. Pl. N.-E. China*, 343 (1955)—Fu in Acad. Sin. Bot., *Ill. Important Chin. Pl. Leguminosae*, 520 (1955)—Kitamura & Murata, *Col. Ill. Herb. Pl.* **2**: 99 (1961)—Kung in *Chinese Journ. Bot.* **1**: 21 (1963)—Hatusima in *Mem. Fac. Agr. Kagoshima Univ.* **6**: 10 (1967)—Acad. Sin. Bot., *Iconogr. Cormophyt. Sin.* **2**: 459, fig. 2648 (1972)—Lee in *Bull. Seoul Nat. Univ. For.* No. 2, 10 (1965); *Ill. Fl. Korea*, 468 (1979)—Ohashi in Satake et al., *Wild Fl. Jap. Herb.* **2**: 205 (1982).

[Plate 12; Fig. 37]

*L. virgata* (non DC.) Sieb. et Zucc. in *Abh. Akad. Muench.* **4**: Abt. 2. 121 (1845), *pro parte, excl. syn.*

*Campylotropis virgata* Miq., *Fl. Ind. Batav.* **1**: 230 (1855)—Maximowicz in *Acta Hort. Petrop.* **2**: 352 (1873).

*L. bicolor* Turcz. f. *microphylla* Miq. in *Ann. Mus. Lugd.-Bat.* **3**: 47 (1867) et *Prol. Fl. Jap.*, 235 (1867), *pro parte*.

*L. bicolor* (non Turcz.) Maxim. in *Acta Hort. Petrop.* **2**: 355 (1873), *pro parte*.

*L. bicolor* β. *intermedia* Maxim. in *Acta. Hort. Petrop.* **2**: 356 (1873), *pro parte*.

*L. cyrtobotrya* Miq. var. *pedunculata* Nakai, *Lesp. Jap. Korea*, 46 (1927).

*L. cyrtobotrya* Miq. var. *longiramea* Nakai, *Lesp. Jap. Korea*, 46 (1927).

*L. cyrtobotrya* Miq. var. *acutifolia* H. Koidzumi in *Journ. Pl. Iwateken* **2**: 77 (1937), *typum non vidi.*

*L. anthobotrya* Ricker in *Amer. Journ. Bot.* **33**: 257 (1946), *typum non vidi.*

*L. cyrtobotrya* Miq. var. *longifolia* Nakai ex T. Lee in *Bull. Seoul Nat. Univ. For.* No. 2, 11 (1965).

*L.×Nakaii* T. Lee in *Bull. Seoul Nat. Univ. For.* No. 2, 24 (1965).

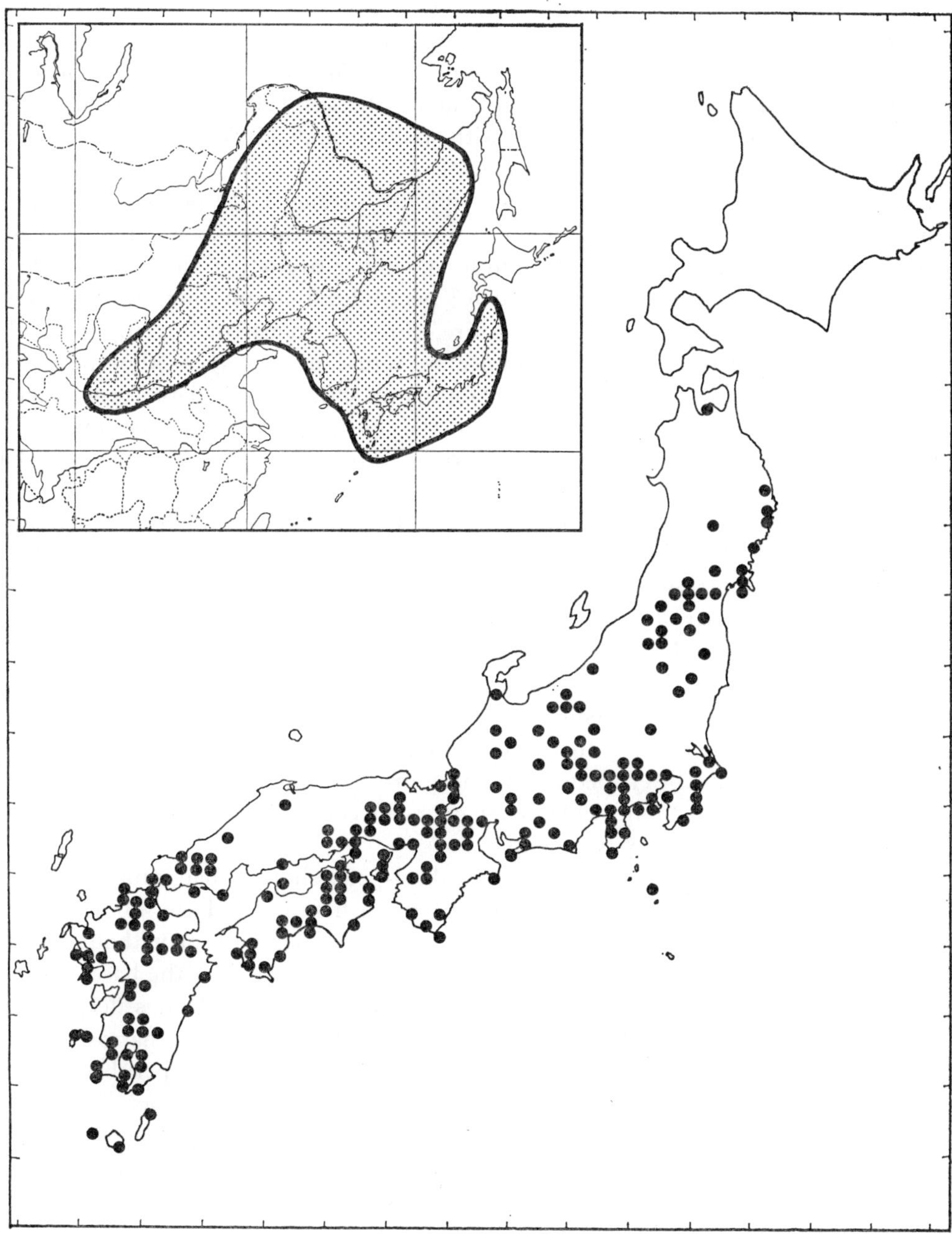

Fig. 37.   Distribution of *L. cyrtobotrya*.

Japanese name. Marubahagi (Makino & Nemoto, 1925), Miyamahagi (Matsumura, 1902).

A perennial plant, 1.5–2.5 m high. Stems ascending, 0.5–1 m high, 1–5 cm in diameter; most of the terrestrial part lives a few years; branched in upper parts (a few branches coming from basal part of the stem). Branches terete, ascending, sometimes dependent later; when young slightly angular with densely ascending appressed-sericeous hairs (hairs 0.3 mm long, whitish), later glabrescent; distal parts die in winter season, proximal parts (and the stem) with axillary and adventitious winter buds (winter buds at lower parts mostly dormant).

Leaves trifoliolate, petiolate, stipulate, spirally arranged. Stipules free, linear-triangular to linear, 2–7 mm long, brownish, persistent. Petioles 0.5–4 cm long, hairy like the young branch. Rachides 5–15 mm long, similar to the petiole. Terminal leaflets (at middle parts of branches) petiolulate; petiolules 1–3 mm long, swollen; lamina thick, 2–5 cm long, 1–3 cm wide, entire, (broadly to narrowly) elliptic to obovate, round or cuneate at the base, obtuse or retuse at the apex (the apex itself with or without a point), upper surface glabrous or sericeous or glabrescent (except along the midrib) (hairs ca. 0.2 mm long) at the anthesis, lower surface appressed-sericeous (hairs 0.3–0.5 mm long); lateral ones similar to terminal but somewhat smaller.

Inflorescence axillary racemous, one at an axil, 1–2 cm long including the peduncle, very compactly 4–10-flowered; peduncles 0.5–1 cm long (rarely elongating up to 3 cm or more), hairy like the young branch. Primary bracts ovate, about 1 mm long, pubescent, persistent; secondary bracts similar to primary ones.

Flowers 8.9–12.6 mm long at anthesis; pedicels 0.8–3 mm long, terete, curved, tomentose. Bracteoles at the base of the calyx broadly to narrowly ovate or obtuse, 0.6–1.2 mm long, pubescent outside, glabrous inside, persistent. Calyx 4.5–6.2 mm long, tubulate, four-lobed more than the middle part, densely appressed pubescent; tube 2.2–2.7 mm long; lobes subequal in length or the lower one longest; lateral ones 2.3–3.5 mm long, 0.9–1.2 mm wide, triangular-lanceolate, acuminate at the apex; upper one occasionally slightly two-clefted.

Standard longer than keel-petal and wings, keel-petal shorter than wings (S>W> K). Standard red-purple inside, paler outside, whitish near the base, obovate, 8.1–12.4 mm long, 4.6–6.8 mm wide, round or retuse at the apex with or without a point attenuated at the base without distinct claw, inflexed-auriculate near the base, spreading in anthesis from the part almost 1/2 the way from the base. Wings deeper red-purple than standard, 8.1–10.7 mm long with distinct claw; lamina narrowly obovate to obovate, 5.0–6.8 mm long, 1.8–2.4 mm wide, auriculate at upper basal part; claw 3.5–4.5 mm long, whitish. Keel-petal 7.6–9.6 mm long with distinct claw; lamina obovate, 4.0–5.1 mm long, 2.1–2.7 mm wide, paler than wings, deepest in the apical part; claw 3.6–5.0 mm long, whitish. Stamens 10, nearly same length, 7.4–9.4 mm long, diadelphous. Anthers elliptic, shallowly retuse at the apex, ca. 0.5 mm long, before anthesis yellow. Pistils 7.5–9.5 mm long; ovary elliptic, 1.5–2 mm long at anthesis, pubescent, subsessile; style 7–8 mm long, pubescent at the basal part, glabrescent at the apical part.

Fruits compressed broadly elliptic, 4.5–5.5 mm long, 3.5–4.5 mm wide, subsessile, densely pubescent.

Seeds reniform, ca. 3 mm long, ca. 2 mm wide.

Distr. Japan (Honshu, Shikoku, Kyushu), Korea, N. China, and Amur.

Voucher and representative Specimens.

**Japan** (Siebold s.n., L 944.161-356).

HONSHU. AOMORI. Higashitsugaru-gun, Kominato-machi, Mt. Yamato-yama (Hosoi, 6 Oct. 1956, KANA)*. IWATE. Kamihei-gun, Kousi-mura, Ogawa (Sasamura 0717, TNS); Shimohei-gun, Tarou-machi, Oohira (Nakaike, 19 Sept. 1977, TI)*; Esashi-shi, Koeji-toge, alt. 600 m (Murata 615, TI); Miyako, Mt. Gassan (Sugawara, 9 Sept. 1966, TUSG); Kesen-gun, Otomo-mura (Toba 455, TI—Lectotype of *L. patens* var. *rotundifolia* Honda)*. MIYAGI. Kamaishi-shi, Heita (Sasamura 0713, TNS); Togatta (Miyabe, 30 Aug. 1893, A); Katta-gun, Shichigashuku-machi, Seki (Suzuki, 18 Sept., TNS)*; Katta-gun, Kawasaki, at the foot of Mt. Zaou-san (Suzuki, 25 Sept., TNS); Kami-gun, Shikama-mura (Suzuki, 17 Sept. 1937, TNS); Daitodake (Ohashi, 2 Sept. 1960, TUSG); Okachi, Kamaya-touge (Sugawara & Tohda, 2 Sept. 1969, TUSG); Shiroishi-shi, Ootakasawa, Misawa (Ueno, 18 Sept. 1973, TNS); Ishinomaki-shi, Magiyama (Naito, 15 Oct. 1972, TUSG); Oshika Peninsula, Makinosaki (Koidzumi, Sept. 1928, KYO); Oshika-gun, Tashiro-jima (Kikuchi, 6 Nov. 1967, TUSG). YAMAGATA. Yamagata (Yuki, 5 Sept. 1936, TNS)*; Yamagata-shi, near Yamadera, alt. 300–400 m (Ohba & Akiyama 2642–2647, TI); Higashine-shi, Sekiyama–Sekiyama-touge, alt. 300–500 m (Ohba & Akiyama 2650–2653, TI); Kitamurayama-gun, Numazawa (Okuyama 24141, TNS); Aranuma (at the foot of Mt. Shirotaka-yama) (Yuki 3948, TNS); Yamakami-mura, Mt. Adzuma, upstream from Kasamatsu-onsen (Koidzumi, Aug. 1929, KYO); Minamiokitama-gun, Tamaniwa-mura (Koidzumi, 23 Aug. 1927, KYO); Nishiokitama-gun, Oguni-machi, near Numasawa, alt. ca. 300 m (Ohba & Akiyama 2675–2679, TI); Yonezawa-shi, Narushima (Midorikawa 1388, 1414, TI). FUKUSHIMA. Date-gun, Moniwa-mura (Nemoto, 3 Sept. ?, TNS); Yama-gun, Yamato-machi, Obusehara, alt. 200–300 m (Ohba & Akiyama, 1918–1928, TI); Fukushima-shi, Iizaka-machi-Nakano, Sekiba–Suginotaira, alt. 200–300 m (Ohba & Akiyama 2668, TI); Fukushima-shi, Iizaka-machi, Nakano, Ootaki, alt. 300–400 m (Ohba & Akiyama 2669–2674, TI); Kawanuma-gun, Yanaizu-machi, Fuji (Ohba & Akiyama 1793–1801, TI); Tamura-gun, Takine-mura, Mt. Otakine (Shimizu, 13 Aug. 1956, KYO); Nishishirakawa-gun, Koseki-mura (Imai, 8 Sept. 1933, TNS); Minamiaidzu-gun, Tajima-machi, Nagano (Ohba & Akiyama, 1947–1950, TI); Minamiaidzu-gun, Shimogou-machi (Ohaba & Akiyama, 1936–1943, TI). GUNMA. Mt. Myoko-san (Nakajima, 4 Aug. 1965, KANA). TOCHIGI. Nasu-gun, Kuroiso-machi, Kamisakitama (Mizushima, 17 Aug. 1948, TI); Shimotsuga-gun, Minakawa-mura, Kashiwagura (Furuse B.Nos. 150, 151, 154, TI). IBARAKI. Kashima-gun, Namino-mura (Tsurumachi 17, TI). SAITAMA. Toda-no-hara (Togasi, 26 Sept. 1937, TNS); Mt. Monomi-yama (Okuyama 17979, TNS); Hanno-shi, Mt. Izugatake (Akiyama 331, TI); Chichibu, Bukozan (Saito & Yashima, 12 Aug. 1971, TUSG); Chichibu-gun, Ootaki-mura, Kawamata–Akazawadeai, alt. 700–1000 m (Murata, Chen, & Takahashi 8516, TI). TOKYO. Mt. Takao-san (Sugimoto 11237, TI); Nishitama-gun, Kori, Ôtaba, alt. 700 m (Yamazaki 233, TI); Nishitama-gun, Hinohara-mura, Ogouchi-touge, alt. ca. 1000 m (Ohba 2161, TI); Itsukaichi, Mt. Sengen-one (Ohba & Akiyama 15, 20, TI); Itsukaichi, Tokisaka (Akiyama 280–288, TI); Mt. Kariyose (Hisauchi, 16 Aug. 1931, TI); Isl. Miyake-jima, Kamitsuki-mura (Hayashi, 12 Sept. 1936, KYO; 35, TI). CHIBA. Matsudo (Shimizu, 1936, TNS); Chosei-gun, Mobara-machi (Yamazaki, 29 Aug. 1942, TI); Yatsumi (Okuyama 09590, TNS); near Jakkou-fudou (Kurata 762, TNS)*; Kimitsu-gun, Akimoto-mura, Mt. Kanou-san (Makino in 1930, MAK); Torami (Wakana 7489, KANA); Choja-machi, seaside of Izuri-ura (Kanai & Ohashi, 6 Nov. 1966, TI; TUSG); Choshi, Nanatsuike (Fukuda, 22 Sept. 1968, TI); Mt. Kiyosumi-yama (Mizushima, 23 Oct. 1949, TI). KANAGAWA. Yokohama (Hisauchi, 21 Sept. 1929, TI); Kanazawa (Hisauchi 2625, TI); Mt. Jinmuji-yama (Hisauchi, Momiyama & Maekawa 4025, TI); Zushi-shi, NE side of Mt. Futagoyama, alt. 209 m (Ohashi & Ohba 713, TI); Kamakura (Momiyama, 30 Oct. 1935, TI); Enoshima (Hayata, 20 Sept.

1925, TI); Nishiikuta (Kazami, 11 Sept. 1959, TUSG); Tsukui-gun, Kamise-machi, Mt. Kage-nobu-yama–Myoou-touge (Kanai, 4 Aug. 1954, TI); Mts. Tanzawa, Toriuchi-mura–Mt. Yake-yama (Sawada, 25 Aug. 1935, TI); Mts. Tanzawa, Mizunashizawa, Tozawarindoo (Midorikawa 206, TI); Ashigara Pass (Hisauchi, 26 Aug. 1916, TI); Hakone, Mt. Souun-san–Oowaku-dani, alt. ca. 800 m (Mizushima, 21 Oct. 1951, TI); Ashigarashimo-gun, Hakone-machi, Miyagino (Ohba & Akiyama 621–629, TI). NIIGATA. Minamiuonuma-gun, Shiozawa-machi, Shimizu, alt. 500 m (Ito 18313, TNS). NAGANO. Motoyama (Jack, 2 Sept. 1905, A); Miegawa (Jack, 2 Sept. 1905, A); Chiisagata-gun, Tobucho, Narahata–Mt. Yunomaru, alt. 1200 m (Murata, 20 Aug. 1962, KYO); Chiisagata-gun, Kakuma-keikoku–Kakuma-touge (Sakakibara, 30 Aug. 1971, TNS); Minamisaku, Kitaaigi-mura, near Shiroiwa (Hara, 11 Aug. 1958, TI); Higashichikuma-gun, Shiga-mura (Oota, 27 Aug. 1967, KANA); Kiso-gun, Ootaki-mura, alt. 900–1300 m (Ozaki, 6 Aug. 1974, KANA); Mt. Ontake, Kurosawa-guchi, 5 goume–1 goume (Ito, 11 Aug. 1891, TNS); Suwa-shi, Kirigamine, alt. 1800 m (Tateishi 330, TI); Suwa-shi, Mt. Moriyasan, alt. 1100–1650 m (Midorikawa 511, 524, 545, 581, 612, 638, TI); Minamiazumi-gun, Hotaka-machi, Ariake–Nakabusa-onsen, alt. 1200 m (Murata 8087, TI); Nagano-shi, Susobana-kyo, alt. ca. 500 m (Ohba & Akiyama 498, TI); Mt. Togakushi-yama (Muramatsu, 7 Aug. 1925, TI); Mt. Hakuba-san (Shimura, 21 Aug. 1904, TNS); Oomachi-shi, Taira, Kurosawa (Ohba & Akiyama 458a–c, TI); Sinanooiwake (Kondo, 2 Sept. 1927, TI); Ooshika-mura (Muramatsu 1180, TNS); Shimoina-gun, Minamiwadamura, Sokoine–Nagoyama (Asano 8117, 13039, TI). YAMANASHI. Daibosatsu-touge (Okuyama 20399a, TNS); Mitsu-touge (Okuyama 24078, TNS); Iwao-mura, Shibotu, near Mt. Fuji (Okamoto, 7 Sept. 1933, KYO); W side of Mt. Fuji (Ohba 69858, TUSG); Mt. Mitake, Shosenkyo (Mizushima, 2 Sept. 1948, TI)*; Mt. Minobu-san (Okuyama 10267, TNS); Mt. Takagawa-yama, alt. ca. 930 m (Tuyama & Tuyama, 8 Dec. 1974, TI); Mt. Yatsuga-dake, Ooizumi (Okuyama 12841, TNS); Mt. Houou-san, Komugawa, alt. 800 m (Yamazaki, 7 Sept. 1954, TI). SHIZUOKA. Mt. Amagi (Sugimoto, 15 Sept. 1925, TI—Type of *L. idzuensis* Sugimoto); Kamikarino, Semo-notaki (Sugimoto 19857, TI); Yugashima (Kanai s.n., TI); Itoo-shi, E side of Mt. Oomuro-yama, alt. ca. 300 m (Murata 7481, 7485, TI); Mts. Ashitaka, Ikenotaira ES, alt. 500 m (Kanai 6083, TI); Mt. Togasa (Sato, 18 Sept. 1963, KYO); Kakegawa-shi, the foot of Mt. Ogasayama (Ohba 3497, TI); Fujinomiya-shi, the foot of Mt. Kenashiyama (Asagiri-kogen), alt. ca. 900 m (Takahashi 1014, KYO; TNS); Inasa-gun, Idaira-mura, Higashikakuroda (Hashimoto, 27 Sept. 1933, TNS); Shuchi-gun, Mikura-mura, Mt. Haruno-yama (Hashimoto, 28 Aug. 1933, TNS). AICHI. Shinshiro-shi (Torii, 14 Sept. 1941, TUSG); Minamishitara-gun, Tsukude-mura (Torii 6122, TUSG); Kitashitara-gun, Shitara-machi, Nagae, alt. 500–600 m (Mimoro, Tsugaru, & Nishiyama 2067, TNS); Ichinomiya-mura, Nagayama (Torii 6234, TUSG); Toyohashi-shi (Torii 6120, TUSG); Higashikasugai-gun, Seto-machi (Shioda, 21 Sept. 1930, KYO). GIFU. Kani-gun, Koizumi-mura (Shiota 5387, A); Kani-gun, Hiromi-machi, Mt. Goryo-yama (Shiota, Sept. 1933, KYO; 7030, A); Kani-gun, Kani-machi, Kukuri, alt. ca. 170 m (Ohba & Akiyama 3065–3071, 3073, TI); Toki-gun, Shimoishi-machi (Shiota, 18 Aug. 1930, KYO; 2098, A)*; Gujyo-gun, Takasu-mura, Hirugano, alt. 890 m (Murata, 27 Aug. 1961, KYO); Ena-gun, Mt. Ena-san (Shiota, 10 Aug. 1932, TI); Ena-gun, Hirukawa-mura, Mt. Kasagi (Shiota 9607, A)*; Yoshida-gun, Kotakari-mura, N. foot of Mt. Ibuse-yama, Kotori Pass–Umehata (Fukuoka 7426, KANA). ISHIKAWA. Kanazawa-shi, Utatsu-yama (Midzuno, 1 Sept. 1944, KANA); Nanao-shi, Nishisankai, Ninomiya (Komaki, 13 Sept. 1964, KANA). FUKUI. Tsuruga-shi, near Ikenokouchi (Kitamura & Murata, 25 Aug. 1961, KYO); Nishisaigou-mura, Hayase (Okamoto 12990, TI). SHIGA. Gamou-gun, Mt. Watamuki-yama (Hashimoto 10121, TNS); Gamou-gun, Minamihitsusa-mura, Kiyota, (Hashimoto 9939, TNS); Kitahitsusa-mura, Yamadera (Hashimoto, 1 Sept. 1950, TNS); Kanzaki-gun, Minamigokasho-mura, Kannonji (Hashimoto 10013, TNS); Iga-gun, Minamisoma-mura, Niihama (Hashimoto 10394, TNS); Taka-

shima-gun, Makino-cho, Mt. Akasaka-yama, alt. 200 m (Murata, 6 Sept. 1964, KYO); Takashima-gun, Shinasahi-cho, alt. ca. 200 m (Murata 22312, KYO); Takashima-gun, Taka-
shima-machi, Kanpuu-touge–Suzu-touge (Nishiyama 2265, TNS); at the foot of Mt. Mikamiyama, Shinohara–Ishibe (Murata, 6 Sept. 1959, KYO); Mt. Hieizan (Murata 17688,
TI); Mt. Yatsuo-yama (Hayashi, 15 Aug. 1940, TI); Kurita-gun, Tanakamiyama, alt. 200 m
(Murata, 29 Sept. 1958, KYO); Ohmihachiman-shi, near Kagami, Kagamiyama Hill, alt.
ca. 200 m (Murata, 30 Aug. 1962, KYO); Kaigake-mura (Hashimoto 2208, KYO); Kouga-
gun, Ishibe-machi Kinshoudou (Hashimoto 11419, SHO). MIE. Komono (Matsumura, 6
Aug. 1883, TI); Mt. Hossaka-yama (Maekawa 2627, TI); Inbe-gun, Mt. Chida-yama (Mago-
fuku 169, TI); Aoyama (Takeuchi, 23 July 1946, TI); Kameyama-shi, at the summit of Mt.
Nonobori–Sakamoto (Fukuoka, 9 Sept. 1962, KYO); Yunoyama-onsen (Togashi 59875, TI);
Minamimuro-gun, Kamishiyama-mura (Koide, 1 Oct. 1957, KYO). NARA. Mt. Takamado-
yama (Koidzumi, 6 Oct. 1932, KYO); Nara-shi, Kasuga–Ninnikusen (Murata, 17 Sept. 1967,
KYO); Katsuragi-mura, Fushimiue, Mt. Kongou-san, alt. 600 m (Okamoto, 14 Oct. 1934,
KYO); Gose-shi, Mt. Kongô–Takama, alt. 1100–500 m (Kurosaki 6930, SHO). WAKAYAMA.
Mt. Ryumon (Kitamura, 15 Aug. 1950, KYO); Arita, Yawata (Okamoto 3829, TI); Mt. Kouya-
san (Nakajima, Aug. 1920, TI); Iwawaki-mura (Ui 8935, TNS); Nishimuro-gun, Susami-mura
(Ui, Oct. 1909, MAK); Nachi, Myohouzisan (Enomoto 424, TI); Higashimuro-gun, Kumano-
gawa-cho, Koguchi–Ohara, along the River Wadagawa (Murata, 15 Oct. 1960, KYO); Higa-
shimuro-gun, Ooshima-mura (Nakajima, 17 Aug. 1928, TI). KYOTO. Atago-gun, Kuta-mura,
Hacchoudaira, Touge-no-tani (Momiyama, 24 Sept. 1934, TI)*; Arashi-yama (Koidzumi, 29
Sept. 1932, KYO); Onogō (Okamoto, 17 Aug. 1933, TNS); Ayabe-shi, Tachi-cho–Iden, alt.
ca. 100 m (Murata, Okubo & Nishida 34002, KYO); Tsuzuki-gun, Ujidawara-cho, Mt. Jubu-
sen (Murata 37096, KYO); Funai-gun, Nishimotoume-mura, Hankokusan (Araki 2023, KYO);
Minamikuwata-gun, Higashibetsuin-mura, Katada-goe (Araki 15063, KYO). OSAKA. Taka-
tsuki-shi, Honzanji (Shimizu, 15 Sept. 1978, SHO); Hontoyonaka-machi (Ui, 9 Sept. 1935,
SHO); summit of Mt. Iwawaki (Okamoto 1451, SHO); Kitashirakawa-gun, Muroike (Yoshino,
2 Sept. 1934, KYO); Mishima-gun, Iwate-mura (Tagawa, 16 Oct. 1935, KYO); Mt. Kongo
(Kongo-san) (Togashi 59824, TI); Toyonaka-shi, Kamishinden–Shiba (Murata, 5 Oct. 1958,
KYO); Kumanoda-mura (Ui 39, TI); Yosikawa-mura (Ui 8936, TNS). HYOGO. Mt. Rokko
(Koidzumi, Oct. 1927, KYO; TNS); Arima-gun (Ui, 26 Oct. 1934, TI)*; Miki-shi, Shimizu-
cho, alt. 100–200 m (Murata & Nishimura 267, KYO); Hikami-gun, Kagura-mura, Mt. Awaga-yama (Araki 13763, KYO); Hikami-gun, Ikugou-mura, Mt. Kenji-san (Hosomi 195(2),
MAK); Wada-mura, Mt. Ishikane-yama (Hosomi 7(2), TI); Kousei-mura, Mt. Takatori-yama,
Fudouue (Adachi, 25 Aug. 1930, KYO); Naka-gun, Hora-touge (Araki 15527, KYO); Asago-
gun, Asago-cho, Okutataragi, alt. 200 m (Fukuoka 1006, KYO); Taki-gun, Taki-cho, Shimo-
sasami, Nakanomori–Kotanaka (Kurosaki 9653, KANA); Shisou-gun, Mikata-mura (Tatebe,
2 Sept. 1937, KYO); Mt. Minou-san (Tashiro, 10 Sept. 1931, TNS)*; Aioi-shi, Isl. Tsuna-
jima, alt. 0–10 m (Fukuoka & Kurosaki 549, KYO); Mt. Shosha-zan (Muroi, 27 Aug. 1934,
KYO); Kanzaki-gun, Ohkawauchi-cho, Tonomine Hills (Mizushima & Koyama, 17 Oct. 1966,
TI); Taki-gun, Shikkawa-mura (Araki 14538, 14541, KYO); Akō, Une-mura (Muroi 373, A)*;
Isl. Awaji, Sumoto (Muroi 3031, A); Isl. Awaji-shima, Tsuna-gun, Hokutan-cho, Ni–Jôryû-
ji (Kurosaki 11823, KANA). OKAYAMA. Wake-gun, Hinase-machi, Sougo (Nanba 9449, KAG);
Mituisi (Muroi 6226, A); Oku-gun, Oku-machi, Shiriume (Nanba 9422, KAG)*; Tamano-shi,
Binwari (Nanba 9442, KAG); Mitsu-gun, Kamogawa-machi, Uedanishi (Nanba 9377, KAG)*;
Kurashiki-shi, Mt. Yuka-yama (Nanba 9451, KAG); Jobou, Nakai (unknown collector, 9 Aug.
1933, KYO); Shitsuki-gun, Takayama (Yoshisawa, 21 Aug. 1962, KANA). HIROSHIMA. Sunami
(Fujii, 16 Aug. 1963, KANA). TOTTORI. Hino-gun, Nichinan-machi, Shimoabire–Chaya, alt.
ca. 500 m (Ohba & Akiyama 2195, TI). SHIMANE. Nogi-gun, Hirose-machi, nr. Nishihida

(Ohba & Akiyama 2121, TI)*; Naka-gun, Haza-mura, Mt. Ohsa-yama (Furuwa, 5 Aug. 1927, KAG); Suzu-no-Ohtani-yama (Okuyama & Utsumi 11628, TNS). YAMAGUCHI. Ohshima-gun, Kuga (Oka 32180, TNS); Kumage-gun, Marifu-mura (Maekawa, 22–23 July 1935, TI); Kuga-gun, Nishiki-machi, Mt. Heike-dake (Miyake, 24 Aug. 1972, KANA); Tsuno-gun, Kano-machi, Kano (Imada 3364, KANA); Tokuyama, Isl. Ootsu-shima (Okamoto, 9 Oct. 1958, KYO); Ogouri, Nanamagari (Oda 3002, TI); Saba-gun, Namerayama-kokuyuurin (Oka 2910, TNS); Yunomura, Namera (Okamoto, 25 Aug. 1949, KYO); Tokusa (Oda 3134, TI); Yama-guchi-shi (Oda 3114, TI); Abu-gun, Atō-chō, Chōmonkyō, alt. ca. 200 m (Tateishi, 19 Sept. 1975, TI); Abu-gun, Tsubaki-mura, Kouchi (Nikai, 31 Aug. 1917, TNS); Hagi-shi, Kasayama (Tasaka, 15 Oct. 1955, KYO); Akiyoshi-dai (Murata 18866, KYO); Shimonoseki-shi, Ozuki (Tateishi 3638, TI); Toyoura-gun, Katsuyama-mura (Murata, 4 Oct. 1936, KYO).

SHIKOKU. KAGAWA. Shoozu(Shozu)-gun (Shoodo-shima), Kankakei (Midorikawa 810914–810920, TI); Kita-gun, Mure-machi, Mt. Goken-yama, alt. 370 m (Midorikawa 810957–810959, TI); Kita-gun, Miki-machi, Nakayama–Ohtaki-ji (Kawasaki, 5 Aug. 1957, TI)*; Kagawa-gun, Shioe-cho, Mt. Ryuozan (Murata 20485, KYO); Takamatsu-shi, Yashima (Midorikawa 810939–810942, TI); Ayauta-gun, Ryonan-machi, Toda (Midorikawa 810902–810906, TI). TOKUSHIMA. Itano-gun, Itano-machi, Mt. Ooasa-yama (Inobe, 1 Sept. 1960, TNS); Tokushima-shi, Bizan (Takatou, 3 Oct. 1965, KYO); Myodou-gun, Kamomyo-mura (Nikai 2569, TI; TNS); Myozai-gun, Shimobungoyama-mura (Tashiro, 5 Aug. 1935, TNS)*; Awa-gun, Awa-machi, Izawa (Takatou, 2 Oct. 1978, KYO); Mima-gun, Waki-machi, Mt. Ootaki-san (Inobe, 20 Sept. 1962, KYO); Miyoshi-gun, Ikeda-machi, Kurosawa (Takatou, 22 Sept. 1978, KYO); Oe-gun, Koyadaira-mura (Nikai, 14 Aug. 1904, TNS); Mt. Tsurugi-san (Takahashi, 9 Aug. 1932, KYO); Anan-shi, Tachibana-machi (Takatou, 8 Nov. 1975, KYO); Kaibe-gun, Shishikui-machi, Takaitamae (Takatou, 10 Oct. 1965, KYO). EHIME. Ochi-gun, Yuge-chô (Isl. Yuge-jima) (Kitagawa, 12 Sept. 1956, KYO); Niihama-shi, Funagi (Ishikawa, 10 Sept. 1963, KYO); Nibukawa (Satou, 29 Sept. 1960, TNS); Uma-gun, Oku-no-in (Tashiro, 24 Sept. 1930, KYO); Mt. Takanawa-san (Tashiro, 28 Sept. 1930, KYO); near Uwajima (Ha-tusima 22075, KAG); Minamiuwa-gun, Uchiumi-mura, Kashiwazaki–Oohama, alt. 0–70 m (Fukuoka 8709, KYO); Minamiuwa-gun, Shirobe-machi (Murata 6869, TI). KOUCHI. Sanri-mura (Makino, 2 Oct. 1892, TNS); Tosayama (Makino in 1913, MAK); Takaoka-gun, Ochi-machi, Yokokura-yama (Shoma, Otomo & Mori, 18 Oct. 1970, TUSG); Nagaoka-gun, Ohsu-gimura, Sugi (Murata, 21 Sept. 1957, KYO); Katsurahama (Kikuchi et al., 19 Oct. 1968, TUSG); Iwata-gun, Saga-machi, Iyoki (Murata 6690A, TI); Mt. Sasayama (Okuyama 16019, TNS); Yata-gun, Higashiyama-mura (Kanematsu, 14 Sept. 1913, MAK)*; Sukumo-shi, Sasa-yama (Okuyama 16019, TNS).

KYUSHU. FUKUOKA. (Nabeshima 14, 15, TI); Moji, Mt. Tonoe-yama (Yoshioka, 10 Oct. 1933, KYO); Kitakyushu-shi, Kokura-ku, Hirao-dai, alt. ca. 400 m (Kurosaki 8603, KYO); Munakata-gun, Ooshima-mura (Nabeshima 7, TI); Kasuya-gun, Shiga-machi, Shikanoshima (Tateishi 3139, TI); Kasuya, Tatara, Najima Takinoura (Tanaka 100109–24 & 26, A); Dazai-fu (Osada, 23 Aug. 1947, TNS); Tagawa-gun, Kawaramachi, Mt. Kawaradake (Shimizu, 20 Aug. 1955, KYO); Kurate-gun, Miyata-mura (Nabeshima, 12 Oct. 1928, KYO); Akidzuki-machi (Nabeshima 23, TI); Asakura-gun, Yasukawa-mura (Nabeshima 3, 4, 8, 12, 13, 16, 17, 21, & 18 Sept. 1926, TI—Syntype of *L. cyrtobotrya* Miq. var. *longiramea* Nakai); Chikushino-shi, Yoshiki, alt. 50–100 m (Ohba & Akiyama 2375–2377, TI); Miike-gun, Ginsui-mura, Kami-uchi (Sugino, 7 Sept. 1930, TI). OITA. Mt. Iwa-dake (Yatabe & Matsumura, 17 July 1882, TI); Handa (Mayebara 2113, TI); Beppu, Mt. Garandake (Naito, 31 July 1965, TUSG); Mt. Taisenzan (Tashiro, 11 Aug. 1933, KYO); Kuzu-gun, Mts. Kuju, Kanjigoku–Sugamori (Ha-shimoto, 13 Oct. 1951, TI); Mt. Onotake (Mayebara 2128, TI). NAGASAKI. (Oldham 331, L 908.118-1064); Iwajajama (Buerger, Oct., L 908.118-1365); Mt. Kunimi-dake (Baba, Aug.

1948, KAG); Mt. Inasa-dake (Tashiro, 19 Oct. 1911, TNS); Nagasaki-shi, Mt. Iwaya-san, alt. 150–470 m (Murata 8380, TI); Nagasaki-shi, Namishi–Tairagou (Ohba & Akiyama 2404, TI); Nishisonogi-gun, Sotome-machi, Kurosaki, near the seashore (Ohba & Akiyama 2405, TI); Nishisonogi-gun, Sotome-machi, Kurosaki–Shitsu, near the seashore (Ohba & Akiyama 2406–2413, TI); Nishisonogi-gun, Sotome-machi, Kounoura–Oonakao (Ohba & Akiyama 2419, 2421–2426, TI); Nishisonogi-gun, Kinkai-machi, Fumyo (Ohba & Akiyama 2462–2466, TI); Nishisonogi-gun, Kinkai-machi, Fumyo–Tanimon (Ohba & Akiyama 2468–2469, 2470–2473, TI); Nishisonogi-gun, Sanwa-machi, Kayaki, near Takero, alt. 0–100 m (Ohba & Akiyama 2474–2475, TI). MIYAZAKI. Higashiusuki-gun, Kitaura-mura, Mikawauchi, Honguchi, alt. 100 m (Murata 10051, TI); Hyuga-shi, Mimitsu, along the Ishinami River, alt. 20–100 m (Tateishi 3446, TI); Kitamoroagata-gun, Takasaki-machi, Takasakishinden (Minamitani, 22 Aug. 1976, KYO). KUMAMOTO. Iinogoe (Mayebara 2115, TI); Kamoto-gun, Dakema-mura (Tashiro, 26 Aug. 1932, TNS); Kikuchi-gun, Suigen-mura (Tokunaga, 22 Aug. 1928, KYO); Aso-gun, Aso-machi, Futae Pass–Mt. Yagoyama, alt. ca. 700 m (Tateishi 3310, TI); Mt. Ryuhou-zan (Kojima, 10 Sept. 1933, KYO); Yatsushiro-gun, Miyaji-mura, Mt. Koroku-san, Kamimiyayama (Kôzuma, 23 Sept. 1032, TNS); Yatsushiro-gun, Izumi-mura, Kurigi, Miya-madani, alt. 600 m (Shimada, 13 Oct. 1963, KYO); Yatsushiro-gun, Taneyama-mura, Koura (Shimada, 18 Sept. 1927, KYO); Ashikita-gun, Taura-mura (Kaneta, 8 Nov. 1936, KYO; TI); Tanoura (Mayebara, 19 Oct. 1926, TI); Hitoyoshi-shi, Yatake, alt. 650 m (Sako 6573, KAG). KAGOSHIMA. Mt. Kirishima, alt. 100–1000 m (Tashiro in 1917, A); Mt. Kurino-dake (Mura-matsu, 23 Sept. 1935, KYO; TI); Mt. Takakuma-yama (Yamazaki, 11 Aug. 1942, TI); Ushine-mura (Ide, 28 Oct. 1934, TNS); Tarumizu-shi, Saruga-jo, alt. 300 m (Hatusima 31987, KAG); Isl. Sakura-jima (Tashiro, 22 Oct. 1916, TNS); Kagoshima-shi, Kasamatsu (Ohba & Akiyama 776, TI); Ibusuki-gun, Yamakawa-cho, Yamakawa–Nagasakibana, seaside (Tateishi 3518, TI); Bonotsu-cho, Mt. Isoma (Sako 3562, KAG); Bonotsu-cho, Imadake (Sako 3563, KAG); Kawanabe-gun, Bounotsu-machi, Kushi–Akime, alt. 0–100 m (Ohba & Akiyama 2592–2595, 2599–2601, TI); Kawanabe-gun, Kasasa-machi, Akaugi, near Sanya, alt. ca. 95 m (Ohba & Akiyama 2618–2619, TI); Satuma-gun, Harimochi (Hatusima & Sako 29804, KAG); Isl. Shi-mokoshiki-jima, Kashima-mura (Hatusima 16646, 16605, KAG); Isl. Tane-ga-shima, Nishino-machi (Hatusima 15590, TI); Isl. Yakushima, Mt. Motchon, alt. 300 m (Hatusima & Sako 29759, KAG); Isls. Tokara, Isl. Kuchierabu-jima, alt. 10 m (Tomari & Kurata, KAG).

**China.**

MANCHURIA (Fukumoto in 1935, MAK). Peiling (Sugiura, Aug. 1923, TI); Mt. Fenghu-ang (Houou) (Kitagawa, 5 Aug. 1931, TI).

**Korea.**

KANGWON (KOGEN). Mt. Gumgang San, Banbutsuso (Soto-Kongo) (Hiratsuka, 22 Aug. 1934, TI); Woon Moon-Ri (Unkanri) (Nakai 5583, TI—Lectotype of *L. cyrtobotrya* var. *pedun-culata* Nakai); Mt. Sulak San (Hara, 13 Sept. 1983, TI). KYONGSANGNAM. Fusan (Sargent, 6 Sept. 1903, A)*; Mts. Jiri San (Chiisan); Mt. Sanpo San (Okamoto, 25 Aug. 1934, TI).

This species is well characterized by calyx-lobes with acuminate apices, wings longer than keel-petals, and inflorescences that are usually shorter than subtending leaves. Some plants, however, have inflorescences longer than the subtending leaves, and are identical to the type of var. *pedunculata* by Nakai (1927). In these, the inflores-cences have longer raches but the internodes of flowers remain short, and the flowers are compactly crowded near the apical part. Nakai also described another variety, var. *longiramea*, which has elongated branches (up to 1–3 m long) without branchlets. As these forms are found sporadically throughout the range and intermediate forms

112

are frequently observed, we are reluctant to treat those varieties above as separate taxa, even as form rank.

Var. *acutifolia* H. Koidz. is characterized by elliptic leaves with an acute to cuneate base and an acute apex, although the shape of leaflets varies from broadly to narrowly elliptic to obovate by individuals. *L. anthobotrya* Ricker is characterized by "nearly sessile, mostly terminal, clusters" and a calyx with "acuminate teeth" which distinguish it from *L. cyrtobotrya*. But its diagnostic characters of calyx and inflorescence agree with those of *L. cyrtobotrya* and the shape of leaflets is variable, so it appears reasonable to recognize it as conspecific with *L. cyrtobotrya* even though we have not been able to examine a type specimen of *L. anthrobotrya*.


Forma **kawachiana** (Nakai) Hatusima in *Mem. Fac. Agr. Kagoshima Univ.* **6:** 10 (1967).

*L. kawachiana* Nakai, *Lesp. Jap. Korea*, 47 (1927).

*L. patens* Nakai var. *rotundifolia* Honda in *Bot. Mag. Tokyo* **45:** 422 (1931), syn. nov.

*L. cyrtobotrya* Miq. var. *patens* H. Koidzumi in *Journ. Pl. Iwateken* **2:** 77 (1937).

*L. cyrtobotrya* var. *kawachiana* (Nakai) Ohwi in *Journ. Jap. Bot.* **26:** 234 (1951).

Japanese name. Kawachihagi (Nakai, 1927).

Stem and branches with patent hairs.

Distributed sporadically in the range of this species.

Representative specimens.

HONSHU. Osaka. Kongosan (Kongō-san, Mt. Kongo) (Tada, 22 Aug. 1897, ti—Lectotype and syntype of *L. kawachiana* Nakai).


Among the specimens cited under *L. cyrtobotrya*, those referrable to f. *kawachiana* are marked with an asterisk.

This form is found sporadically in the distribution range of *L. cyrtobotrya*. The offspring of an individual of f. *kawachiana* have also branches and inflorescence axes with patent hairs, so it is considered that the direction of hairs on branches and inflorescence axes is determined hereditarily.


**Lespedeza melanantha** Nakai

In *Bot. Mag. Tokyo* **37:** 78 (1923); *Lesp. Jap. Korea*, 57 (1927).

[Plate 13; Figs. 38 & 39]

*L. bicolor* Turcz. var. *purpurea* Nakai [ex Mori, *Enum. Cor. Pl.*, 217 (1922), *nom. nud.*—Nakai in *Bull. Nat. Sci. Mus.* No. 31, **65** (1952), *pro syn. L. melanantha* Nakai].

*L. melanantha* Nakai var. *longifolia* Uyeki in *Journ. Chosen Nat. Hist. Soc.* **17:** 54 (1934), *typum non vidi*.

*L. melanantha* Nakai var. *densa* Nakai in *Bot. Mag. Tokyo* **49:** 349 (1935).

*L. homoloba* Nakai var. *higoensis* T. Shimizu in *Journ. Fac. Textile Sci. Technol. Shinshu Univ.* **36**, ser. A. **12:** 42 (1963); in *Acta Phytotax. Geobot.* **21:** 27 (1964).

*L. bicolor* Turcz. var. *higoensis* (T. Shimizu) Murata in *Acta Phytotax. Geobot.* **29:** 95 (1978).

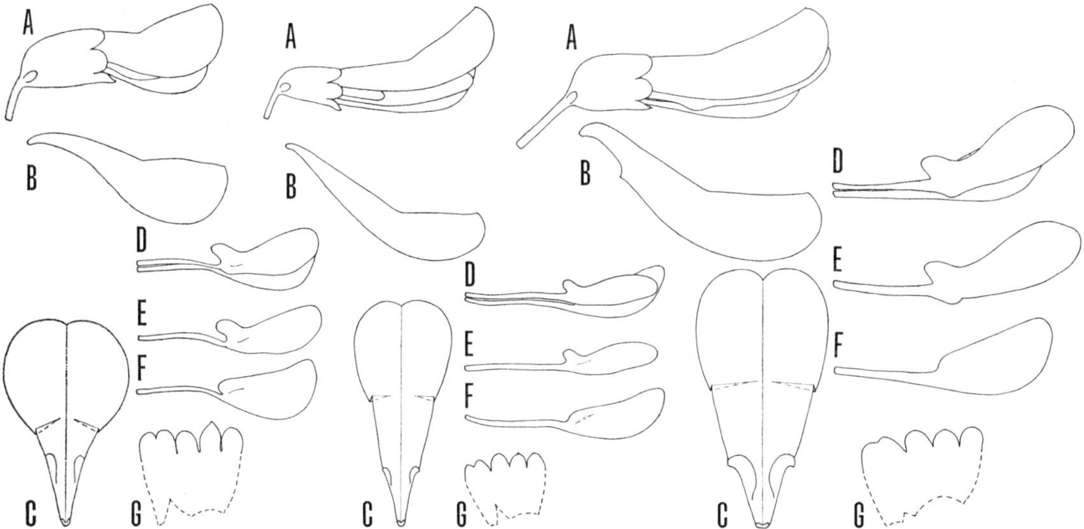

Fig. 38.  Flowers of *L. melanantha*.
A, flowers, lateral view;  B, standards, lateral view;  C, standards, opened;
D, wings and keel-petals;  E, wings;  F, keel-petals;  G, calyces, dissected. Left. Korea, Mt. Chiisan (Nakai 7, TI—Lectotype of *L. melanantha* Nakai). Center. Kyushu, Kumamoto (Shimizu 5102, KYO—Holotype of *L. homoloba* var. *higoensis* T. Shimizu). Right. Cultivated in Tokyo (ex Tsushima Isl.) (Akiyama, 1 Sept. 1984, TI).  All×3.3.

*L. bicolor* Turcz. var. *melanantha* (Nakai) S. Akiyama et H. Ohba in *Journ. Jap. Bot.* **59**: 191 (1984).

Japanese name. Kurobana-kihagi (Nakai, 1923).

A perennial plant, 0.3–0.5 m high. Stems ascending, 0.5–0.8 cm in diameter; branched in upper parts (a few branches coming from basal part of the stem). Branches terete, ascending or spreading; when young slightly angular with sparsely ascending appressed-sericeous hairs (hairs 0.1–0.2 mm long), later glabrescent; distal parts die in winter season, proximal parts (and the stem) with axillary and adventitious winter buds (winter buds at lower parts mostly dormant).

Leaves trifoliolate, petiolate, stipulate, spirally arranged. Stipules free, lineartriangular to linear, 2–5 mm long, brownish, persistent. Petioles 2–4 cm long, nearly glabrous. Rachides 5–10 mm long, similar to the petiole. Terminal leaflets (at middle parts of branches) petiolulate; petiolules 1–2 mm long, swollen; lamina 1.5–2 cm long, 1–2 cm wide, entire, (broadly) elliptic to obovate, round at the base, obtuse or retuse at the apex (the apex itself with or without a point), upper surface nearly glabrous, lower surface appressed-sericeous (hairs 0.1–0.2 mm long); lateral ones similar to terminal ones but somewhat smaller.

Inflorescence axillary racemous, one per axil, 2–5 cm long including the peduncle, loosely 2–6-flowered, peduncles 1–3 cm long, nearly glabrous. Primary bracts ovate,

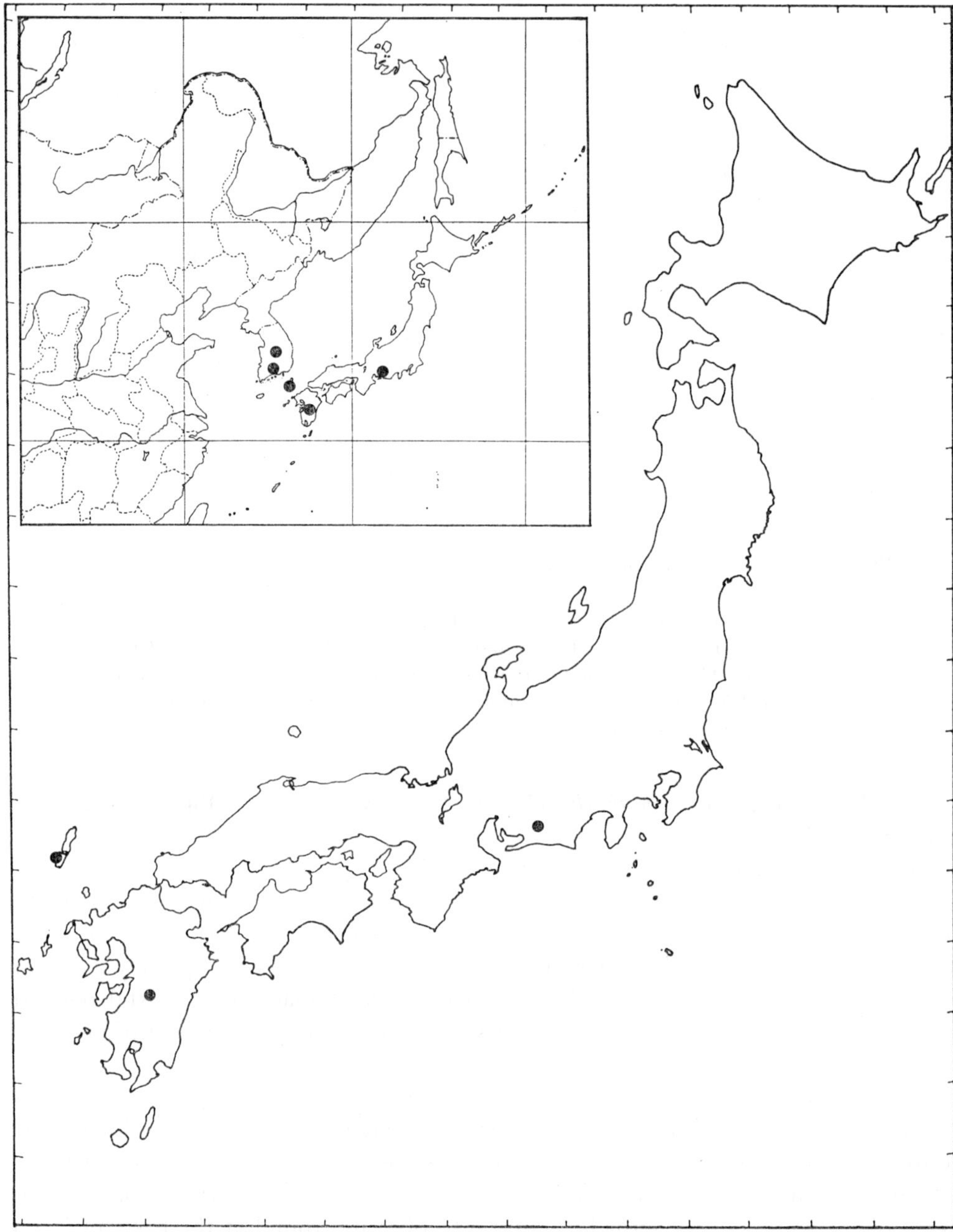

Fig. 39.   Distribution of *L. melanantha*.

0.7 mm long, nearly glabrous, persistent; secondary bracts similar to primary ones.
Flowers ca. 10 mm long at anthesis; pedicels 1.5–3 mm long terete, curved, nearly glabrous. Bracteoles at the base of the calyx broadly ovate to obtuse, 0.2–0.3 mm long, sparsely pubescent outside, glabrous inside, persistent. Calyx 3.0–3.5 mm long,

tubulate, slightly four-lobed, sparsely appressed-pubescent; tube ca. 2.5 mm long; lobes subequal in length; lateral ones ca. 0.7 mm long, ca. 0.8 mm wide, semicircular, round at the apex; upper one slightly two-clefted.

Standard longer than keel-petal and wings (or nearly equal to wings), keel-petal shorter than (or nearly eqnal to) wings (S≧W>K). Standard dark red-purple inside, paler outside, whitish near the base, obovate, ca. 10 mm long, ca. 5 mm wide, retuse at the apex with or without a point, attenuate at the base without distinct claw, inflexed-auriculate near the base, spreading in anthesis from the part almost 1/2 the way from the base. Wings deeper dark red-purple than standard, ca. 10 mm long with distinct claw: lamina narrowly obovate, ca. 6 mm long, ca. 2.3 mm wide, auriculate at upper basal part, slightly auriculate or tapering at lower basal part; claw ca. 4.0 mm long, whitish. Keel-petal ca. 9 mm long with distinct claw; lamina obovate, ca. 5 mm long, ca. 2.5 mm wide, paler than wings, deepest in the apical part; claw ca. 4 mm long, whitish. Stamens 10, nearly the same length, ca. 8 mm long, diadelphous. Anthers elliptic, shallowly retuse at the apex, ca. 0.5 mm long, before anthesis yellow. Pistils ca. 8.5 mm long; ovary elliptic, ca. 1 mm long at anthesis, sparsely pubescent, stalked; stalk ca. 1 mm long; style ca. 6 mm long, sparsely pubescent at the basal part, glabrescent at the apical part.

Fruits compressed, broadly elliptic, ca. 5 mm long, ca. 4 mm wide, nearly glabrous.

Seeds renifrom, ca. 3 mm long, ca. 2 mm wide.

Distr. Japan (Honshu, Kyushu, Tsushima Isl.) and Korea.

Representative specimens.

**Japan.**

HONSHU. AICHI. Kitashitara-gun, Miwa-mura (Inami, 21 Aug. 1941, TNS; Torii, 26 July 1942, TNS); Miwa-mura, Kawai (unknown collector, 28 Aug. 1940, TNS; Yamamoto in 1943, TUSG).

KYUSHU. KUMAMOTO. Kuma-gun, Itsugi-mura, Toji–Tenguiwa (Shimizu 5102, KYO— Holotype of *L. homoloba* Nakai var. *higoensis* T. Shimizu); Itsugi-mura, Itagi, Tenguiwa (Otomatsu 10491, 10492, TI). NAGASAKI. Isl. Tsushima, Mt. Shiratake (cultivated in Tokyo) (Akiyama, 1 Sept. 1984, TI).

**Korea.**

KYONGSANGNAM. Mt. Jiri San (Chiisan) (Nakai 7, TI—Syntype of *L. melanantha* Nakai). CHOLLANAM (ZENNAN). Mt. Pakwoon San (Hakuunzan) (Nakai, 21 Aug. 1934, TI—Holotype of *L. melanantha* Nakai var. *densa* Nakai).

This species, described from Mt. Jiri San (Chiisan), Korea, is reported from three isolated localities in Japan: Mt. Shiratake (Tsushima Island), Itsuki (Kumamoto Pref., Kyushu), and Mt. Horaiji (Aichi Pref., Central Honshu). It is remarkable that the range of morphological variation among the different localities is very narrow. This species has been regarded as conspecific with *L. bicolor* Turcz. and often treated as one of its varieties or forms. But it is apparently distinguishable from *L. bicolor* by the wings, which are longer than the keel-petal, and semicircular calyx-lobes.

Forma **rosea** Nakai in *Journ. Jap. Bot.* **15**: 680 (1939).

*L. melanantha* Nakai var. *rosea* Nakai in *Tyosen-Sanrin-Kaiho* No. 122, 25 & 32 (1935), *cum diagnos. Jap., nom. illeg.*

*L. bicolor* Turcz. var. *melanantha* (Nakai) S. Akiyama et H. Ohba f. *rosea* (Nakai) S. Akiyama et H. Ohba in *Journ. Jap. Bot.* **59**: 191 (1984).

Japanese name. Beni-kurobana-kihagi (Nakai, 1935), Yakushimahagi (Hort).

Corolla red-purple.

Ditsr. Sporadic in the range of this species.

Specimens examined.

**Korea.**

CHUNGCHONGPUK (CHUHOKU). Mt. Zokuri-san (Sackri San) (Nakai 15000, TI—Lectotype of *L. melanantha* Nakai f. *rosea* Nakai).

**Japan.**

Cultivated in Botanical Gardens, Nikko (Ohba & Akiyama 1038, TI).

The typical f. *melanantha* grows 1 m high or more in Tokyo but this is a dwarf form. This form is cultivated as an ornamental plant in Japan.

A plant with dark red-purplish flowers from Mt. Shiratake, Tsushima Islands, was transplanted to Tokyo. The next year the color of the flowers became paler, therefore the flower color is affected by habitat.


**Series Formosae** S. Akiyama et H. Ohba, *ser. nov.*

Vexilla unguiculata. Et hibernaculorum squamae et folia spiraliter disposita.
Typus: *L. formosa* (Vogel) Koehne.

Key to the species of the series *Formosae*
1. Calyx-lobe triangular, acuminate at the apex. Branches distinctly angular. Leaflet broadly ovate ....................................................................................... *L. Davidii*
1. Calyx-lobe triangular or elliptic, acute or obtuse at the apex. Branches terete. Leaflet elliptic to ovate.
    2. Auricles of standard well developed, reniform. Upper surface of leaves glabrous except along midrib .......................................................... *L. homoloba*
    2. Auricle of standard narrowly lunate to lunate
        3. Standard slightly longer than or nearly equal to keel-petal. Wing narrowly ob-ovate. Upper surface of leaves pubescent or glabrous. Aerial parts of stem and branches perennial .......................................................... *L. formosa*
        3. Standard shorter than keel-petal; claw very short. Wing obovate, ascending at anthesis. Uppersurface of leaves glabrous except along midrib or sometimes pubescent. Most of aerial parts of stem and branches annual ............... *L. patens*


**Lespedeza homoloba** Nakai

In *Bot. Mag. Tokyo* **37**: 76 (1923); *Lesp. Jap. Korea*, 55 (1927)—Makino & Nemoto, *Fl. Jap.*, 735 (1925)—Ohwi, *Fl. Jap.*, 578 (1953); *Fl. Jap.* rev. ed., 790 (1965); *Fl. Jap.* Eng. ed., 559 (1965); *Fl. Jap.* new ed., 790 (1975)—Kitamura & Murata, *Col. Ill. Herb. Pl.* **2**: 100 (1961)—Hatusima in *Mem. Fac. Agr. Kagoshima Univ.* **6**: 10 (1967) —Ohashi in Satake et al., *Wild Fl. Jap.* **2**: 205 (1982).

[Plate 14; Fig. 40]

*L. nikkoensis* Nakai, *Lesp. Jap. Korea*, 49 (1927).
*L. retusa* Nakai, *Lesp. Jap. Korea*, 51 (1927).

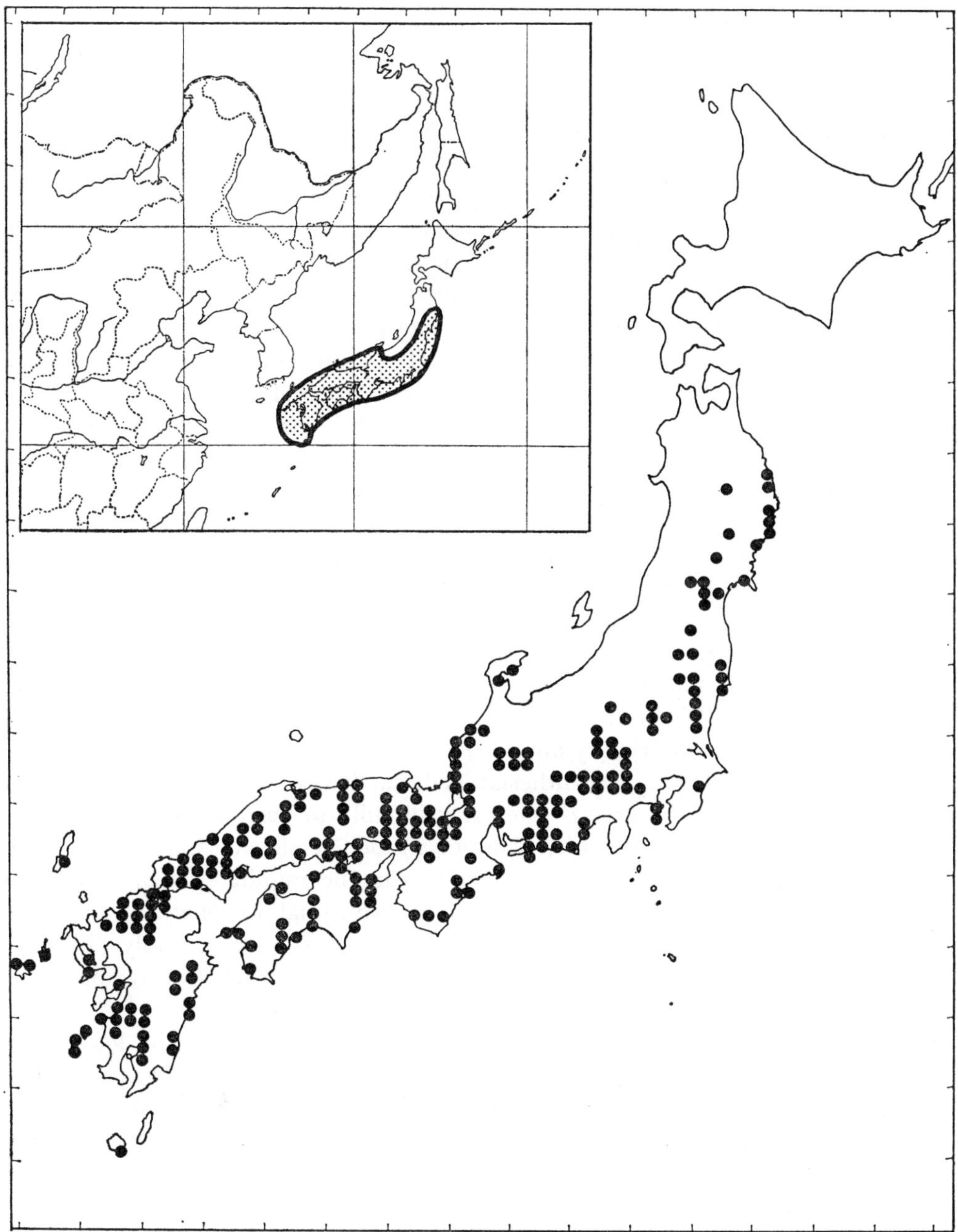

Fig. 40.   Distribution of *L. homoloba*.

*L. rotundiloba* Nakai, *Lesp. Jap. Korea*, 70 (1927).

*L. sendaica* Nakai in *Bot. Mag. Tokyo* **43**: 449 (1929).

*L. homoloba* Nakai var. *emarginata* H. Koidzumi in *Journ. Pl. Iwateken* **2**: 78 (1937), *typum non vidi*.

118

*L. homoloba* Nakai var. *longiloba* H. Koidzumi in *Journ. Pl. Iwateken* **2**: 78 (1937), *typum non vidi.*

*L. bicolor* (non Turcz.) var. *japonica* (non Nakai) Masamune in *Sci. Rep. Kanazawa Univ.* **3**: 131 (1955).

Japanese name. Tsukushihagi (Nakai, 1927), Nikko-shirahagi (Nakai, 1927).

A perennial plant, 1.5–2.5 m high. Stems ascending, 0.5–1 m high, 1–3(–5) cm in diameter; branched in upper parts (a few branches coming from basal part of the stem). Branches terete, ascending, sometimes dependent later; when young slightly angular with ascending appressed-sericeous hairs (hairs 0.2–0.3 mm long, whitish), later glabrescent; distal parts die in winter, proximal parts (and stem) with axillary and adventitious winter buds (winter buds at lower parts mostly dormant).

Leaves trifoliolate, petiolate, stipulate, spirally arranged. Stipules free, linear-triangular to linear, 2–7 mm long, brownish, persistent, Petioles 0.5–4 cm long, almost glabrous. Rachides 5–15 mm long, similar to the petiole. Terminal leaflets (at middle parts of branches) petiolulate; petiolules 1–3 mm long, swollen; lamina sometimes reddish, 2–6 cm long, 1.5–3.5 cm wide, entire, (broadly) elliptic to (broadly) ovate, round at the base, retuse or obtuse at the apex (the apex itself with or without a point), upper surface glabrous (except along the midrib), lower surface appressed-sericeous (hairs 0.1–(0.15) mm long); lateral ones similar to terminal but somewhat smaller.

Inflorescence axillary racemous, one (rarely 2–3) per axil, (2–)4–10(–15) cm long including the peduncle, somewhat loosely 4–12-flowered; peduncles 1–4 cm long, hairy like the young branch. Primary bracts ovate, about 1 mm long, pubescent, persistent; secondary bracts similar to primary ones.

Flowers 9.4–14.4 mm long at anthesis; pedicels 0.8–3 mm long terete, curved, tomentose. Bracteoles at the base of the calyx broadly to narrowly ovate or obtuse, 0.6–1.2 mm long, pubescent outside, glabrous inside, persistent. Calyx 3.6–4.4 mm long, tubulate, four- or sometimes five-lobed near the middle part or above, appressed-pubescent; tube 1.7–2.4 mm long; lobes subequal in length or the lower one longest; lateral ones 1.7–2.5 mm long, 1.0–1.6 mm wide, broadly elliptic to elliptic, obtuse or acute at the apex but not acuminate; when four-lobed upper ones slightly two-clefted.

Standard slightly shorter than (or sometimes equal to) keel-petal and longer than wings, keel-petal longer than wings (K≧S>W). Standard pale red-purple inside, whitish outside, 8.4–13.1 mm long with distinct claw, 5.4–7.7 mm wide; lamina (broadly to narrowly) elliptic to ovate, round or retuse at the apex with or without a point, inflexed-auriculate near the base, reflexing in anthesis from the part near the base; auricles reniform, well developed; claw whitish, (1.4–)1.7–2.9 mm long. Wings deeper red-purple than standard, 6.8–11.0 mm long with distinct claw; lamina narrowly oblong, 4.5–7.1 mm long, 1.6–2.9 mm wide, auriculate at upper basal part, slightly or minutely auriculate or tapering at the lower basal part; claw 1.6–2.9 mm long, whitish. Keel-petal 8.5–13.5 mm long with distinct claw; lamina narrowly obovate, 6.6–9.7 mm long, 2.7–3.8 mm wide, paler than wings and standard, deepest in the apical part; claw 2.3–3.9 mm long, whitish. Stamens 10, nearly the same length, 8.2–13.2 mm long, diadelphous. Anthers elliptic, shallowly retuse at the apex, ca.

0.5 mm long, before anthesis yellow. Pistils 8.4–13.4 mm long; ovary elliptic, ca. 1.5 mm long at anthesis, pubescent, subsessile; style 7–8 mm long, pubescent at the basal part, glabrescent at the apical part.

Fruits compressed round or broadly elliptic, 5–7 mm long, 4–5 mm wide, subsessile, slightly to densely pubescent or nearly glabrous.

Seeds reniform, ca. 3.5 mm long, ca. 2.5 mm wide.

Distr. Japan (Honshu, Shikoku, Kyushu).

Voucher and representative specimens.

**Japan.**

HONSHU.  IWATE. Morioka (Toba, 5 Sept. 1932, TI); Shimohei-gun, Tanohata-mura, Raga (Kikuchi, 29 Sept. 1967, TNS); Shimohei-gun, Tarou-machi, Oohira (Nakaike, 19 Sept. 1977, TI); Kamihei-gun, Oodzuchi-machi, Kirikiri (Kanai, 3 Sept. 1959, TI); Kamaishi-shi, Heita (Sasamura 0714, TNS); Okkirai, Rasho-touge (Koidzumi, Sept. 1928, KYO); Higashi-iwai-gun, Mt. Tabashine-yama (Kikuchi, 17 Sept. 1967, TNS); Mt. Iwate (Arimoto, 16 July 1903, GH).  MIYAGI. Tamatsukuri-gun, Kawatabi (Sugawara, 30 Sept. 1964, TUSG); Ishino-maki-shi, Magiyama (Naito, 15 Oct. 1972, TUSG); Ojika-gun, Onagawa-cho, Urajuku (Kikuchi, 21 Oct. 1967, TNS); at the foot of Mt. Funagata-yama (Suzuki, 26 July 1937, TNS); Sendai (Ichihashi, Sept. 1928, TI—Holotype of *L. sendaica* Nakai); Sendai-shi, Kawauchi (Ohba & Akiyama 569, 570, TI); Shibata-gun, Kawasaki-machi, Kamafusa (Ohba & Akiyama 578–581, TI); Shiogama (Jack, 27 Aug. 1905 (three sheets), GH); Matsushima (Wilson 7052, A). YAMAGATA. Higashine-shi, Sekiyama–Sekiyama-touge, alt. 300–500 m (Ohba & Akiyama 2654–2661, TI).  FUKUSHIMA. Fukushima-shi, Iizaka-machi-Nakano, Sekiba–Suginotaira, alt. 200–300 m (Ohba & Akiyama 2663–2667, TI); Yama-gun, Inawashiro-machi, near Tago-numa, alt. 500–600 m (Ohba & Akiyama 2746–2751, 2752, TI); Tamura-gun, Miharu-machi, Kaiyama, alt. 300–400 m (Ohba & Akiyama 2776–2778, 2780–2783, TI); Futaba-gun, Hirono-machi (Tateishi 705, TI); Asaka-gun, Mitsumori-touge (Kurosawa, 31 July 1957, TI); Shira-kawa-machi (Saito, 10 Sept. 1936, TI); Nishishirakawa-gun, Koseki-mura (Imai 44, TI); Taira (Sugawatari in 1931, TI); Iwaki-shi, Tairanumanouchi (Midorikawa 255, TI); Ishiki-gun, Akai-mura, Mt. Aaki-dake, alt. 150 m (Ikegami 27522, KANA); Yunomata-onsen (Tobita, 10 Aug. 1936, TI).  GUNMA. Mt. Asamayama (Faurie 6990, A); Tone-gun, Shirasawa-mura, Namae, alt. ca. 550 m (Ohba & Akiyama 1524–1527, TI); Tone-gun, Shirasawa-mura, Taka-hira, alt. ca. 500 m (Ohba & Akiyama 1528–1538, TI); Tone-gun, Minakami-machi, Yubiso, alt. ca. 600 m (Ohba & Akiyama 1562–1574, TI); Mt. Yagasaki-yama (Hara & Kurosawa, 3 Aug. 1977, TI); Kanraku-gun, Myoogi-machi, Mt. Myoogi, alt. 700–1080 m (Furuse, 16 Aug. 1957, A); Kanraku-gun, Minamimaki-mura, Mt. Kurotaki-yama (Furuse 12482, KYO); Ma-tsuida-machi (Wakana, 26 Sept. 1954, TNS); Onga, Takaiwa (Ishii, 12 Sept. 1965, TNS); Tano-gun, Nakazato-mura, Otsuku, alt. 500–700 m (Murata 8159, TI); Tano-gun, Ueno-mura, Kawawa–Fujidoo, alt. 500–700 m (Murata 8171–8182, TI).  TOCHIGI. Nikko, Uma-gaeshi (Matsumura, 12 Aug. 1885, TI—Holotype of *L. nikkoensis* Nakai); Nikko-shi, Hanaishi, alt. 700–800 m (Ohashi, Akiyama, & Iijima 1528, 1529, 1531, 1533, 1534, TI); Nikko-shi, Hanaishi-machi (Ohba & Akiyama 1422–1429, TI); Nikko to Lake Chuzenji (Sargent, 3 Sept. 1892 & 8 Sept. 1892, A); Shimotsuga-gun, Itaga, Mt. Kawabake-yama (Maekawa 80248, TI); Shimotsuga-gun, Minakawa-mura, Mt. Oohira-yama (Furuse B155, TI); Mt. Kogashi-yama (Kaimoto, 26 Sept. 1937, TNS); Kawachi-gun, Ohya (Furuse, 23 Sept. 1952, A).  IBARAKI. Kuji-gun, Mt. Nabeashi-san (Tsurumachi, 29 Sept. 1935, KYO); Kuji-gun, Fukuroda-mura (Honda, 7 Sept. 1937, TI); Mt. Nishikanasa-san (Okuyama 12578, TNS); Taishi-machi, Fuku-roda (Honda s.n., TI); Mito (Ando 84, TI—Syntype of *L. homoloba* Nakai).  SAITAMA. Mt. Bukou-san (Hisauchi 1351, TI); Chichibu, Kamabuse-touge (Okuyama, 13 Oct. 1940, TNS);

120

Hannou-shi, Mt. Izugatake (Akiyama 81-13–30, TI). TOKYO. Ongata-mura (Satow, 23 Sept. 1934, TI); Nishitama-gun, Mita-mura, Mt. Sougaku-san (Kanai 3721, TI); Mt. Takamizu-yama (Hisauchi, 18 Sept. 1932, TI); Mt. Mitou-san (Okazaki, 14 Aug. 1954, TNS); Mt. Takao, near the summit, alt. ca. 500 m (Tateishi 496, TI); Mt. Jinba (Suzuki, 7 Oct. 1939, A); Minamitama-gun, Yui-mura (Mizushima, 3 Oct. 1951, A; TI); Cultivated in Botanical Gardens, Koishikawa (unknown collector, 10 Oct. 1926, TI—Holotype of *L. rotundiloba* Nakai). CHIBA. Sanbu (Wakana, 6 Sept. 1963, TNS). KANAGAWA. Yokosuka (Kouno, 7 Oct. 1895, TI); Misaki, Aburatsubo (Momiyama, 4 Oct. 1927, TI); Miura-shi, Hatsune-cho, Wada Beach (Makino, KYO). NAGANO. Karuizawa, Mt. Yagasaki-yama (Kanai, 30 Aug. 1955, TI); Saku-shi, Uchiyama, Matsui (Sato 3155, TI); Mt. Ontake (Hatusima 15392, TUSG); Mt. Komagatake (Yabe, 19 Aug. 1903, TI—Syntype of *L. homoloba* Nakai); at the foot of Mt. Senjo-dake, Todai, alt. 1500 m (Uematsu, 24 Sept. 1949, TI); Shimoina-gun, Tenryû-kyô, alt. 400 m (Shimizu, 19 Sept. 1959, KYO); Shimoina-gun, Yamamotomura, Maniwadaira (Asano, 19 Sept. 1963, TI); Shimoina-gun, Yasuoka-mura (Asano 21635, TI). YAMANASHI. Kitatsuru-gun, Tabayama-mura, Saora-touge–Taba (Kanai, 10 Sept. 1955, TI); Daibosatsu-tohge (Okuyama 12737, TNS); Kita-tsuru-gun, Sasago-mura, Seihachi-tooge (Furuse 21925, KAG); Higashiyamanashi-gun, Mitomi-mura, Mt. Kentoku (Furuse, 14 Sept. 1957, A); Mitake-Shosenkyo, Senga-taki (Mizushima, 2 Sept. 1949, TI); Nishiyatsushiro-gun, Sebire-ko–Ochii (Oomura 18557, TNS); Mt. Kushigata-yama, alt. 900 m (Matsuda, 24 Sept. 1955, TI); Mt. Kaikoma, alt. 900 m (Uematsu, 22 Sept. 1949, TI); Mt. Houou-san (Uematsu, 1 Oct. 1949, TI). SHIZUOKA. Shizuoka, Yawata (Sugimoto, 2 Oct. 1928, TNS); Fujieda-shi (Shimizu 337, TI); Abe-gun, Miwa-mura, Yuyama–Aizawa (Mizushima 1661, TI); Abe-gun, Ikawa-mura, Dam–Umikubo (Kurata, 12 Sept. 1974, TNS); Ogasa-gun, Mt. Ogasa-yama (Shimizu, 13 Oct. 1929, TI); Iwata-gun, Shikiji-mura (Hashimoto, 30 Sept. 1933, TNS); Mt. Akiba-san (Okuda, 1 Oct. 1960, TNS); Hamamatsu-shi, Miyakoda (Furuse, 8 Oct. 1959, A; TNS). AICHI. Kitashitara-gun, Tomiyama-mura, Yamanaka-mura, Maiki (Torii, 30 Sept. 1951, TI); Kitashitara-gun, Inabu-machi, Kamigou (Torii, 11 Sept. 1960, TI); Kitashitara-gun, Toyone-mura, Mt. Chausu-yama, alt. 900–1100 m (Mimoro, Tsugaru, & Nishiyama 2073, TNS); Hourai-machi, Tsugeno (Torii, 12 Oct. 1958, KYO); Minamishitara-gun, Mt. Honguu-san (Torii, 3 Nov. 1953, TI); north of Futagawa (Koyama 7116, TI); Toyohashi-shi, Iwasaki (Satake, 27 Sept. 1938, TNS); Atsumi-gun, Tahara-machi, Mikawatahara–Zaoyama, alt. 5–253 m (Ohba & Akiyama 3247–3263, TI); Atsumi-gun, Atsumi-machi, Konkayama (Furuse, 21 Oct. 1061, A); Higashikasugai-gun, Kanko-ji (Shiota 7029, A); Nagoya, Higashiyama (Maekawa 369, TI). GIFU. Nakatsugawa (Jack, 6 Sept. 1905, A); Takayama-shi, NE foot of Mt. Genji-dake, Chishima–Matsukura-kannon, alt. 600–900 m (Fukuoka 7474, KYO; TNS); Oono-gun, Kuguno-cho, Dangumi–Mt. Funa-yama, alt. 900 m (Fukuoka 7501, KYO); Mt. Kurai-yama (Tashiro, 14 Sept. 1929, KYO); Nakayamashichiri (Sugimoto, 23 Sept. 1933, KYO); Gujyo-gun, Takasu-mura, Hirugano, alt. 890 m (Murata, 27 Aug. 1961, KYO); Toki-gun, Hiyoshi-mura (Shiota, 17 Sept. 1932, TI); Kani-gun, Kani-machi, Kukuri, alt. ca. 170 m (Ohba & Akiyama 3055, 3073–3095, TI); alt. 170–300 m (Ohba & Akiyama 3097–3105, 3114–3119, 3127, TI); Toki-gun, Hiyoshi-mura (Shiota 7248, A); Ena-gun, Toumachi (Hasegawa, 1 Aug. 1935, A). ISHIKAWA. Ogi (Satomi, 21 Sept. 1957, KANA); Noto-jima, inter Mukôda (Masamune 14030, KANA); Isl. Noto-jima, Suso (Satomi s.n., KYO); Nomi-gun, Tatsunokuchi-machi (Satomi, 20 Sept. 1959, KYO); Yamashiro-onsen (Satomi, 13 Nov. 1955, MAK). FUKUI. Sakai-gun, Awara-cho, Namimatsu (Fukuoka 3222, KYO); Itai-gun, Oshima-mura (Tashiro, 19 Sept. 1933, KYO; TNS); Nishiyasui, Mt. Kunimi-yama (unknown collector, Oct. 1955, TNS); Takefu-shi, Ikenoue-machi (Yamada, 16 Aug. 1968, KANA); Nanjo-gun, Imajo-machi, Magotani (Ohba & Akiyama 1960–1963, TI); Tsuruga-shi, Kehi-no-matsubara (Kodama, 21 July 1962, TNS); Saburi (Hori, Oct. 1955, TNS). SHIGA.

Ika-gun, Suino, SE foot of Mt. Yokoyama-dake (Fukuoka, 14 Sept. 1963, KYO); Ika-gun, Yogo-machi, Nakanogo–Shimoniu (Ohba & Akiyama 1994–2028, TI); Higashiasai-gun, Asai-machi, Takeyama, south of Mt. Kanakuso, Futamata-bashi–Myoto-taki (Fukuoka, 15 Sept. 1962, KYO); Sakata-gun, Santô-chô, Nagaoka–Samegai (Kurosaki 5808, KYO); Mt. Ibuki-yama (Hisauchi 1225, 1248, TI); Yasu-gun, Yasu-machi, Mt. Myokouji-yama (Hashimoto 10500, TNS); at the foot of Mt. Mikamiyama, Shinohara–Ishibe (Murata, 6 Sept. 1959, KYO); Gamou-gun, Ichinobe-mura, Ichinobe (Hashimoto 10175, 10184, TNS); Kurita-gun, Kanekatsu-mura, Higashizaka, Ishibemichi (Hashimoto 11430, SHO); Mt. Hieizan (Murata 17689, TI).    MIE. Isshi-gun, Nakagawa-mura, Tengeiji (Maekawa 2577, 2681, TI); Mt. Asama-yama (Nakai, Oct. 1932, TI); Kuki (Okamoto, 11 Sept. 1957, KYO); Owase (Satake & Okuyama, 10 Oct. 1942, TNS).    NARA. Nara-shi, Mizukami-ike (Murata 20876, TI); Yoshino-gun, Kawakami-mura (Koidzumi, Sept. 1924, KYO).    WAKAYAMA. Hidaka-gun, Kamiminabe-mura, Tamano-tani (unknown collector, 11 Oct. 1931, TI); Nishimuro-gun, Asso-mura (Nakajima, Sept. 1924, TI); Nishimuro-gun, Shirahama-cho (Murata, 18 Oct. 1960, KYO); Shinguu (Chihomine) (Satake & Okuyama 17614, TNS); Nachi, Myohozisan (Enomoto, 4 Oct. 1971, TI).    KYOTO. Yosa-gun, Higatani-mura (Horie, 23 Sept. 1933, TNS); Maizuru-shi, mon. Aoba-yama (Furuike 64471, KANA); Ikaruga-gun, Okukanbayashi-mura, Kimioyama (Murata, 15 Sept. 1956, KYO); Fukutiyama, Mutobe (Togasi, 21 Sept. 1949, TI); Ayabe-shi, Tachicho–Iden (Murata, Okubo & Nishida 34005, KYO); Kitakuwada-gun, Kita-machi, Koshiokamino-machi, alt. 380 m (Kuwashima 18791, TNS); Mt. Kurama (Hiroe 12.892, TNS); Kyoto-shi, Sakyo-ku, Shugakuin (Koidzumi, 30 Oct. 1932, KYO); Sakyo, Iwakura (Aragi, 23 July 1985, A); Arashiyama (Koidzumi, 29 Sept. 1932, TI); Kameoka-shi, Asahi, Mimata-kyo, alt. 200 m (Murata 19658, TI; TNS); Nansou-gun, Umaji-mura, near Ikejiri (Yamamoto, 8 Sept. 1940, TNS); Otokuni-gun, Modzume (Yamamoto 4471, TNS); Kuze-gun, Kumiyama-machi, on the bank of Ujigawa (Sato, 22 July 1975, TI); Tsuzuki-gun, Tanabe-tyō (Tanabe-machi) (Murata, 23 Sept. 1954, KYO).    OSAKA. Toyonou-gun, Kumanoda-mura (Ui 34, TI); Higashi-toyonaka (Ui 8963, TNS); Kawabe-gun, Yamamoto (Togashi, 21 Sept. 1953, GH; TI); south of Sakai, Shinodayama (Seto, 5 Oct. 1952, KANA).    HYOGO. Mikata-gun, Hamasaka-machi, Futsukaichi (Tanaka, 10 Nov. 1974, KYO); Nishihama-mura, Moroyose (Hiratsuka 23, TI); Asako-gun, Awaga-mura, Mt. Awaga-yama (Araki 13728, KYO); Hikami-gun, Mt. Awagamine, alt. 400 m (Tagawa & Iwatsuki 3983, TNS); Taki-gun, Imada-machi, Kamaya–Aidashinden (Shimizu, 18 Sept. 1980, KANA); Miki-shi, Shizimi-cho, alt. 100–200 m (Murata & Nishimura 228, TNS); Miki-shi, Hosokawa-cho, alt. 50–150 m (Murata & Nishimura 324, KYO; TNS); Ono-shi, Kawainishi-machi, Aonogahara (Shimizu, 24 Sept. 1978, SHO); Tasai-shi, Kohokke (Shimizu, 26 June 1977, SHO); Nakayamadera (Togasi, 6 Nov. 1948, TI); Takaraduka (Kuwashima, 22 Oct. 1933, TNS); Arima-gun, Namase (Ui 45, TI); Nishinomiya-shi, Koyoen–Jyurinji, alt. ca. 200 m (Murata, 22 July 1964, TI); Kobe-shi, Kita-ku, Yamadacho, Tenkatuji, Tanigami–Ooike, alt. 250–450 m (Fukuoka & Kurosaki 1181, TI); Kobe, Suzuran-dai (Okamoto, 30 Sept. 1940, TI); near Kobe, Sakae (Muroi 6214, A); Ashiya-shi, Mt. Rokkô, near Okuike, Imori-ike Pond (Seto 19771, KANA); Akashi-shi, Tsuchiyama (Shimizu, 21 Sept. 1976, SHO); Takasago-shi, Ooishi, Ishinohoden (Kurosaki 12925, SHO); Takino, Mt. Koumyoji-yama (Tashiro, 23 Sept. 1932, KYO); Mt. Shosya-zan (Shimizu, 23 Nov. 1977, SHO); Shikama-gun, Yumesaki-machi, Suginouchi (Uchiumi U167, SHO); Shikama-gun, Yumesaki-machi, Kotsubo–Hisasaka (Iwatani, 15 Sept. 1972, SHO); Akô-shi, Port of Akô, alt. 0–4 m (Kurosaki 8662, KYO).    OKAYAMA. Bizen-shi, Funasaka, Fukaya-no-taki, alt. 150 m (Kurosaki & Kato 183, SHO); Wake-gun, Hinase-machi, Sougo (Nanba 9450, KAG); Oku-gun, Ushimado, Nishiwaki (Nanba 9452, KAG); Okayama-shi (Yamazaki 7831, TI); Kojima-gun, Kogushi (Nishihara, 30 Aug. 1935, TNS); Tsukubo-gun, Hayashima-cho, Yao, alt. 200 m (Murata 30307, KYO; TI); Mitsu-gun, Mitsu-machi, Ichiba (Ohba & Akiyama 683–687,

TI); Jobou-gun, Kayou-machi, Tatsuchi (Nanba 9415, KAG); Katsuta-gun, Kitayoshino-mura (Hiratsuka 18, TI); Tomada-gun, Kamo-machi, Kurami (Nanba 9438, KAG); Atetsu-gun, Nohara (Nishihara, 16 Sept. 1950, TNS); Tsukubo-gun, Asabara, Sugō-son (Uno, 10 Oct. 1950, A). HIROSHIMA. Fukayasu-gun, Kamo-machi, Yamano (Enmei-kyo) (Matsumoto, 15 Sept. 1971, KYO); Kamiishi-gun, Taishakukyô, near Sakurabashi, alt. 400 m (Shimizu, 23 Sept. 1955, KYO); Hiroshima-shi, Kabe-machi, Nabarakyo, alt. ca. 300 m (Shimoda, 2 Oct. 1981, TI); Miyoshi-shi, Takasugi-machi, Kata (Fukuoka, Kurosaki & Itô 2801, SHO); Takada-gun, Mt. Otsuchi, Yoshidaguchi–the summit, alt. ca. 400 m (Terabayashi 700, KYO); Yamagata-gun, Geihoku-machi, Yawata-kougen (Shimizu, 13 Oct. 1979, SHO); Saeki-gun, Oono-mura, Matsugahara (Oka 38190, KANA); Itsukushima (Sato, 18 Sept. 1932, KANA). TOTTORI. Tottori (Hiratsuka 8, TI); Yadzu-gun, Kouge-machi, Yamashidani (Shimizu, 7 Oct. 1977, KANA); Iwami-gun, Kokufu-machi, Mitani (Tanaka, 26 Sept. 1975, KYO); Omokage-mura, Mt. Omokage-yama (Hiratsuka 43, TI); Daisen-mura (Maruyama 36, TI); Hino-gun, Nichinan-machi, nr. Kamihagiyama (Ohba & Akiyama 2150–2152, TI); Hino-gun, Nichinan-machi, nr. Abire, alt. ca. 600 m (Ohba & Akiyama 2155–2161, TI); Hino-gun, Nichinan-machi, nr. Shimoabire, alt. 500–550 m (Ohba & Akiyama 2162–2184, TI); Hino-gun, Nichinan-machi, Shimoabire–Chaya, alt. ca. 500 m (Ohba & Akiyama 2196, TI); Hino-gun, Hino-machi, near Kamisuge, alt. 200–300 m (Ohba & Akiyama 2213–2214, TI). SHIMANE. Yasugi-shi, Wada (Ohba & Akiyama 2047–2048, TI); Nogi-gun, Hirose-machi, Sugawara (Ohba & Akiyama 2067–2069, 2073–2074, TI); Nogi-gun, Hirose-machi, Fube (Ohba & Akiyama 2075, 2077, 2078–2079, 2080–2086, 2088–2098, TI); Nogi-gun, Hirose-machi, Fube, nr. Fube Dam (Ohba & Akiyama 2099–2109b, TI); Nogi-gun, Hirose-machi, nr. Nishihida (Ohba & Akiyama 2110–2111, 2112–2114, 2115–2119, TI); Nita-gun, Yokota-machi, near Nakamura, alt. 600–700 m (Ohba & Akiyama 2138–2143, TI); alt. ca. 500 m (Ohba & Akiyama 2135–2136, TI); Nita-gun, Mt. Sentuu-zan (Furuumi, 17 July 1922, TI); Mt. Sanbe-san (Satomi, 12 Aug. 1952, KANA); Mt. Shinzouji-yama (Tashiro, 12 Sept. 1936, TNS); Ouchi-gun, Kawamoto-machi (Maruyama, 12 Sept. 1935, KYO); Naka-gun, Haza-mura, Oosayama (Furuwa, 6 Aug. 1918, KAG); Mt. Gori-san (Maruyama, 24 Oct. 1937, TNS); Masuda-shi, Masuda (Murata 22150, KYO); Kanoashi-gun, Tsuwano-machi, alt. 250 m (Tateishi 3058, TI). YAMAGUCHI. Kuga-gun, Takane-mura, Mt. Jakuchi-san (Migo, 17 Oct. 1953, KANA); Tsuno-gun, Kano, Oohira (Migo, 5 Sept. 1952, KANA); Abu-gun, Atō-chō, Midoohara–Hosono (Tateishi 3011, TI); Hagi, Mt. Shitsu-yama (Nikai, 5 Oct. 1930, TNS); Tokuyama-shi, Toda, Kuwabara (Migo, 11 Oct. 1953, KANA); Yoshiki-gun, Ouchi-mura, Hikami (Nikai 287, TI—Syntype of *L. homoloba* Nakai; TNS); Yoshiki-gun, Sayama (Oda 3120, TI); Miyano, Ooyamaji (Furuya, 13 Sept. 1932, TI); Yoshiki-gun, Natajima-mura, Shima (Oda 3152, TI); Ogouri (Oda 3140, TI); Yamaguchi-shi, Yuda, Mt. Gongen-yama (Oda 3124, TI); Mine-gun, Akiyoshi (Oka 19289, TNS); Mine-gun, Mitou-machi, Naganobori–Kaerimizu (Miyake, 2 Sept. 1969, KANA); Toyoura-gun, Toyoda-machi, Ichinomata (Imada 1656, KANA); Shimonoseki-shi, Ozuki Tateishi 3627, TI).

SHIKOKU. KAGAWA. Shozu-gun, Kobe (Satomi, 25 Oct. 1962, TI); Shozu-gun, Dosho-machi, Choshi-taki (Oka 40152, TI); Shoozu-gun (Shoodo-shima), Futagoura (Ohashi, Ohba & Murata, 10 Nov. 1975, TI); Mitoyo-gun, Nio-machi, Tsutajima (Kawasaki, 7 Aug. 1957, TI). TOKUSHIMA. Naruto-shi, Ootsu-machi, Nakayama (Takatou, 27 Aug. 1965, KYO); Itano-gun, Ooasa-machi, Mt. Ooasa-yama (Takatou, 28 Sept. 1965, KYO); Itano-gun, Kamisaka-machi, Kanyake, Ooyama-dera (Takatou, 30 Sept. 1978, KYO); Myodou-gun, Kamomyo-mura (Nikai 2570, 2571, 2572, 2573, TI—Syntypes of *L. homoloba* Nakai; TNS; 2574, TI—Lectotype of *L. homoloba* Nakai; TNS); Myozai-gun, Ano-mura, Agawa (Inobe 412, TNS); Myozai-gun, Ishi-machi, Fujinoki (Takatou, 13 July 1968, KYO); Myozai-gun, Kamiyama-machi, Orono (Takatou, 28 Sept. 1978, KYO); Mt. Tairyuji-san (Okuyama, 3 Oct. 1950, TNS).

EHIME. Uma-gun, Sankaku-ji, Horikiri–Oku-no-in (Araki, 24 Sept. 1930, KYO); Imabari-shi, Sakurai, Hebiike (Yamamoto, 16 Oct. 1977, KYO); Matsuyama-jo (Satow, 28 Sept. 1960, TNS); Shusou-gun (Akutagawa, 2 Oct. 1934, TI); Yawatahama-shi (Kikuchi, 4 Oct. 1969, TUSG); Sada-misaki (Ninatsu–Natori) (Satomi, 8 Oct. 1952, KANA); Nishiuwa-gun, Ikata-mura, Minatoura (Nomura, 30 Sept. 1954, KYO); Uwajima, Naraoku–Narukawa (Okuyama 21804, TNS). KOUCHI. Kannoura (Togasi, Oct. 1957, TI; TNS); Mitani, Higashi-no-tani (Na-gaoka-gun, Ootoyo-mura) (Makino, 6 Oct. 1892, MAK); Nangoku-shi, Shiraki-dani, alt. 100 m (Ohashi, 20 June 1973, TI); Mt. Chokou-san (unknown collector, 18 Oct. 1943, TI); Tosa-shi, Syoryu-ji (Sohma, Otomo & Mori, 19 Oct. 1970, TUSG); Kaminokae-mura (Watanabe, Oct. 1885, TNS); Takaoka-gun, Hayama-mura, Kainokawa, alt. 300 m (Murata, Nishimura & Takahashi, 8 Nov. 1969, KYO); Takaoka-gun, Higashitunoyama-mura, Mt. Irazu-yama (Tagawa 2973, KYO); Takaoka-gun, Notsu-mura, Nishikiyama (Momiyama 48, TI); Sukumo-shi, Katashima (Kitamura & Murata, 25 Nov. 1966, KYO); Aki-shi, between Ohwi and Ka-wanaro, along the upper reaches of the Ioki-gawa River (Wood et al. 4683, A).

KYUSHU. FUKUOKA. Moji, Shiranoe (Hashimoto, Oct. 1952, TI); Moji, Mt. Tonokami-yama (Yoshioka, 10 Oct. 1933, TNS); Noogata-shi, Ueki (Sugiyama, 10 July 1974, TI); Kasu-ya-gun, Mt. Inunaki-yama (Nakajima, 17 Sept. 1933, TI); Kasuya-gun, Shikanoshima (Naka-jima, 11 Oct. 1936, TI); Kasuya, Tatara, Najima (Ichikawa 200856-2, A); Munakata-gun, Tsuyazaki-machi (Nakajima 23, TI); Fukuoka-shi, Noma (Nakajima, 15 Sept. 1933, TI); Fukuoka-shi, Mt. Aburayama, alt. 50–200 m (Tateishi 3064, TI); Fukuoka-shi, Yamada, alt. ca. 100 m (Ohashi & Tateishi 958, TI); Mt. Houman-zan (Nabeshima, 9 Oct. 1927, KYO); Amagi-shi, Bodaiji, alt. 50–100 m (Ohba & Akiyama 2393–2397, TI); Asakura-gun (ut "Hon-gun"), Shiwa-mura (ut "oppido Shido") (Nabeshima 22, 6 Sept. 1925, TI—Holotype of *L. retusa* Nakai); Asakura-gun, Yasukawa-mura (Nabeshima, 12 Sept. 1930, TI); Asakura-gun, Aki-dzuki-machi (Nabeshima 8, TI); Asakura-gun, Koishiwara-mura, Okuhata, alt. ca. 450 m (Ohba & Akiyama 756, TI); Mt. Gozen-dake (Matsumura 17, KAG); Kaho-gun, Ootani-mura (Naka-jima, 1 Oct. 1938, TI). SAGA. Saga-gun, Mt. Kinryu-zan (Hashimoto, 22 Sept. 1951, TI); Ogi-gun, Tenzan (Hashimoto, 2 Oct. 1951, TI). NAGASAKI. Oomura (Toyama, 1 Oct. 1949, TNS); Ohmura-shi (Toyama, 4 Oct. 1952, A); Nagasaki-shi (Makino, Sept. 1908, KYO); Nishi-sonogi-gun, Kinkai-machi, near Fumyo (Ohba & Akiyama 2448–2451, TI); Nishisonogi-gun, Kinkai-machi, Fumyo (Ohba & Akiyama 2452–2461, TI); Isl. Gotou, Narao (unknown col-lector 4, KAG); Fukue-shi, Oomagari–Oosakatoge (Fukuoka 11848, SHO); Isl. Tsushima, Izu-hara (Makino in 1904, MAK). OITA. Oono-gun, Hakusan-mura (Nakajima, 27 Aug. 1933, TI). KUMAMOTO. Amakusa-gun, Ooyano-machi, Mt. Hidake (Yamashiro, 20 Nov. 1966, TNS); Amakusa-gun, Isl. Takamoku-jima (Yamashiro, 7 June 1964, TNS); Amakusa-gun, Ariake-machi, Kamikawatsu (Shimada, 9 Oct. 1966, KYO); Hitoyoshi-shi, Yatake (Sako 6563, TNS); Aida (Mayebara 2110, TI); Kuma-gun, Kuma-mura, Isshôchi, alt. ca. 100 m (Tateishi 3607, TI); Ashikita-gun, Tsunagi-mura (Kaneta, 4 Oct. 1936, KYO); Minamata-shi, Mt. Oni-dake, alt. 400 m (Shimada, 19 Sept. 1965, KYO); Minamata (Kaneta, 15 Sept. 1935, TNS). MIYA-ZAKI. Mt. Sobo-san (Sako, 11 Oct. 1955, KAG); Hōi–Mt. Ōkue (Hatusima & Sako 24910, KAG); Takachiho (Momiyama, 8 Oct. 1958, TI); Nishiusuki-gun, Hinokage-machi, near Furu-zono, alt. 150–250 m (Tateishi 3348, TI); Higashiusuki-gun, Tougou-machi, Yamage, alt. 40 m (Kanai, 18 Oct. 1958, TI); Hyuuga-shi, Nishikawauchi (Murata 8568, TI); Koyu-gun, Kawaminami-machi, Nanuki, alt. 40 m (Tateishi 3440, TI); Miyazaki-shi, Shimokita (Miyazawa 32, TI); Awosima-mura (Miyazawa 38, TI); Minaminaka-gun, Kitagô-machi, Inohae (Tatei-shi 1090, TNS). KAGOSHIMA. Honjo-mura (Muramatsu, 1 Oct. 1936, TI); Yamano (Mura-matsu, 9 Sept. 1934, TI); Isa-gun, Yamano (unknown collector s.n., KAG); Miyanojo, Shibi (Cult. Tokyo University) (Hara, 3 Oct. 1959, TI); Yunoo–Yoshimatsu, alt. 300 m (Hatu-sima & Sako 25499, KAG); Mt. Kurino-dake (Muramatsu, 23 Sept. 1935, TI); Kirishima,

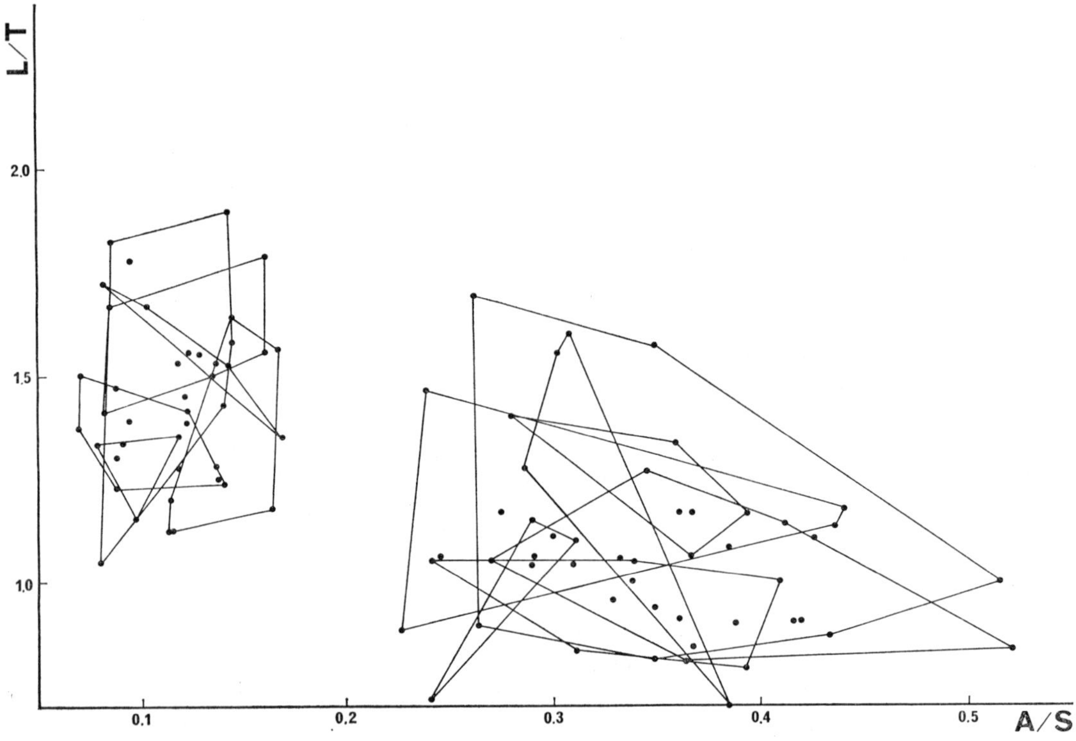

Fig. 41. Scatter diagram showing the variation of *L. homoloba* and *L. formosa* subsp. *velutina*.

A/S, ratio of the width of auricle to one half the width of standard; L/T, ratio of the length of calyx-lobe to the length of calyx-tube.

Maruo (Sako 4273, KAG); Aira-gun, Kitanagamoda (Hatusima 31286, KAG); So-gun, Iwa-gawa-cho, Ootori-kyo, alt. 100 m (Hatusima & Sako 32043, KAG); Akune-shi, Orikuchi (Sako 6483, KAG); Sinkawa-keikoku (Hatusima 16962, KAG); Isl. Koshikijima (Hatusima 16563, A); Isl. Shimokoshiki-jima, Kashima-mura (Hatusima 16607, TI); Isls. Koshiki-jima, Isl. Kami-koshiki (Sato), Arahitozaki–Naka-koshiki (Imae, 22 Oct. 1960, KANA); Isl. Yaku-shima, Onoaida (Hatusima 14852, A; KAG); Is. Yakushima, SE side, Nakabase, alt. 50–100 m (Yama-zaki, Ohba, Murata & Akiyama 2156, TI).

It is interesting that the distributions of *L. cyrtobotrya*, *L. Buergeri* and this species are similar in Japan but quite different over their entire range (Figs. 37, 40, 46). This species, endemic to Japan, is characterized by well-developed reniform auricles of standards and glabrous upper surface of leaves. Nakai (1927, 29) recognized *L. homoloba* Nakai, *L. nikkoensis* Nakai, *L. retusa* Nakai, *L. rotundiloba* Nakai, and *L. sendaica* Nakai as distinct species based on the shape of the calyx-lobes and the length of the inflorescences. However, these characters are variable even in a single site. Koidzumi (1937) described two varieties within this species, var. *emarginata* and var. *longiloba*, based on the shape of the sitpules, calyx-lobes, bracts and bracteoles, and the length of the calyx. These characters are also quite variable.

This species is distinguishable from *L. formosa* subsp. *velutina* (Nakai) S. Akiyama et H. Ohba by its well-developed auricles of the standard, shorter calyx-lobes, and glabrous upper surface of leaves (Plate 8, 9; Fig. 41). In western Japan where these two species grow together, various intermediate forms are found and are considered to be putative natural hybrids between them (Akiyama & Ohba, 1983a).

Forma **luteiflora** Akasawa in *Bull. Kochi Women's Univ. Ser. Nat. Sci.* **22**: 1 (1974).

*L. homoloba* Nakai f. *albiflora* Sugimoto, *New Keys Woody Pl. Jap.*, 241 (1972), *nom. nud.*

*L. homoloba* Nakai f. *albiflora* Inobe in *Journ. Geobot. (Kanazawa)* **21**: 31 (1973), *nom. nud.*

Japanese name. Shirobana-tsukushihagi (Sugimoto, 1973), Kibana-tsukushihagi (Akasawa, 1974).

Corolla white.

Distr. Sporadic in the range of this species.

Specimens examined.

HONSHU.  SHIGA. Gamou-gun, Ichinobe-mura (Hashimoto, 24 Sept. 1934, TI).

## Lespedeza formosa (Vogel) Koehne

*Deutsche Dendrol.*, 343 (1893), *excl. syn. Desmodium racemosum* Sieb. et Zucc., *L. Sieboldii* Miq.—Schindler in Engl., *Bot. Jahrb.* **49**: 580 (1913), *excl. syn. Desmodium racemosum* Sieb. et Zucc., *L. racemosa* Sieb. *herb. ex* Miq., *L. Sieboldii* Miq., *L. bicolor* (non Turcz.) Maxim. var. *typica* (*p.p.*), *L. bicolor* Turcz. var. *Sieboldii* (Miq.) Maxim., *L. cyrtobotrya* Miq. (*p.p.*), *Desmodium penduliflorum* Oudem. et var. *albiflora* Schindl.; in Sargent, *Pl. Wils.* **2**: 107 (1914), *pro parte*—Fu in Acad. Sin. Bot., *Ill. Important Chin. Pl. Leguminosae*, 519 (1955), *pro parte, excl. syn. L. Thunbergii* (DC.) Nakai—Hatusima in *Mem. Fac. Agr. Kagoshima Univ.* **6**: 7 (1967), *pro parte, excl.* f. *sericea* Hatusima, f. *versicolor* (Nakai) Hatusima, f. *albiflora* (Sieb. *in sched* ex Miq.) Hatusima, var. *shiroyamensis* (Hatusima) Hatusima et *syn. L. japonica* L. H. Bailey, *L. nipponica* Nakai—Acad. Sin. Bot., *Iconogr. Cormophyt. Sin.* **2**: 459, fig. 2647 (1972), *pro parte, excl. syn. L. Thunbergii* (DC.) Nakai—Akiyama & Ohba in Ohba & Malla, *Himal. Pl.* **1**: 225 (1988).

A perennial plant, 0.5–2 m high. Stems ascending or not distinct, branched in upper part or many branches coming from the basal parts of stem and underground parts. Branches terete, ascending, sometimes spreading, when young slightly angular with densely ascending appressed-sericeous or spreading hairs (hairs 0.2–0.3 mm long, whitish), later glabrescent; distal parts die in winter season, proximal parts (and stem) with axillary and adventitious winter buds.

Leaves trifoliolate, petiolate, stipulate, spirally arranged. Stipules free, linear-triangular to linear, 2–7 mm long, brownish, persistent. Petioles 1–5 cm long, hairy like the young branch. Rachides 5–15 mm long, similar to the petiole. Terminal leaf-lets (at middle parts of branches) petiolulate; petiolules 1–3 mm long, swollen; lamina 2–6 cm long, 1.5–3.5 cm wide, entire, elliptic to ovate, round at the base, acute or obtuse at the apex (the apex itself with or without a point), upper surface glabrous or pubescent, lower surface appressed-sericeous (hairs 0.2–0.3 mm long); lateral ones similar to terminal but somewhat smaller.

126

Inflorescence axillary racemous, one (rarely 2–3) per axil, 2–15 cm long including the peduncle, loosely or compactly 4–14-flowered; peduncles 1–2 cm long, hairy like the young branch. Primary bracts ovate, about 1 mm long, pubescent, persistent; secondary bracts similar to primary ones.

Flowers 10–15 mm long at anthesis; pedicels 1–3 mm long, terete, curved, tomentose. Bracteoles at the base of the calyx broadly to narrowly ovate or obtuse, ca. 1 mm long, pubescent outside, glabrous inside, persistent. Calyx 3.5–6 mm long, campanulate, four-lobed above or near the middle part, appressed pubescent; lateral one triangular with acute apex; upper one slightly two-clefted.

Standard nearly equal to, longer or shorter than keel-petal, keel-petal longer than wings. Standard red-purple inside, paler outside, 9.5–13.5 mm long with distinct claw; lamina (broadly to narrowly) elliptic, round or retuse at the apex with or without a point, inflexed-auriculate near the base, reflexing in anthesis from the part near the base; claw whitish. Wings deeper red-purple than standard, 8.5–10.5 mm long with distinct claw; lamina narrowly (to broadly) obovate to oblong, auriculate at upper basal part, slightly auriculate or tapering at the lower basal part; claw whitish. Keel-petal 10.5–13.5 mm long with distinct claw; lamina narrowly obovate, paler than wings, deepest in the apical part; claw whitish.

Ovary elliptic, ca. 1.5 mm long at anthesis, pubescent, stalked; stalk ca. 1.5 mm long.

Fruits compressed (broadly) elliptic, 7–12 mm long, 4–5 mm wide, densely to slightly pubescent (or nearly glabrous), stalked; stalk 1.5–2 mm long.

Distr. E. Himalayas, China, Korea, Taiwan, and Japan.

Our previous paper deals with this species (Akiyama & Ohba, 1988). This species has the widest range in sect. *Macrolespedeza* (Japan to E. Himalayas through Korea, Taiwan, and China) and shows remarkable geographical variation in the shape of calyx-lobes, and in the relative length of corolla to calyx and calyx-lobe to calyx-tube. Three subspecies are recognized: subsp. *formosa* from S.E. China, Hong Kong, Macao, and Taiwan; subsp. *elliptica* from S.W. China and Assam; and subsp. *velutina* from Japan, Korea, and E. China.


Key to subspecies and varieties
1.  Corolla is three to four times longer than calyx. Calyx-lobe is nearly equal to or slightly shorter than calyx-tube. Upper surface of leaves glabrous or pubescent. Distributed in Hong Kong, Taiwan and S.E. China ..................................... subsp. *formosa*
1.  Corolla is two to three times longer than calyx. Upper surface of leaves pubescent, very rarely glabrous.
    2.  Calyx-lobe is one and one-half to three times longer than calyx-tube. Distributed in S.W. China and Assam..................................................... subsp. *elliptica*
    2.  Calyx-lobe is slightly longer or one and one-half times longer than calyx-tube. Distributed in Japan, Korea, and E. China ............................ subsp. *velutina*, 3
        3.  Plant shrubby. Hairs on stem and branches usually appressed. Distributed in Japn (W. Hoshu & N. Kyushu), Korea and E. China .................... var. *velutina*
        3.  Plant dwarffy, less than 1 m high, with many branches from underground portion of stem and branches. Hairs on stem and branches usually spreading. Endemic to S. Kyushu (Satsuma Penin. & Isls. Danjo) .................. var. *satsumensis*

Subsp. **formosa**: S. Akiyama & H. Ohba in Ohba & Malla, *Himal. Pl.* **1**: 225 (1988).

Japanse name. Taiwanhagi (Hatusima, 1969).

Further synonyms and specimens examined are cited in Akiyama & Ohba (1988).

Distr. S.E. China, Hong Kong, Macao, and Taiwan.


Subup. **elliptica** (Benth. ex Maxim.) S. Akiyama et H. Ohba in H. Ohba & Malla, *Himal. Pl.* **1**: 226 (1988).

Distr. S.W. China and Assam.

Further synonyms and specimens examined are cited in Akiyama & Ohba (1988).


Subsp. **velutina** (Nakai) S. Akiyama et H. Ohba in H. Ohba & Malla, *Himal. Pl.* **1**: 227 (1988).                                        [Plate 15–17; Fig. 42]

*L. bicolor* Turcz. var. *velutina* Nakai in *Bot. Mag. Tokyo* **37**: 74 (1923).

*L. kiusiana* Nakai, *Lesp. Jap. Korea*, 26 (1927)—Akiyama & Ohba in *Journ. Jap. Bot.* **61**: 96 (1986).

*L. intermedia* Nakai, *Veg. Mt. Chirisan* 56 (1915); in *Bot. Mag. Tokyo* **37**: 76 (1923), non Britt. (1893).

*L. intermedia* Nakai var. *angustifolia* Nakai in *Bot. Mag. Tokyo* **37**: 77 (1923).

*L. maritima* Nakai in *Bot. Mag. Tokyo* **37**: 78 (1923).

*L. japonica* L. H. Bailey var. *intermedia* (Nakai) Nakai, *Lesp. Jap. Korea*, 23 (1927) —Lee in *Bull. Seoul Nat. Univ. For.* No. 2, 15 (1965).

*L. japonica* L. H. Bailey var. *angustifolia* (Nakai) Nakai, *Lesp. Jap. Korea*, 25 (1927).

*L. japonica* L. H. Bailey var. *retusa* Nakai, *Lesp. Jap. Korea*, 26 (1927).

*L. Uekii* Nakai in *Bot. Mag. Tokyo* **42**: 457 (1928).

*L. japonica* L. H. Bailey var. *spicata* Nakai in *Journ. Jap. Bot.* **15**: 533 (1939).

*L. tetraloba* Nakai in *Journ. Jap. Bot.* **15**: 680 (1939).

*L. penduliflora* Oudem. var. *albiflora* (Sieb. ex Miq.) Ohwi f. *angustifolia* (Nakai) Ohwi, *Fl. Jap.*, 679 (1953), *nom. nud.*

*L. Thunbergii* (DC.) Nakai var. *albiflora* (Schneid.) Ohwi f. *angustifolia* (Nakai) Ohwi, *Fl. Jap.* rev. ed., (790) 1438 (1965); *Fl. Jap.* Eng. ed., 559 (1965), ut "*L. Thunbergii* (DC.) Nakai f. *angustifolia* (Nakai) Ohwi"; *Fl. Jap.* new ed., 790 (1975).

*L. japonica* L. H. Bailey var. *intermedia* Nakai f. *retusa* (Nakai) T. Lee in *Bull. Seoul Nat. Univ. For.* No. 2, 17 (1965).

*L. maritina* Nakai f. *Uekii* (Nakai) Hatus. in *Mem. Fac. Agr. Kagoshima Univ.* **6**: 10 (1967).

*L. Thunbergii* (DC.) Nakai var. *intermedia* (Nakai) T. Lee in *Bull. Seoul Nat. Univ. For.* No. 6, 56 (1969).

*L. Thunbergii* (DC.) Nakai var. *intermedia* (Nakai) T. Lee f. *retusa* (Nakai) T. Lee in *Bull. Seoul Nat. Univ. For.* No. 6, 56 (1969).

*L. japonica* L. H. Bailey f. *angustifolia* (Nakai) Murata in *Acta Phytotax. Geobot.* **29**: 100 (1978)—Ohashi in Satake et al., *Wild Fl. Jap.* **2**: 205 (1982).

Japanese name. Bitchu-yamahagi (Nakai, 1923), Chosen-yamahagi (Nakai, 1923).

A perennial plant, 1.5–2 m high. Stems ascending, 0.5–1 m high, 1–3 cm in diameter; most of the terrestrial part lives only a few years; branched in upper parts (a few

128

branches coming from the basal part of the stem). Branches terete, ascending, sometimes dependent later; when young slightly angular with densely ascending appressed-sericeous hairs (hairs 0.2–0.3 mm long, whitish), later glabrescent; distal parts die in winter, proximal parts (and stem) with axillary and adventitious winter buds (winter buds at lower parts mostly dormant).

Leaves trifoliolate, petiolate, stipulate, spirally arranged. Stipules free, linear-triangular to linear, 2–7 mm long, brownish, persistent. Petioles 1–5 cm long, hairy like the young branch. Rachides 5–15 mm long, similar to the petiole. Terminal leaflets (at middle parts of branches) petiolulate; petiolules 1–3 mm long, swollen; lamina 2–6 cm long, 1.5–3.5 cm wide, entire, elliptic to ovate, cuneate, or round at the base, acute or obtuse at the apex (the apex itself with or without a point), upper surface pubescent (hairs 0.1–0.15 mm long, usually more than 40/mm²), lower surface appressed-sericeous (hairs 0.2–0.3 mm long); lateral ones similar to terminal but somewhat smaller.

Inflorescence axillary racemous, one (rarely 2–3) per axil, (2–)4–10(–15) cm long including the peduncle, somewhat loosely to loosely 6–14-flowered; peduncles 1–4 cm long, hairy like the young branch. Primary bracts ovate about 1 mm long, pubescent, persistent; secondary bracts similar to primary ones.

Flowers in middle September to middle October, 11.7–12.8 mm long at anthesis; pedicels 1.5–3 mm long terete, curved, tomentose. Bracteoles at the base of the calyx broadly to narrowly ovate or obtuse, ca. 1 mm long, pubescent outside, glabrous inside, persistent. Calyx 3.9–4.4 mm long, campanulate, four-lobed above the middle part, appressed pubescent; tube 1.6–1.7 mm long; the lower lobe longest; lateral ones 2.2–2.7 mm long, 0.8–1.1 mm wide, triangular-lanceolate, acute at the apex but not acuminate; upper ones slightly two-clefted.

Standard nearly equal to keel-petal, keel-petal longer than wings (S≥K>W). Standard red-purple inside, paler outside, 10.7–12.4 mm long with distinct claw, 6.0–7.2 mm wide; lamina (broadly to narrowly) elliptic to obovate, round or retuse at the apex with or without a point, inflexed-auriculate near the base, reflexing in anthesis from the part near the base; claw whitish, 2.0–2.4 mm long. Wings deeper red-purple than standard, 8.6–10.1 mm long with distinct claw: lamina narrowly obovate to oblong, 6.0–7.3 mm long, 2.0–3.2 mm wide, auriculate at upper basal part, slightly auriculate or tapering at the lower basal part; claw 2.9–3.4 mm long, whitish. Keel-petal 10.5–12.0 mm long with distinct claw; lamina narrowly obovate, 7.4–8.6 mm long, 3.2–3.7 mm wide, paler than wings, deepest in the apical part; claw 3.1–3.6 mm long, whitish. Stamens 10, nearly the same length, 10–11.5 mm long, diadelphous. Anthers elliptic, shallowly retuse at the apex, ca. 0.6 mm long; before anthesis yellow. Pistils 11.5–12 mm long; ovary elliptic, ca. 1.5 mm long at anthesis, pubescent, stalked; stalk ca. 1.5 mm long; style 7–8.5 mm long, pubescent at the basal part, glabrescent at the apical part.

Fruits compressed (broadly) elliptic, 7–8(–9) mm long, 4–5 mm wide, densely to slightly pubescent (or nearly glabrous), stalked; stalk 1.5–2 mm long.

Seeds reniform, 3–3.5 mm long, 2–2.5 mm wide.

Distr. Japan (W. Honshu and N. Kyushu), Korea, and E. China.

Voucher and representative specimens.

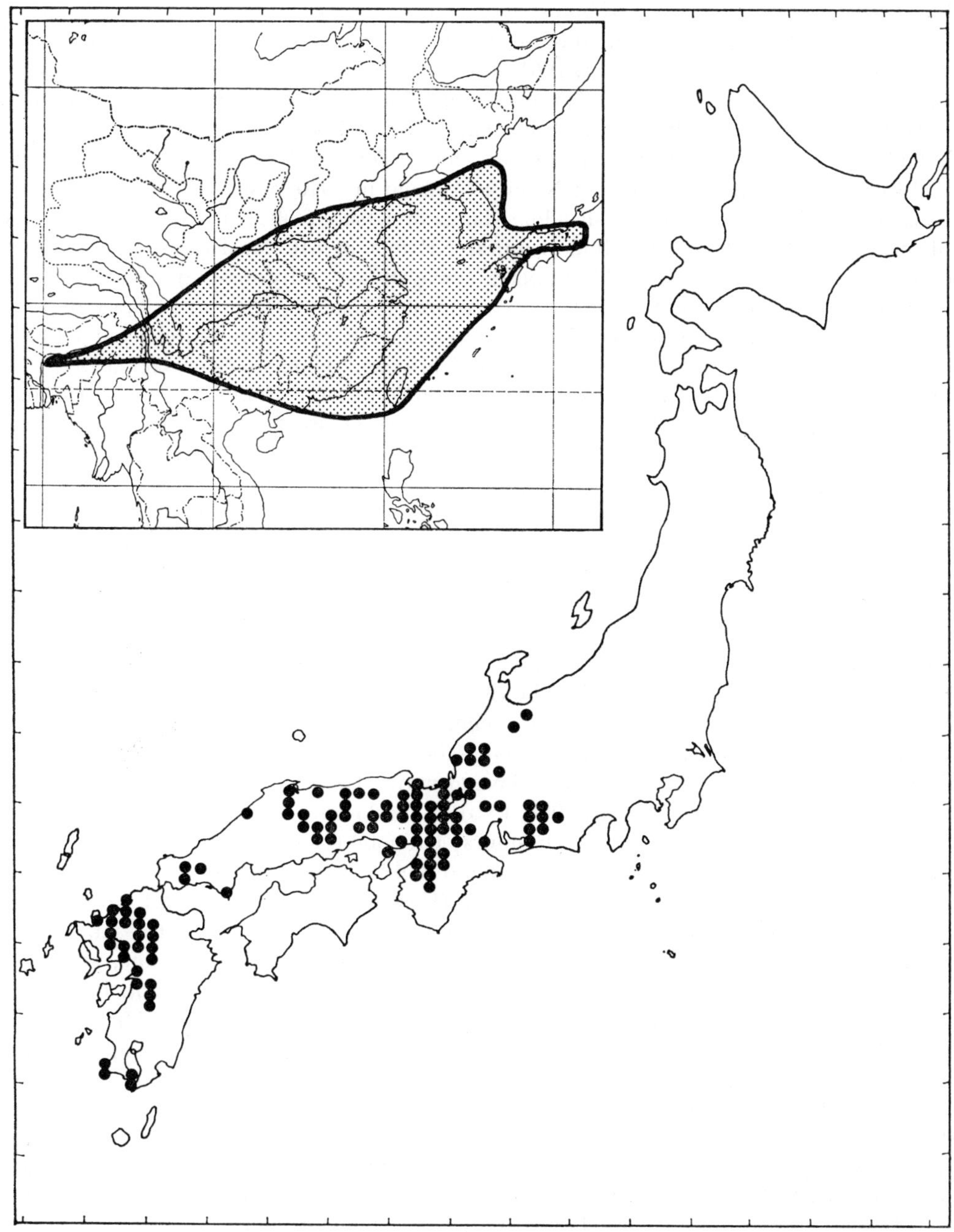

Fig. 42.   Distribution of *L. formosa*.

## Japan.

HONSHU.   SHIZUOKA. Iwata-gun, Sakuma-mura, Urakawa–Kawakami (Torii 14312 &
28 Sept. 1958, KYO); Mt. Ogasa-yama (Shimizu, 13 Oct. 1929, KYO).   AICHI. Miwamura,
Ikebasaka (Torii, 6 Sept. 1959, KYO); Goten-mura (Torii 6156, KYO); Noda-mura (Torii

130

617, TUSG); Kamitsugu-mura, Orimoto-touge (Torii 6110, TUSG); Inabu-machi, Natsuyaki (Torii, 10 Oct. 1967, KYO); Kitashitara-gun, Danmine-mura, Damine (Torii 9897, TNS); Shitara-machi, Kiyosaki (Torii, 17 Sept. 1961, KYO); Mt. Dando (Murata 6635, KYO); Minamishitara-gun, Hourai-machi, Kadoya (Torii, 22 Sept. 1963, TI); Houraiji-mura, Onbara (Torii 12403, TNS); Horaizizan, alt. 350 m (Takeuchi, 10 Sept. 1946, A; TI); Minamishitara-gun, Tsukude-mura (Torii s.n., TI); Shinshiro-shi, Oomi (Torii, 1 Oct. 1950, TUSG); Toyohashi-shi, Iwahi-machi (Torii 6212, TUSG); Nagoya (Makino in 1946, MAK). GIFU. Yoshiki-gun, Miyagawa-mura, along Mannami-Rindo, Mannami–Utsubo, alt. ca. 500 m (Iwatsuki & Fukuoka 370, KANA); Mugi-gun, Itadori-mura, NE foot of Mt. Kooka-san, Kabe–Takagatani-goe (Fukuoka & Yamashita 227, KANA); Yamagata-gun, Miyama-mura, Kamisaki, Mt. Funabuse-yama (Nakano, 3 Oct. 1964, KANA); Yourou (Okamoto, 12 Sept. 1938, KYO); Takayama-shi (Nagase 7111-1–8, 7133-1–5, TI). TOYAMA. Nei-gun, Mt. Kureha-yama (Takagi, 7 Oct. 1934, TNS); Nakaniikawa-gun, Kamiichi-machi, Hinotani (Kurosaki, 3 Oct. 1966, KANA); Nakaniikawa-gun, Iwamuro-taki (Satomi 26433, KYO). FUKUI. Oono-shi, Heisen-ji (Satomi 24518, KYO); Oono-shi, Mt. Arashima-dake (Watanabe 60, TOFO); Oono-shi, Nakayasumi–lower elevation of Mt. Arashima-dake (Fukuoka 8245, KYO); Oono-shi, Fukui–Mt. Iifuri-yama (Fukuoka 8262, KYO); Asuwa-gun, Takada–Eihei-ji (Watanabe, 25 Oct. 1964, TOFO); Imatate-gun, Itagakisaka (Okamoto-mura) (Hosoi, 5 Oct. 1936, KYO); Takefu-shi, E. foot of Mt. Hino-san (Kurosaki 8004, TI); Katsuyama-shi, Rokuroshi (Watanabe, 13 Sept. 1964, TOFO); Nanjô-gun, Kôno-mura, Jindo, alt. 150–250 m (Kurosaki 11013, KANA); Tsuruga-shi, Iwagomori (Watanabe, 20 Sept. 1964, TOFO); Mikata-machi, Mikata–Kiyama (Watanabe 121, TOFO); Mt. Sanjusangen-yama (Watanabe 12690, TOFO); Mt. Aoba-yama (Mimura, 2 Oct. 1960, TOFO); Saburi (Hori, Oct. 1955, TNS). SHIGA. Mt. Handoji (Kitamura, 2 Sept. 1952, A); Ohtsu-shi, Kokubo (Makino in 1933, KYO); Mt. Yatsuo-yama (Hayashi, 15 Aug. 1940, TI); Ika-gun, Kinomoto-cho, S. of Mt. Tsuchigura, Tsuchigura–Hassō Pass (Fukuoka 6168, KYO); Ika-gun, Kinomoto-cho, Suino, SE foot of Mt. Yokoyama-dake (Fukuoka 6121, KYO); Higashiasai-gun, Hayami-mura, Umawatashi (Hashimoto, 21 Sept. 1941, KYO); Ibuki (Tashiro, 2 Sept. 1928, KYO); Kanzaki-gun, Mt. Kannonji-yama (Hashimoto 10014, TNS); Gamou-gun, Ichinobe-mura, Ichinobe (Hashimoto 10176, TNS); Kouga-gun, Ishibe-machi, Kanekatsumichi (Hashimoto 11421, KYO); Suzukatoge (Kitamura, 17 Sept. 1931, KYO); Kurita-gun, Kanekatsu-mura (Hashimoto 17774, TNS); Mt. Iwama-yama (Sono, 1 Sept. 1907, TNS); Kokubu-mura (Makino in 1933, MAK); Kusatsu-shi, Yamada (Makino, Sept. 1933, KYO); Takashima-gun, Makino-cho, Mt. Akasaka-yama, alt. 200 m (Murata, 6 Sept. 1964, KYO); Takashima-gun, Imadzu-machi (Hashimoto, Sept. 1941, KYO); Shiga-gun (Katata-cho), Hanaore-toge, alt. 590 m (Murata 19030, TI); Mt. Hira (Tashiro, 9 Oct. 1927, KYO); Mt. Hiei-zan, Yokokawa-chudou (Okamoto, 30 Sept. 1933, TI). MIE. Ichishi-gun, Yamato-mura, alt. 200 m (Maekawa, Sept. 1945, TI); Tsu, Nishihorihata (Umemura, Sept. 1894, MAK); Ueno-shi, Iwakura-kyo (Shimizu, 5 Nov. 1978, SHO). NARA. Mt. Monju-yama (Hara, 23 Oct. 1932, KYO); Mt. Ikoma (Kageyama, 8 July 1938, TNS); Ikoma-gun, Katagiri-tyo, Matsuo-yama (Murata, 28 Oct. 1956, KYO); Ikoma-gun, Ikoma-cho, Narukawa (Murata, 10 Oct. 1965, KYO); Nara-shi, Yagyu (Okubo & Nishida 15, KYO); Yamabe-gun, Tsuge-mura (Hara, 29 Aug. 1956, KYO); Sakurai-shi (Shimizu, 10 Oct. 1982, SHO); Takaichi-gun, Asuka-mura, Imotoge, alt. 400–500 m (Mimoro, Tsugaru & Fujimoto 4243, KYO); Takatori-machi, Tsubosaka-touge (Okamoto, 3 Nov. 1933, TI); Gose-shi, Mt. Katsuragi-yama, alt. 300–900 m (Mimoro, Tsugaru, & Mizoguchi, 22 Sept. 1975, TNS); Katsuragi-mura, Fushimi (Okamoto, 14 Oct. 1934, KYO); Gose-shi, the summit of Mt. Kongo-san–Nagara, alt. 1100–300 m (Kurosaki 6444, SHO); Yoshino-gun, Yoshino-machi, Mt. Yoshino-yama (Yamanouchi, 1 Nov. 1960, TNS). WAKAYAMA. Hashimoto-shi, Niu-gawa-tsuji (Ogawa, 17 Sept. 1962, TNS); Kouya-san (Fukuoka, 1 Nov. 1959, SHO); Naga-gun, Minaminokamimachi (Yamamoto, 27 Sept. 1941, TNS); Arita-gun, Yawata-mura, Chikai (Okamoto, 22 Sept. 1938, KYO). KYOTO. Kyoto, Shishi-ga-dani (Ohwi 9084, TI); Kyoto, Ichizyôzi (Yamazaki, Oct. 1962, TI); Mt. Yosida (Matsuda, 24 Oct. 1979, KYO); Sakyo-ku, Iwakura (Okamoto, 17 Sept. 1938, KYO); Kyoto-shi, Yase (Koidzumi, 30 Sept. 1932, TNS); Kyoto-shi, Saga, Arashi-yama (Koidzumi, 29 Sept. 1932, KYO); Mt. Hiei-zan (Makino, Sept. 1932, MAK); Takao (Tan in 1926, KYO); Kitakuwada-gun, Chii-mura, Ashiu, Sugo (Okamoto, 9 Sept. 1936, KYO); Oohara-mura, Aogi-touge (Okamoto 7125, TI); Mt. Nyoi-ga-dake (Okamoto, 1 Oct. 1949, TNS); Mt. Daimonji (Hiroe 16464, KYO); Kasa-gun, Ooe-machi, Arita, Jonodan (Kuwashima, 24 Sept. 1961, TNS); Kasa-gun, Kasai-mura, Tadehara (Araki 8762, KYO); Fukuchiyama, Mt. Jinnan-yama (Araki, 2 Sept. 1928, KYO); Fukuchiyama-shi, Mt. Tori-ga-take (Nagai, 15 Sept. 1973, KYO); Amata-gun, Nakayakuno-mura (Okamoto, 22 Sept. 1949, KYO); Ayabe-shi, Tachi-cho–Iden, alt. ca. 100 m (Murata, Okubo & Nishida 34006, KYO); Ikaruga-gun, Okukanbayashi-mura, Kimioyama (Murata, 15 Sept. 1956, KYO); Mt. Chorogadake, Hodo-su–the summit, alt. 550 m (Murata, 23 Sept. 1961, KYO); Minamikuwata-gun, Nishibe-tsuin-mura, Yuzuhara (Araki 8810, KYO). OSAKA. Mt. Myoken (Togashi, 11 Sept. 1950, TI); Mishima-gun, Takatsuki-machi (Kuwashima, 1 Oct. 1933, TNS); Takatsuki, Nariai–Kawa-kubo (Tagawa, 30 Oct. 1954, KYO); Katsuo-ji (Ui 42, 43, TI); Toyono-gun, Kanno-mura (Ui, 30 Oct. 1934, TI); Ikeda (Murata 5958, KYO); Hirakata-shi, Saka (Okada, Nov. 1962, KYO); Kisaichi–Shijonawate (Kitamura, 2 Oct. 1949, KYO); Yao-shi, Takayasu, near Juso-toge, alt. 300 m (Murata, 11 Oct. 1959, KYO); Minamikawachi-gun, Taishi-chô, Mt. Nijô, Nijô-dani, alt. 100–350 m (Seto, 9 Oct. 1975, KANA); Kongo (Ui 8928, TNS); Kawachinagano-shi, Taki-hata–Zaou-touge (Shimizu, 23 Sept. 1976, SHO); Sennan-gun, Higashitottori-mura, around Ogawa-onsen (Nakazima, 22 Sept. 1957, KYO). HYOGO. Mt. Kurosawa (Muroi 379, A); Taki-gun, Jotou-machi, Furusaka-touge (Hosomi, 20 Sept. 1967, KYO); Hikami-gun, Ichi-jima-machi, Kamitakeda (Hosomi, 14 Sept. 1937, KYO); Takeda-mura (Hosomi, 30 Sept. 1936, TNS); Hikami-gun, Sakiyama-mura (Hosomi, 14 Sept. 1937, TI); Kamikuge-mura, Akusa (Hosomi, 26 Sept. 1937, TI); Mt. Rokko, twenty-cross (Muroi 6732, A); Mt. Hachi-buse, Okubo–Takamaru (Murata 21138, KYO); Mt. Hyo-no-sen (Iwatani, 18 Aug. 1964, TNS); Mikata-gun, Wakasugi (Muroi 5641, A); Mikata-gun, Kumatugi (Muroi 5540, A); Mikata-gun, Sekinomiya (Muroi 5679, A); Mikata-gun, Yohka (Muroi 5574, A); Asako-gun, Takeda (Muroi 6204, A); Asako-gun, Awaga-mura, Mt. Awaga-yama (Araki 23720, KYO); Asako-gun, Santou-machi (Ishimura, 8 Aug. 1963, KANA); Asako-gun, Ikuno-cho (Makino, Oct. 1939, MAK); Asago-gun, Asago-cho, Okutataragi, alt. 200 m (Fukuoka 10007, KYO); Kanzaki-gun, Ookouchi-cho, Kamioda, Mineyama-kougen, alt. 940–1040 m (Kurosaki 13930, SHO); Shisou-gun, Yasutomi-machi, Anji (Uchiumi U193, SHO); Shikama-gun, Yumesaki-machi, Mt. Myojin-yama (Uchiumi U390, SHO); Yumesaki-machi, Suginouchi (Iwatani I72233, SHO); Mt. Minou-san (Tashiro, 10 Sept. 1932, KYO); Akou-gun, Ashiankyo-mura (Tashiro, 8 Sept. 1934, KYO); Akou-gun, Takada-mura (Muroi, 18 Sept. 1934, KYO); Mt. Nabebutayama (Muroi 6703, A); Isl. Awaji-shima, Tsuna-gun, Hokutan-cho, Nojima-sho-nyudo–Todoroki (Kurosaki, 10 Sept. 1983, SHO). OKAYAMA. Tomada-gun, Awa-mura, Oosugi (Nanba, 18 Sept. 1968, KAG); Katsuta-gun, Shôboku-cho, Hiroto (unknown collector, 23 Sept. 1908, KYO); Katsuta-gun, Kitayoshino-mura (Hiratsuka, 20 Oct. 1938, TI); Tsu-yama-shi, Fukuda (Nanba 9424, KAG); Mitsu-gun, Kamogawa-machi, Takao (Ohba & Aki-yama 681, TI); Mitsu-gun, Mitsu-machi, Kubo, alt. 50–100 m (Ohba & Akiyama 650–657, TI); Mitsu-gun, Mitsu-machi, Masera, alt. 50–100 m (Ohba & Akiyama 662–667, TI); Jōbō-gun, Kawamo (Nanba, 10 Sept. 1949, A); Tsukubo-gun, Kamo-mura (Tsuboi, 18 Sept. 1929, KYO); Tsukubo-gun, Asabara (Uno, 26 May 1946, A); Kibi-gun, Makane-mura, Itakura (Nikai 1072, TI—Lectotype of *L. bicolor* Turcz. var. *velutina* Nakai and lectotype of *L. kiusiana* Nakai; TNS—Isolectotype); Kibi-gun, Takamatsu-mura, Okayama Prefectural Ag-

132

ricultural School (cult.) (Nikai 1142, 1143, TI—Syntypes of *L. bicolor* Turcz. var. *velutina* Nakai and *L. kiusiana* Nakai); Takamatsu-machi (cult.) (Koidzumi, 20 Sept. 1942, KYO); Takahashi (Nishihara, 10 Sept. 1901, MAK); Atetsu-gun, Miyoshi-mura, Kuraida (Shimizu, 14 Oct. 1954, KYO); Atetsu-gun, Tetta-cho, Hanaki, alt. 240–300 m (Fukuoka & Kurosaki 1949, KYO). HIROSHIMA. Miino (Yatsuhoko-mura) (Maruyama, 11 Aug. 1946, TI). TOTTORI. Wakasa-machi, Mt. Hyo-no-sen (Tanaka 21214, KYO); Tottori-shi (Nakada, 24 Sept. 1938, TI—Holotype of *L. japonica* var. *spicata* Nakai); Hino-gun, Nichinan-machi, near Kasagi, alt. ca. 500 m (Ohba & Akiyama 2201–2204, TI). SHIMANE. Ugasho (Maruyama, 23 Sept. 1947, TNS); Yatsuka-gun, Iwasaka-mura (Makino, 24 Sept. 1906, MAK); Nima-gun, Nima-machi (Tanaka, 20 Sept. 1936, KYO). YAMAGUCHI. Kumage-gun, Tabuse-machi, Seto (Minami, 5 Sept. 1977, TNS); Abu-gun, Atō-chō, Midoohara–Hosono (Tateishi 3026, TI); Yamaguchi-shi, Imamichi (Oda 3129, TI); Yoshiki-gun, Iseki (Oda 3128, TI).

KYUSHU. FUKUOKA. Itoshima-gun, Sakurai (Yahara, 14 Sept. 1969, KAG); Itoshima-gun, Maebara-cho, Shiraito, alt. ca. 250 m (Tateishi 3731, TI); Itoshima-gun, Fukuyoshi-mura (Hatusima, 13 Sept. 1963, KAG); Mt. Rai-zan (Nakajima 10350, TNS); Itoshima-gun, Mt. Ibara-yama (Yahara, 15 Sept. 1969, KAG); Itoshima-gun, Mt. Tamamaru-yama (Oouchi 11, KAG); Tsukushi-gun, Nakagawa-mura, Minamihata, Saruyama-rindou (Kurata 1304, TOFO); Mt. Kusenbu-yama (Oouchi, 29 Sept. 1970, KAG); Tsukushi-gun, Kasuga-mura (Hatusima 10211, KAG); Amagi-shi, Akizuki, Notori, alt. ca. 200 m (Ohba & Akiyama 718, 719, 722–729, TI); Asakura-gun, Asakura-machi (Ohba & Akiyama 2242, TI); Asakura-gun, Haki-machi, Shiwa, Kouyama, alt. ca. 150 m (Ohba & Akiyama 2238, 2253–2270, TI); Amagi-shi, Kurokawa, Motonome–Nishihara, alt. ca. 180 m (Ohba & Akiyama 2283–2290, TI); Asakura-gun, Yasukawa-mura (Nabeshima 5, TI—Syntype of *L. japonica* L. H. Bailey var. *gracilis* Nakai); Ukiha-gun, Yoshii-mura (Nabeshima, 10 Sept. 1930, KYO); Kurume-shi (Oouchi, 17 Sept. 1970, KAG); Mt. Miike-yama (Tashiro, 3 Sept. 1932, KYO); Miike-gun, Tamakawa-mura, Kunugino (Sugino, 7 Sept. 1932, TI); Miike-gun, Tamagawa-mura, Adatatsu (Sugino, 22 Sept. 1930, TI). SAGA. Mt. Seburi (Baba 17, KAG); Mt. Ten-zan (Tashiro, 6 Sept. 1936, KYO); Ogi-machi, Syakutai (Baba 21, KAG); Fujitsu-gun, Shiota-machi (Baba, 19 Sept. 1958, TNS). NAGASAKI. (Makino, Sept. 1908, MAK). KUMAMOTO. Kamoto-gun, Uchida-mura (Yamashiro, 9 Oct. 1960, TNS); Arao-shi, Hirayamashita (Arao, 20 Aug. 1931, TNS); Tamana-gun, Fumoto-mura (unknown collector 28, TI); Kikuchi-gun, Nishigoushi-machi, Ooike (Shimada, 6 Oct. 1966, KYO); Kumamoto-shi, Mt. Honmyouji-yama (Yamashiro, 23 Oct. 1955, TNS); Kumamoto-shi, Mt. Tatsuta-yama (Yamashiro, 23 Sept. 1957, TNS); Houtaku-gun, Mt. Kohagi-yama (Shimada 7845, KYO); Houtaku-gun, Mt. Kinpou-zan (Kôduma 308, KANA); Shimomasaki-gun, Samata (Koyama, 14 Sept. 1963, KAG); Yatsu-shiro-gun, Mt. Ryuuhouzan (Toshima, 10 Sept. 1935, KYO); Yatsushiro-gun, Mt. Ryûhô (Shimizu, 4 Sept. 1959, KYO); Hitoyoshi-shi, Kuma-gawa (Yamashiro, 10 Sept. 1966, TNS).

**Korea.**
KYONGGI. Yangju (Nakai 2313 (ut "259" in Nakai 1927), TI—Lectotype of *L. intermedia* var. *angustifolia* Nakai). CHUNGCHONGPUK. Mt. Sokri San (Nakai 15009, TI—Holotype of *L. tetraloba* Nakai). In littore insulae Hokitsutô (Nakai 9821, TI—Holotype of *L. maritima* Nakai). KYONGSANNAN (KEINAN). In littore circa Torai, (Ueki s.n., TI—Holotype of *L. Uekii* Nakai).

Other specimens outside Japan are cited in Akiyama & Ohba (1988).


Forma **alba** (Nakai) S. Akiyama et H. Ohba, *comb. nov.*

*L. intermedia* Nakai var. *alba* Nakai in *Bot. Mag. Tokyo* **37**: 77 (1923).

*L. japonica* L. H. Bailey var. *albiflora* (Sieb. ex Miq.) Nakai, *Lesp. Jap. Korea*, 25 (1927), *pro parte*.

*L. japonica* L. H. Bailey var. *intermedia* Nakai f. *alba* (Nakai) T. Lee in *Bull. Seoul Nat. Univ. For.* No. 2, 16 (1965).

*L. Thunbergii* (DC.) Nakai var. *intermedia* (Nakai) T. Lee f. *alba* (Nakai) T. Lee in *Bull. Seoul Nat. Univ. For.* No. 6, 56 (1969).

Japanese name. Shirobana-bitchu-yamahagi (nov.), Shirobana-chosen-yamahagi (Nakai, 1923).

Corolla white.

Distr. Korea.

Specimen examined.

**Korea.**

KANGWON. Jangjun (Chōzen) (Nakai 5576, TI—Holotype of *L. intermedia* var. *alba* Nakai).


For the white-flowered *Lespedeza*, many names have been described under different species as varieties or forms, although most are cultivated plants. In this paper we treat cultivated white-flowered *Lespedeza* as different species. Native white-flowered form of this species is known from Korea only and described by Nakai as the variety of *L. intermedia* (non Britt.) Nakai. The specimen of white-flowered *Lespedeza* (Kurokawa TI) has a label on which it is written that this plant is wild (Maekawa, 1938). It is doubtful whether this plant is truly wild or naturalized.


Var. **satsumensis** (Nakai) S. Akiyama et H. Ohba in Ohba & Malla, *Himal. Pl.* **1**: 228 (1988). [Plate 18]

*L. satsumensis* Nakai in *Bot. Mag. Tokyo* **42**: 456 (1928)—Hatusima in *Mem. Fac. Agr. Kagoshima Univ.* **6**: 11 (1967)—Akiyama & Ohba in *Journ. Jap. Bot.* **58**: 135 (143) (1983)—Ohba in *Journ. Jap. Bot.* **60**: 78 (1985).

*L. nipponica* Nakai var. *satsumensis* (Nakai) Murata in Kitamura & Murata, *Col. Ill. Herb. Pl. Jap.* **2**: 99 (1961), *nom. nud.*; in *Acta Phytotax. Geobot.* **20**: 198 (1962).

*L. Thunbergii* (DC.) Nakai var. *satsumensis* (Nakai) Ohwi, *Fl. Jap.* rev. ed., (791) 1438 (1965); *Fl. Jap.* new ed., 791 (1975)—Murata in *Acta Phytotax. Geobot.* **29**: 105 (1978)—Ohashi in Satake et al., *Wild Fl. Jap. Herb.* **2**: 205 (1982).

*L. formosa* var. *australis* Hatusima in *Mem. Fac. Agr. Kagoshima Univ.* **6**: 8 (1967).

*L. japonica* L. H. Bailey var. *australis* (Hatusima) Murata in *Acta Phytotax. Geobot.* **29**: 101 (1978)—Ohashi in Satake et al., *Wild Fl. Jap. Herb.* **2**: 205 (1982).

Japanese name. Satsumahagi (Nakai, 1928), Nangoku-chosen-yamahagi (Hatusima, 1967).

A prennial plant, 0.5–1 m high. Stem not distinct, many branches coming from maily underground parts even in late autum. Branches with spreading or ascending appressed-sericeous hair. Receme 2–4 cm long, nearly equal to subtending leaf. Flowers in middle October to late November, 10–12 mm long at anthesis. Standard nearly equal to or shorter than keel-petal (K≧S>W).

Distr. Japan (Satsuma Peninsula, S. Kyushu, and Meshima, Isls. Danjo).

Representative specimens.

KYUSHU. NAGASAKI. Danjo Islands, Meshima Island (Saito, Oct. 1984, TI). KAGOSHIMA. Yamagawacho, Takeyama, alt. 60 m (Hatusima 20917 & 23 June 1957, KAG); Nagasakibana (Hatusima 16731 (cultivated at Kagoshima), KAG; Hatusima, Sako & Kawanabe 22177,

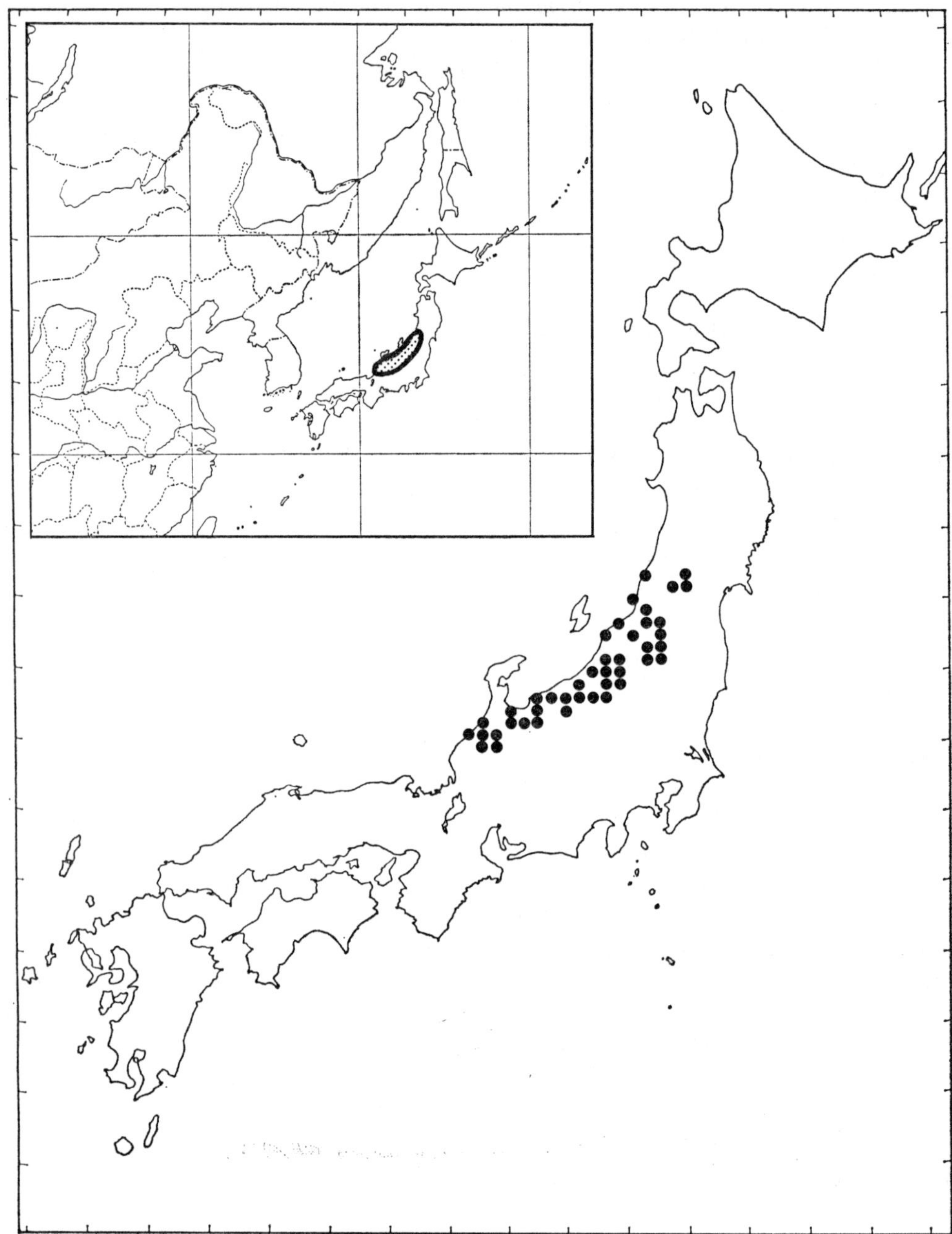

Fig. 43.   Distribution of *L. patens*.

22186, KAG); Ibusuki-gun, Yamakawa-machi, Nagasakibana, alt. 0–50 m (Ohba & Akiyama 806–809, TI); Mt. Isoma-yama (Doi, 23 Oct. 1927, TI—Holotype of *L. satsumensis* Nakai; KYO—Isotype; Nov. 1928, TI); Nomaike (Sugimoto 35591, KYO); Kawanabe-gun, Kasasa-machi, Gotou–Kurose, alt. 50–100 m (Ohba & Akiyama 2602–2604, TI); Kawanabe-gun, Kasasacho, Kurose, alt. ca. 150 m (Ohba & Akiyama 2608–2617, TI; Sakata, 6 Oct. 1980, 25 Oct. 1980, TI); Mt. Nomadake (Hatusima 16433, KAG—Holotype of *L. formosa* (Vogel) Koehre var. *australis* Hatus.); Bonotsu-cho, Imadake (Sako 3564, KAG).

This variety is restricted to the southern part of Satsuma Peninsula and Meshima Island (Danjo Islands), and is slightly different from var. *velutine* (Akiyama & Ohba, 1983b, 88).

**Lespedeza patens** Nakai (f. *patens*)

In *Bot. Mag. Tokyo* **37**: 79 (1923); *Lesp. Jap. Korea*, 20 (1927)—Makino & Nemoto, *Fl. Jap.*, 735 (1925)—Hatusima in *Mem. Fac. Agr. Kagoshima Univ.* **6**: 12 (1967).

[Plate 19; Fig. 43]

*L. bicolor* Turcz. var. *Sieboldii* (Miq.) Maxim. f. *sericea* Matsum. in *Bot. Mag. Tokyo* **16**: 70 (1902), *pro parte, incl. quoad specim.* Togakushi, Aidzu.

*L. patens* Nakai var. *acutifolia* Nakai in *Bot. Mag. Tokyo* **37**: 79 (1923); *Lesp. Jap. Korea*, 20 (1927).

*L. patens* Nakai var. *obtusifolia* Nakai in *Bot. Mag. Tokyo* **37**: 79 (1923); *Lesp. Jap. Korea*, 20 (1927).

*L. grandiflora* H. Koidzumi in *Journ. Pl. Iwateken* **2**: 75 (1937), *typum non vidi.*

*L. penduliflora* Oudem. var. *sericea* (Matsum.) Ohwi in *Journ. Jap. Bot.* **26**: 234 (1951), *pro parte; Fl. Jap.*, 679 (1953), *pro parte*—Kitamura & Murata, *Col. Ill. Herb. Pl. Jap.* **2**: 100 (1961).

*L. penduliflora* Oudem. var. *sericea* (Matsum.) Ohwi f. *pilosella* Ohwi, *Fl. Jap.*, 679 (1953), *nom. nud.*

*L. Thunbergii* (DC.) Nakai var. *obtusifolia* (Nakai) Ohwi, *Fl. Jap.* rev. ed., (790) 1438 (1965); *Fl. Jap.* Eng. ed., 559 (1965); *Fl. Jap.*, bew ed., 790 (1975).

*L. Thunbergii* (DC.) Nakai var. *obtusifolia* (Nakai) Ohwi f. *pilosella* Ohwi, *Fl. Jap.* rev. ed., 791 (1965); *Fl. Jap.* Eng. ed., 559 (1965); *Fl. Jap.* new ed., 791 (1975), *nom. nud.*

*L. formosa* (Vogel) Koehne f. *sericea* Hatus. in *Mem. Fac. Agr. Kagoshima Univ.* **6**: 8 (1967).

*L. Thunbergii* (DC.) Nakai f. *sericea* Hatus. in *Mem. Fac. Agr. Kagoshima Univ.* **6**: 12 (1967), syn. nov.

*L. Thunbergii* (DC.) Nakai var. *acutifolia* (Nakai) Hiyama ex Murata in *Acta Phytotax. Geobot.* **29**: 104 (1978).

*L. Thunbergii* (DC.) Nakai f. *macrantha* (Honda) Murata *sensu.* Ohashi in *Journ. Jap. Bot.* **56**: 242 (1981), *pro parte, excl. syn. L. cyrtobotrya* Miq. var. *macrantha* Honda, *L. patens* Nakai var. *rotundifolia* Honda, *L. grandis* Koidz., *L. penduliflora* Oudem. var. *sericea* (Matsum.) Ohwi.

*L. Thunbergii* (DC.) Nakai var. *patens* (Nakai) Ohashi in *Journ. Jap. Bot.* **61**: 124 (1986).

Japanese name. Kehagi (Nakai, 1923).

A perennial plant, 1–1.5 m high in late summer and early autumn. Stems ascending or spreading, sometimes dependent later, 0.3–0.5 m high in the first year, rarely branched, terete, slightly angular with densely spreading hairs (hairs 0.3–0.6 mm long, whitish); aerial part (except the basal part 5–10 cm from the ground level) annual;

136

basal part with axillary and adventitious winter buds. In the second year a few branches coming from the basal part of the stem, 1–1.5(–2) m long, similar to the stem of the first year, but branched or rarely branched in upper part; most of the aerial part annual; the basal parts (near ground level) with axillary and adventitious winter buds. Branches in subsequent years (5–20 or more) coming from the basal parts of the stem and branches every year and the number of branches increases according to the size of the stock.

Leaves trifoliolate, petiolate, stipulate, spirally arranged. Stipules free, linear-triangular to linear, 2–7 mm long, brownish, persistent. Petioles 1–6 cm long, hairy like the young branch. Rachides 5–30 mm long, similar to the petiole. Terminal leaflets (at middle parts of branches) petiolulate; petiolules 1–3 mm long, swollen; lamina 3.5–9 cm long, 2–5 cm wide, entire, broadly to narrowly elliptic to ovate, round or cuneate at the base, obtuse or retuse at the apex (the apex itself with or without a point), upper surface glabrous or glabrescent (except along the midrib) or sometimes sericeous (hairs 0.1–0.5 mm long) at the anthesis, lower surface appressed-sericeous (hairs 0.5 mm long); lateral ones similar to terminal but somewhat smaller.

Inflorescence axillary racemous, one per axil, 2–5(–10) cm long including the peduncle, somewhat compactly or loosely 4–12-flowered; peduncles 1–2(–4) cm long, hairy like the young branch. Primary bracts ovate, about 1 mm long, pubescent, persistent; secondary bracts similar to primary ones.

Flowers 12.5–15.4 mm long at anthesis; pedicels 2–5 mm long terete, curved, tomentose. Bracteoles at the base of the calyx narrowly elliptic to ovate, 1–1.5 mm long, pubescent outside, glabrous inside, persistent. Calyx 5.3–5.4 mm long, campanulate four-lobed above the middle part, densely appressed pubescent; tube 1.8–1.9 mm long; the lower lobe longest; lateral ones 3.4–3.6 mm long, 1.1–1.2 mm wide, triangular-lanceolate, acute at the apex but not acuminate; upper one slightly two-clefted or not.

Standard shorter than keel-petal and longer than wings, 7.0–8.1 mm wide; lamina obovate to elliptic, acute or round at the apex with or without a point, slightly inflexed-auriculate near the base, reflexed in anthesis from near the base; claw 1.5–1.7 mm long, whitish. Wings deeper red-purple than standard, 6.9–9.4 mm long with distinct claw; lamina obovate, 5.1–7.4 mm long, 2.6–3.4 mm wide, auriculate at upper basal part, tapering at lower basal part; claw 2.4–2.8 mm long, whitish. Keel-petal 12.4–14.6 mm long with distinct claw; lamina narrowly obovate, 9.8–12.3 mm long, 3.7–4.4 mm wide, paler than wings, deepest in the apical part; claw 2.7–3.0 mm long, whitish. Stamens 10, nearly the same length, 12.4–14.5 mm long, diadelphous. Anthers elliptic, shallowly retuse at the apex, ca. 0.7 mm long; before anthesis yellow. Pistils 12.5–14.6 mm long; ovary elliptic, ca. 2 mm long at anthesis, pubescent, stalked; stalk ca. 0.8 mm long; style pubescent at the basal part, glabrescent at the apical part.

Fruits compressed elliptic, 8–10 mm long, 4–5 mm wide, slightly to densely pubescent or nearly glabrous, stalked; stalk ca. 1.5 mm long.

Seeds reniform, 5 mm long, 3 mm wide.

Distr. Japan (Japan Sea side of C. and N. Honshu).

Voucher and representative specimens.

HONSHU. YAMAGATA. Nishitagawa-gun, Atsumi-machi, Koshizawa (Mori, 14 Aug. 1958, MAK)*; Nishimurayama-gun, Tateoka–Tominami (Koidzumi, 12 Sept. 1929, KYO); Nishimurayama-gun, Aterazawa-machi, Aterazawa (Okuyama, 19 Aug. 1910, MAK); Kitamurayama-gun, Ooishida (Yuki, Aug. 1937, TI); Minamioguni-mura (Yuki, Aug. 1933, KYO)*; Kosaka (Kato, 17 Aug. 1934, KYO)*; Nishiokitama-gun, Oguni-machi, Taruguchi–Taruguchi-touge, alt. ca. 300 m (Ohba & Akiyama 2690–2698, TI). FUKUSHIMA. Oonuma-gun, Kaneyama-machi, Honna (Saitou 14997, TI)*; Oonuma-gun, Akasawa-mura, Mt. Akaruyama (Satou, 7 Aug. 1910, MAK); Yama-gun, Nishiaidzu-machi, Karusawa (Ohba & Akiyama 1813–1816, TI); Yama-gun, Yamato-machi, Aikawa–Ichinoki (Ohba & Akiyama 1839–1841, 1843–1851, TI); Aidzu (Yatabe & Matsumura, Aug. 1879, TI—Syntype of *L. bicolor* Turcz. var. *Sieboldii* (Miq.) Maxim. f. *sericea* Matsum. and holotype of *L. patens* Nakai var. *acutifolia* Nakai); Rokujuri-goe (Hoshi, 26 July 1935, KYO). NIIGATA. Mt. Komaga-take (Okuyama 4580, TNS)*; Murakami-shi, Izumi-machi (Miomote-gawara) (Kudou, 8 July 1962, TNS)*; Iwafune-gun, Sekikawa-mura, Takanosu-onsen (Togashi, 8 Oct. 1960, TI)*; Iwahune-gun, Sekigawa-mura, Takanosu, alt. 100 m (Togashi & Yamazaki, 21 July 1965, TI); Shibata-shi, W. foot of Mt. Iide, Higashiakatani–Yunohira spa (Fukuoka & Konta 56, KYO)*; Tsugawa (Togasi, 16 Sept. 1957, A; TI); Higashikanbara-gun, Tsugawa-machi, Kiyokawa (Ohba & Akiyama 1790–1791, TI); Higashikanbara-gun, Mikawa-mura, Koshitori (Ohba & Akiyama 1764–1786, TI); Mt. Hakusan (Ikegami, 7 Sept. 1947, TNS)*; Mt. Yahiko-yama (Ikegami, 19 Sept. 1948, TNS); Minamikanbara-gun, Morimachi-mura, Nagashima (Ikegami, 6 Oct. 1951, TNS)*; Tochio-shi, Hanzogane (Ohba & Akiyama 1698–1738, TI); Kitauonuma-gun, Suhara (Kurata 791-b, TOFO)*; Mt. Hakkoku-san (Ikegami, 23 Sept. 1944, TNS)*; Kitauonuma-gun, Irihirose-mura, Ooshirakawa, alt. 400 m (Yamazaki, 8 Aug. 1962, A; TI)*; Kitauonuma-gun, Kawaguchi-machi, Aikawa–Araya, alt. 150–200 m (Ohba & Akiyama 1631, TI); Koshi-gun, Yamakoshi-mura, Takezawa (Ohba & Akiyama 1646–1661, TI); Koshi-gun, Yamakoshi-mura (Ohba & Akiyama 1689, 1961–1697 TI); Ojiya-shi, near Utogi, alt. ca. 150 m (Ohba & Akiyama 1633–1638, TI); Ojiya-shi, Minaminigoro, Utogi–Nigoro, alt. ca. 200 m (Ohba & Akiyama 1640–1645, TI); Muika-machi–Yatsuga-touge (Ito 12038, TNS); Minami-uonuma-gun, Mt. Koma-ga-dake (Furukawa 119, TNS)*; Mt. Hakkai-zan (Ikegami 11062, TNS); Juni-touge (Hisauchi, 17 Oct. 1947, TNS)*; Nakauonuma-gun, Mimata-mura (Mitsumata), Kiyotsu-kyo (Mizushima, 30 Aug. 1955, A; MAK)*; Tsunan-machi (Okuyama, 10 Aug. 1969, TNS); Kashiwazaki (Abe, 30 Aug. 1905, MAK); Takada-shi, dry river bed near agricultural school (Kurata 261, TOFO–Holotype of *L. formosa* (Vogel) Koehne f. *sericea* Hatus.; KAG—Isotype); Takada-shi, Shichi-touge (Kurata, Oct. 1945, TOFO); Arai-shi, Nozokido, alt. 100–150 m (Ohba & Akiyama 469, 470–471, TI); Arai-shi, Nagasawahara, alt. ca. 200 m (Ohba & Akiyama 484–487, TI); Nakakubiki-gun, Kanaya-mura, Ushirotani (Yoshikawa 116, TNS)*; Sasa-ga-mine (Ikegami, 4 Aug. 1948, TNS); Suginosawa-mura (Togashi, 15 Oct. 1955, A; TI); Itoigawa-shi, Nakagawara, alt. ca. 250 m (Ohba & Akiyama 463–466, TI); Kotaki (Ikegami 795, TNS); Nishikubiki-gun, Mt. Kurohime-yama (Yoshikawa, 7 Sept. 1941, MAK); Oyashirazu (Shinno, 25 Sept. 1935, TNS). NAGANO. Shimominochi-gun, Tozama-mura–Oota(Ohta)-mura (Mizushima, 13 Aug. 1953, A; TI); Shimominochi-gun, Okuyama-mura, Numa-no-ike (Yokouchi, 18 Sept. 1954, MAK); Iiyama-shi, Tomikura, alt. ca. 400 m (Ohba & Akiyama 489, 490–492, TI); Sasagamine, Bokuji–Mt. Kurohime (Nitta, 24–25 Sept. 1963, KYO); Mt. Togakushi (Watanabe, 20 Aug. 1893, TI—Syntype of *L. bicolor* Turcz. var. *Sieboldii* (Miq.) Maxim. f. *sericea* Matsum., holotype of *L. patens* Nakai var. *obtusifolia* Nakai and lectotype of *L. patens* Nakai; MAK; TNS); Kamiminochi-gun, Togakushi-mura, Tochihara, alt. 600–800 m (Ohba & Akiyama 506–511, TI); Kamiminochi-gun, Togakushi-mura, Mt. Togakushi, Okusha–Happonirami (Midorikawa 1213, TI); Kitaazumi-gun, Otari-mura, Kawauchi, alt. ca. 600 m (Ohba & Akiyama 459, 460, 461, 462, TI). TOYAMA.

Shimoniikawa-gun, Asahi-machi, alt. 200 m (Kanai, 29 July 1958, TI)*; Shimoniikawa-gun, Aimoto-mura, River Kuronagi-gawa (Kanai, 5 Aug. 1958, TI)*; Shimoniikawa-gun, Aoki-mura, Mekawa (Kotou, 18 Oct. 1936, KYO)*; Uotsu-gun, Mt. Kekachi-yama (Nagai, 23 Aug. 1956, TI); Kanetsuri (Tashiro, 2 Sept. 1929, KYO)*; River Satsuki-gawa, Kikuishishita (Hara & Kurosawa, 8 Oct. 1976, TI)*; Tateyama-onsen (Maekawa 5978, TI); Toyama (Makino in 1935, MAK)*; Oosawano-machi, Ushigamashi (Oota, 21 Sept. 1981, KANA); Kaminiikawa-gun, Shimoyuu-mura, Teratsu (Sinno, 29 Sept. 1937, TNS)*; Nei-gun, Sugihara-mura, Shirooi (Kotou, 6 Oct. 1940, TNS)*; Higashitonami-gun, Oomaki-onsen (Satomi 22151, TI); Higashi-tonami-gun, Kamitaira, Katsura, Kaitsutani (Kurata 1008, TOFO)*. ISHIKAWA. Kanazawa-shi, Shimotatsumi (Matsuda 2695, KANA); Kanazawa-shi, Mt. Agehara-yama (S.T., 23 Aug. 1963, KANA); Kanazawa-shi, Shimooshihara (Matsuda 127, KANA)*; Kanazawa-shi, Saigawa (Matsuda, 4 Sept. 1955, KANA); Tatsunokuchi-machi, Tedori-gawara (Akiyama 1513–1525, TI); Chuguu-onsen–Mitsumata (Yoshimura & Shirosaki, 25 Sept. 1955, KANA); Oguchi-mura, the foot of Mt. Hakusan, Mekko-dani, alt. 450 m (Shimizu, 7 Aug. 1959, KYO); Ishikawa-gun, Iwama-onsen (Satomi, 23 Aug. 1960, KYO)*; Jyadani (Oyadani-no-yu), alt. 800–1300 m (Hashimoto 6484, KANA)*; Mt. Hakusan, Mt. Shiramine-yama (Karizumi, 20 Aug. 1953, TNS). FUKUI. (Makino, in 1938, MAK); Nisitanaka, Asaki-tyō, Nyu (Maeda 619, KANA)*.

This species is distributed through the Japan Sea side of the central and northern parts of Honshu where heavy snow falls in winter. It has many annual branches coming from the basal parts of the stem and branches of previous years. This species is related to *L. formosa*, especially subsp. *velutina*, although the latter has stem and branches which live a few years or more and axillary buds emerge on their aerial parts.

The typification of *L. patens* Nakai and *L. bicolor* Turcz. var. *Sieboldii* (Miq.) Maxim. f. *sericea* Matsum. is controversial (Murata, 1978, 83; Ohashi 1981, 86). Matsumura (1902) described this species as a form (f. *sericea*) of *L. bicolor* var. *Sieboldii* (=*L. Thunbergii*) based on three specimens collected from Aizu, Tateyama (Tateyamashita on label, in Japanese) and Togakushiyama. These specimens are the syntypes of *L. bicolor* var. *Sieboldii* f. *sericea* as pointed out by Ohashi (1981). When Nakai published *L. patens* as a new species he cited only two of these three syntypes: one is the specimen from Aizu and determined as var. *angustifolia*, and the other is the specimen from Togakushiyama and determined as var. *obtusifolia*. This treatment is not allowed today by the International Codes of Botanical Nomenclature (1983). If the two are distinguished at variety rank in *L. petens*, one repeats the specific epithet, var. *patens*, and the other receives a new name of variety rank. Later (1927) he cited a photograph of the type of var. *obtusifolia* as the type of *L. patens*, and he did not use var. *patens* but var. *obtusifolia* again. In the same monograph he considered the syntype from Tateyama was *L. cyrtobotrya*, but he did not cite *L. bicolor* var. *Sieboldii* f. *sericea* as the synonym of *L. cyrtobotrya*, as pointed out by Ohashi (1986). His treatment remains problematic. When Maekawa described *L. patens* var. *macrantha* he cited *L. bicolor* f. *sericea* [*pro minoribus partibus*] as a synonym. Murata (1978) recognized that the type of var. *obtusifolia* is the type of *L. patens* and var. *obtusifolia* is a superfluous name of var. *patens*. He adopted var. *acutifolia* as the valid name of the variety rank of *L. patens*. Ohashi (1981) considered that *L. patens* was not a new species but a new name of a species. According to him, when lectotypification of *L. patens* was done by Nakai himself in 1927, the lectotypification of *L. bicolor* var. *Sieboldii* f. *sericea* was also done at the same time. Ohashi did not refer to the specimen from Tateyama. On the other hand, Murata (1983) considered that *L. patens* is described as a new species distinguished from *L. bicolor* var. *Sieboldii* f. *sericea*, which is represented by the specimen from Tateyama, and that var. *obtusifolia* is a superfluous name of var. *patens* only under *L. patens* and valid under other species such as *L. Thunbergii*. Ohashi (1986) mentioned that one of three syntypes, Tateyama, is in serious conflict with the original description of f. *sericea* because this specimen has branches with appressed hairs and Nakai excluded this specimen. However, the original description of f. *sericea* is as follows: "Hairs on the stems often spreading or reflexed . . . ." The specimen Tateyama therefore agrees with the original description like as the other two species, Aizu and Togakushiyama, which have branches with spreading and reflexed hairs respectively. From the original description it is possible to choose any three specimens as the lectotype.

In conclusion, it is difficult to decide whether Nakai considered *L. patens* as a new species or a new name. In the case of *L. kiusiana* he wrote it as the name of the species rank for *L. bicolor* var. *velutina*, *L. velutina*, which was confused with another species, *L. velutina* Dunn. But in the case of *L. patens* he did not write such notes. We therefore regard *L. patens* to be described as a new species and adopt Murata's treatment of it. The specimen from Tateyama is the lectotype of *L. bicolor* var. *Sieboldii* f. *sericea*, and the lectotype of *L. patens* is the specimen from Togakushi. Var. *obtusifolia* is a superfluous name of var. *patens* under *L. patens*. And under another species, such as *L. Thunbergii*, there can be two treatments: 1) one is that var. *obtusifolia* is a superfluous name of *L. patens* var. *patens*, so if a new combination is made under other species, var. *patens* should be used as a vallid name; and 2) the other is that var. *obtusifolia* is only a superfluous name of var. *patens* under *L. patens* but if the variety name, var. *obtusifolia*, is considered to be validly published, it can be used as a name of taxon. The present "International Codes for Botanical Nomenclature" (1983) do not restrict this case. From the concept of the Articles concerning autonomy, however, we consider the former treatment is preferable to avoid confusion.

Forma **sericea** (Matsum.) S. Akiyama et H. Ohba, *comb. nov.*

*L. bicolor* Turcz. var. *Sieboldii* (Miq.) Maxim. f. *sericea* Matsum. in *Bot. Mag. Tokyo* **16**: 69 (1902), *excl. quoad specim.* Aizu et Togakushiyama.

*L. cyrtobotrya* Miq. var. *macrantha* Honda in *Bot. Mag. Tokyo* **44**: 671 (1930).

*L. patens* Nakai var. *macrantha* (Honda) F. Maekawa in *Bot. Mag. Tokyo* **48**: 51 (1934).

*L. grandis* Koidz. in *Acta Phytotax. Geobot.* **4**: 159 (1935).

*L. patens* Nakai f. *macrantha* (Honda) Hatusima in *Mem. Fac. Agr. Kagoshima Univ.* **6**: 13 (1967).

*L. Thunbergii* (DC.) Nakai var. *acutifolia* (Nakai) Hiyama ex Murata f. *macrantha* (Honda) Murata in *Acta Phytotax. Geobot.* **29**: 104 (1978)—Ohashi in *Journ. Jap. Bot.* **61**: 124 (1986), ut '*L. Thunbergii* (DC.) Nakai var. *patens* (Nakai) Ohashi f. *macrantha* (Honda) Murata.'

Japanese name. Tateyamahagi (Maekawa, 1934).

Stem and branches with appressed hairs.

140

Distributed sporadically throughout the range of the species.
Representative specimens.
HONSHU TOYAMA. Mt. Tateyama (prov. Etchu) (Matsumura s.n., TI—Lectotype of *L. bicolor* Turcz. var. *Sieboldii* (Miq.) Maxim. f. *sericea* Matsum. decided by Maekawa in 1934); Keyaki-daira–Mt. Mita-sennin-dake (Takenaka, Aug. 1930, TI—Holotype of *L. cyrtobotrya* Miq. var. *macrantha* Honda).
Among the specimens cited under *L. patens*, those referable to f. *sericea* are marked with an asterisk.

Sometimes brances have ascending but not entirely appressed hairs. In these cases it is difficult to decide whether this plant belongs to f. *patens* or f. *sericea*. But tentatively we recognize f. *sericea* based on the hairs of young branches.

Forma **nivea** S. Akiyama et H. Ohba in *Journ. Jap. Bot.* **59**: 128 (1984).
*L. Thunbergii* (DC.) Nakai var. *patens* (Nakai) Ohashi f. *nivea* (S. Akiyama et H. Ohba) Ohashi in *Journ. Jap. Bot.* **61**: 124 (1986).
Japanese name. Yukihagi (Yamazaki in sched. ex S. Akiyama & H. Ohba, 1984).
Corolla white.
Distr. Japan (Niigata Pref.).
Specimens examined.
HONSHU. NIIGATA. Ojiya (Yamazaki, 20 Sept. 1966, TI—Holotype and isotype of *L. patens* Nakai f. *nivea* S. Akiyama et H. Ohba); Koshi-gun, Yamakoshi-mura, (cult.) (Ohba & Akiyama 1688, TI).

*L. Thunbergii* var. *acutifolia* f. *leucantha* Murata does not fall in the variation range of *L. patens*. This is considered to be *L. japonica* with glabrous upper leaf-surfaces. The white-flowered, appressed form of *L. patens* has not been previously described.

## Lespedeza Davidii Franch.

*Pl. David.* **1**: 94, t. 13 (1884)—Forbes & Hemsley in *Journ. Linn. Soc. Bot.* **23**: 180 (1887)—Schindler in Engl., *Bot. Jarhb.* **49**: 580 (1913); in Sargent, *Pl. Wils.* **2**: 107 (1914)—Hand.-Mazz., *Symb. Sin.* **7**(3): 572 (1933)—Fu in Acad. Sin. Bot., *Ill. Important Chin. Pl. Leguminosae*, 522, fig. 516 (1955)—Acad. Sin. Bot., *Iconogr. Cormophyt. Sin.* **2**: 460, fig. 2650 (1972).                    [Plate 20; Figs. 44 & 45]
Japanese name. Okushimohagi (Maekawa, 1966).
A perennial plant, 2 m high. Stems ascending, most of terrestrial parts live a few years; branched in upper parts. Branches winged, ascending; when young with densely ascending or patent hairs. Distal parts of branch die in winter season, proximal parts (and stem) with axillary and winter buds.
Leaves trifoliolate, petiolate, stipulate, spirally arranged. Stipules free, linear, 5 mm long, brownish, persistent. Petioles 2–5 cm long, with patent hairs. Terminal leaflets (at middle parts of branches) petiolulate; 3–8 cm long, 1.8–5 cm wide, entire, (broadly) elliptic, round at the base, retuse or obtuse at the apex (the apex itself with or without a point), upper surface densely pubescent, lower surface appressed-sericeous; lateral ones similar to terminal but somewhat smaller.
Inflorescence axillary racemous, one per axil, 3–12 cm long including the peduncle,

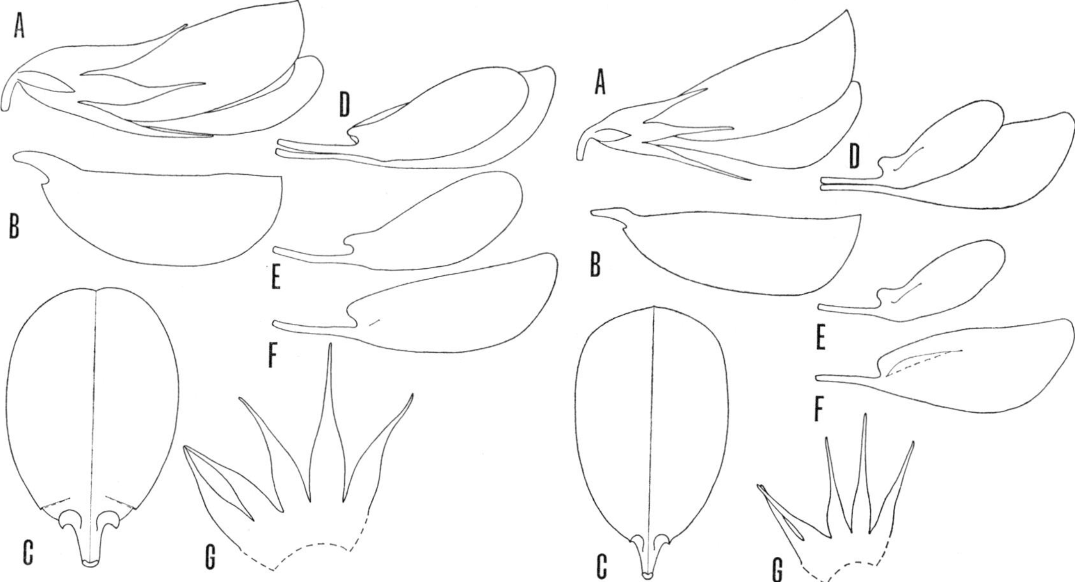

Fig. 44.   Flower of *L. Davidii*.

     A, flowers, lateral view;    B, standards, lateral view;    C, standards, opened; D, wings and keel-petals;    E, wings;    F, keel-petals;    G, calyces, dissected.    Left. China, Kangsi (Chang 206, TI). Right. Cultivated in Tokyo (ex China, Shensi) (Maekawa 179A91, TI).    All×3.3.

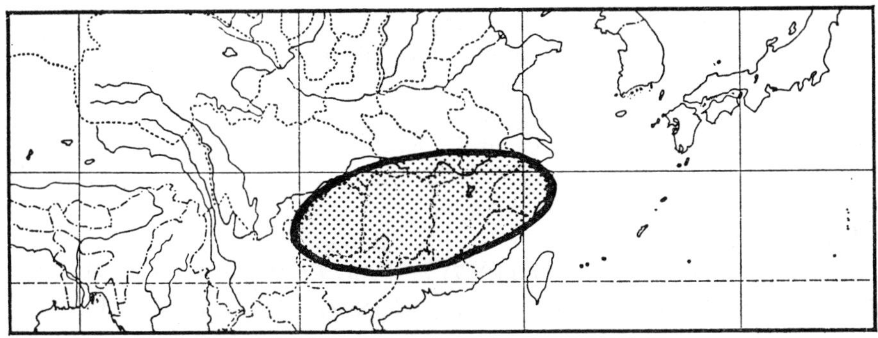

Fig. 45.   Distribution of *L. Davidii*.

6–20-flowered. Primary bracts ovate, about 2–3 mm long, persistent; secondary bracts similar to primary ones.

     Flowers ca. 12 mm long at anthesis; pedicellate. Bracteoles at the base of the calyx narrowly ovate with acuminate apex, ca. 2 mm long, persistent. Calyx ca. 7.5 mm long, campanulate, deeply four-lobed, appressed pubescent; the upper calyx-lobe also deeply two-clefted as if calyx five-lobed; tube ca. 2.5 mm long; the lower lobes somewhat longest; lateral ones ca. 5 mm long, ca. 1 mm wide, triangular with acuminate apex. Standard subequal to (sometimes slightly shorter than) keel-petal and longer than wings, keel-petal longer than wings (K$\geq$S$>$W). Standard ca. 11 mm long with distinct claw, 7 mm wide; lamina elliptic, round or retuse, inflexed-auriculate near the base; auricles broadly lunate; claw ca. 2 mm long. Wings ca. 10 mm long with distinct claw; lamina narrowly oblong, ca. 7 mm long, ca. 3 mm wide, auriculate at upper basal part; claw ca. 3 mm long. Keel-petal ca. 11.5 mm long with distinct claw; lamina narrowly obovate, ca. 8.5 mm long, ca. 3 mm wide; claw ca. 3 mm long. Fruits compressed, broadly elliptic, ca. 8 mm long, 6 mm wide, subsessile, densely pubescent.

Distr. China (Kiangsi, Chechiang, Kwangtung).

Specimens examined.

**China.**

Kuei-chou (Torii, Oct. 6, 1902, TI). KIANGSI (Chianghsi). Lushan (M. Takahashi 24, TI; Cheng 206, TI); Tuchang (F. Maekawa in 1941, TI). KWANGTUNG. Yang Shan Distr., south of Linchow, Yang Shan, and vicinity (T. M. Tsui 598, TI). CHECHIANG. Hangchou (K. Kimura s.n., TI).

This species is characterized by its large, roundish leaflets with dense hairs, distinctly branches, and triangular calyx-lobes with acuminate apices. This is a distinct species in the series *Macrolespedeza*.

### Series Heterolespedeza (Nakai) S. Akiyama et H. Ohba, *stat. nov.*

*Lespedeza* sect. *Heterolespedeza* Nakai in *Jourm. Jap. Bot.* **15**: 531 (1939).
Small shrub; scales of winter-buds and also foliage leaves arranged distichously.
Type species: *L. Buergeri* Miquel (Nakai, 1939).

Key to the species of the series *Heterolespedeza*
1. Standard slightly shorter than or nearly equal to wing, yellow with purple patches near the base. Keel-petal pale yellow. Calyx-lobe with obtuse or acute apex .................................................................................................. *L. Buergeri*
1. Standard longer than wing, red-purple. Keel-petal pale red-purple. Calyx-lobe with acuminate apex ................................................................. *L. Maximowiczii*

### Lespedeza Buergeri Miquel

In *Ann. Mus. Lugd.-Bat.* **3**: 47 (1867) et *Prol. Fl. Jap.*, 235 (1867)—Maxim. in *Acta Hort. Petrop.* **2**: 353 (1873), *excl. quoad specim.* Wilford (cf. Schindler, 1913), Oldham 333—Franchet & Savatier, *Enum. Pl. Jap.* **1**: 101 (1875)—Forbes & Hemsley in *Journ. Linn. Soc. Bot.* **23**: 179 (1887), *pro parte*—Diels in Engl., *Bot. Jahrb.* **29**: 415 (1901)—Matsumura in *Bot. Mag. Tokyo* **16**: 70 (1902); *Ind. Pl. Jap.* **2**: 267 (1912) —Schneider, *Ill. Handb. Laubholz.* **2**: 113, fig. 70k. (1907)—Pampanini in *Nuov. Giorn. Bot. Ital.* n. ser. **17**: 398 (1910); **18**: 123 (1911)—Schindler in Engl., *Bot. Jahrb.* **49**: 580 (1913); in Sargent, *Pl. Wils.* **2**: 106 (1914)—Makino & Nemoto, *Fl. Jap.*, 734 (1925)—Rehder in *Journ. Arn. Arb.* **7**: 171 (1926)—Momiyama in *Journ. Jap. Bot.* **9**: 127 (1933)—Ohwi, *Fl. Jap.*, 789 (1953); *Fl. Jap.* rev. ed., 678 (1965); *Fl. Jap.* Eng. ed., 559 (1965); *Fl. Jap.* new ed., 789 (1975)—Fu in Acad. Sin. Bot., *Ill. Important Chin. Pl. Leguminosae*, 521, fig. 514 (1955)—Kitamura & Murata, *Col. Ill. Herb. Pl. Jap.* **2**: 98 (1961)—Kung in *Chinese Journ. Bot.* **1**: 20 (1963)—Hatusima

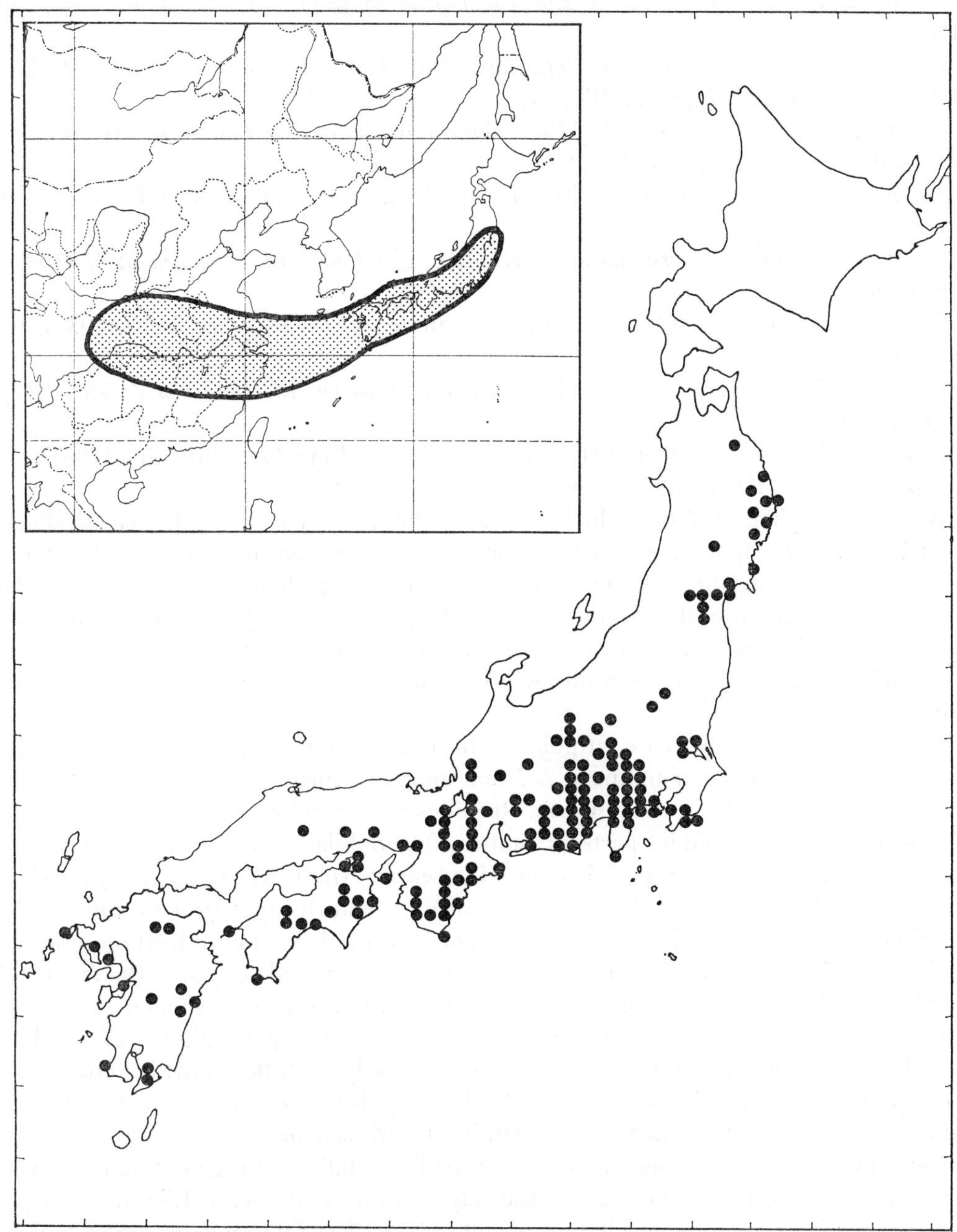

Fig. 46.   Distribution of *L. Buergeri*.

in *Mem. Fac. Agr. Kagoshima Univ.* **6:** 6 (1967), *excl.* ssp. *praecox* (Nakai) Hatusima, ssp. *tricolor* (Nakai) Hatusima—Acad. Sin. Bot., *Iconogr. Cormophyt. Sin.* **2:** 460, fig. 2649 (1972)—Ohashi in Satake et al., *Wild Fl. Jap. Herb.* **2:** 205 (1982).

[Plates 21 & 22; Fig. 46]

*L. Oldhamii* Miq. in *Ann. Mus. Bot. Lugd.-Bat.* **3**: 48 (1867) et *Prol. Fl. Jap.*, 236 (1867).

*L. Sieboldii* Miq. in *Ann. Mus. Bot. Lugd.-Bat.* **3**: 47 (1867) et *Prol. Fl. Jap.*, 235 (1867), *pro parte, quoad specim.*, Textor.

*L. Buergeri* Miq. var. *Oldhamii* (Miq.) Maxim. in *Acta Hort. Petrop.* **2**: 354 (1873)— Matsumura, *Ind. Pl. Jap.* **2**: 267 (1912).

*L. bicolor* Turcz. β. *intermedia* Maxim. in *Acta. Hort. Petrop.* **2**: 356 (1873), *pro parte*.

*L. Buergeri* Miq. f. *longiracemosa* H. Koidzumi in *Journ. Pl. Iwateken* **2**: 77 (1937), *typum non vidi*.

*L. Buergeri* Miq. f. *subbiloba* H. Koidzumi in *Journ. Pl. Iwateken* **2**: 77 (1937), *typum non vidi*.

*L. Buergeri* Miq. var. *tetraloba* H. Koidzumi in *Journ. Pl. Iwateken* **2**: 76 (1937), *typum non vidi*.

*L. Buergeri* Miq. f. *Oldhamii* (Miq.) Sigimoto, *New Keys Jap. Trees*, 463 (1961).
Japanese name, Kihagi (Matsumura, 1912).

A perennial plant, 1–3 m high. Stems ascending, 0.5–1 m high, 1–5 cm in diameter; branched in upper parts (a few branches coming from basal part of the stem). Branches terete, spreading or sometimes dependent later; when young slightly angular with ascending appressed-sericeous or ascending hairs (hairs 0.3–0.6 mm long, yellowish), later glabrescent; most distal parts die in winter season, proximal parts (and stem) with axillary and adventitious winter buds (winter buds at lower parts mostly dormant).

Leaves trifoliolate, petiolate, stipulate, alternately arranged (phyllotaxy 1/2). Stipules free, linear-triangular to linear, 2–7 mm long, brownish, persistent. Petioles 0.5– 4 cm long, similar to the young branch. Rachides 5–15 mm long, similar to the petiole. Terminal leaflets (at middle parts of branches) petiolulate; petiolules 1–2 mm long, swollen; lamina 1–5 cm long, 0.5–3 cm wide, entire, (narrowly to broadly) ovate or elliptic, round or cuneate at the base, acute or obtuse at the apex (the apex itself with or without a point), upper surface glabrescent (except along the midrib) or sericeous (hairs ca. 0.5 mm long) at anthesis, lower surface appressed-sericeous (hairs ca. 0.5 mm long, yellowish); lateral ones similar to terminal but somewhat smaller.

Inflorescence axillary racemous, one (rarely two to three) per axil, 2–4(–7) cm long including the peduncle, somewhat compactly to loosely 4–10-flowered; peduncles 0.5– 1 cm long, hairy like the young branch. Primary bracts ovate, about 1 mm long, pubescent, persistent; secondary bracts similar to primary ones.

Flowers 8.5–10.3 mm long at anthesis; pedicels 0.8–2 mm long terete, curved, tomentose. Bracteoles at the base of the calyx round to elliptic, 1–1.5 mm long, pubescent outside, glabrous inside, persistent. Calyx 1.8–2.9 mm long, campanulate, four-lobed near the middle part, appressed pubescent; tube 1.1–1.4 mm long; lobes subequal in length or the lower one longest; lateral ones 0.7–1.7 mm long, 0.7–0.9 mm wide, broadly ovate or triangular-lanceolate, obtuse or acute at the apex but not acuminate; upper one slightly two-clefted.

Standard shorter than keel-petal and longer than (or sometimes as long as or shorter than) wings, keel-petal longer than wings (K>S≧W). Standard pale yellow with purple patches near the base inside, whitish outside, 7.3–8.8 mm long with distinct claw, 5.0–6.0 mm wide; lamina round to elliptic or broadly obovate, round or retuse at the apex with or without a point, inflexed-auriculate near the base, spreading in anthesis from the part near the base; auricles well developed; claw whitish, 1.1–1.5 mm long. Wings purple or deep red-purple, 7.3–8.6 mm long with distinct claw; lamina narrowly obovate to oblong, 5.0–6.4 mm long, 2.0–2.5 mm wide, auriculate at upper basal part, slightly or minutely auriculate or tapering at the lower basal part; claw 2.2–2.7 mm long, whitish. Keel-petal 8.0–10.1 mm long with distinct claw; lamina obovate, 5.7–7.7 mm long, 2.7–3.5 mm wide, pale yellow; claw 2.4–2.8 mm long, whitish. Stamens 10, nearly the same length, 7.8–9.9 mm long, diadelphous. Anthers elliptic shallowly, retuse at the apex, ca. 0.5 mm long; before anthesis yellow. Pistils 7.9–10.0 mm long; ovary narrowly elliptic, ca. 1.8 mm long, ca. 0.5 mm wide at anthesis, pubescent; style 6–8 mm long, pubescent at the basal part, glabrescent at the apical part, stalked; stalk ca. 1.2 mm long.

Fruits compressed, narrowly elliptic, 11–15 mm long, 5–5.5 mm wide, stalked; stalk 1.5–2 mm long, slightly to densely pubescent or nearly glabrous.

Seeds reniform, 3.5–4 mm long, 2–2.5 mm wide.

Distr. Japan (Hoshu, Shikoku, Kyushu), E. China.

Voucher and representative specimens.

**Japan** (Buerger s.n., GH; L 951326 132); (Textor s.n., L 908.118-2364—Syntype of *L. Sieboldii* Miq.); (Oldham 335, L 908.118-1398—Holotype of *L. Oldhamii* Miq., ut "Japan & Korea Arch. coll."); *ad littus maris* . . . (Siebold s.n., L 908.118-1368—Syntype of *L. Buergeri* Miq.).

HONSHU. IWATE. Kuji-shi, Ookawame (Murata & Tabata, 28 July 1967, KYO); Ryusenkutsu (Shimizu, 1 Aug. 1956, KYO); Shimohei-gun, Iwaizumi-cho, Mt. Ureira (Shimizu, 11 Aug. 1956, KYO); Shimohei-gun, Tanohata-mura, Naraiga (Kikuchi, 28 Sept. 1967, TUSG); Shimohei-gun, Kariya-mura, Oshikado-touge, alt. 300–400 m (Kanai, 2 Sept. 1959, TI); Shimohei-gun, Akka-mura (Hosoi, 15 Sept. 1952, A); Kamihei-gun, Oodzuchi-machi, Kirikiri (Kanai, 3 Sept. 1959, TI); Kamihei-gun, Oohashi (Toba 15, TI); Miyako-shi, Omoe (Yamamoto, 28 July 1971, TNS); Kesen-gun, Takada (Toba, 538, TI). MIYAGI. Jogi-onsen (Kamio, Aug. 1934, KYO); Mt. Kurikoma (Okuyama, 21 Aug. 1952, TNS); Ojika-gun, Onagawa-cho, Urajuku (Kikuchi, 21 Oct. 1967, TNS); Mono-gun, Naruse-cho, Miyato-jima (Naito, 19 Aug. 1973, TUSG); Daitodake (Ôhashi, 2 Sept. 1960, TUSG); Kawasaki, Kamafusa, along River Goishi, alt. 150 m (Sohma 1237, TUSG); Zaou, Aone (Satake & Ito, 17 Oct. 1958, TNS); Tougatta, Torii (Momiyama, 12 July 1936, TI); Sendai, Idohama (Kikuchi & Okazaki, 17 Oct. 1969, TUSG); Sendai, Aobayama (Kimura, Sugaya, Saito & Mori, 10 July 1959, TI). YAMAGATA. Yamaike (Doi in 1966, TNS). FUKUSHIMA. Date-gun, Kunimi-machi, Kaida (Sato, Kaneko & Hayashi, 3 Aug. 1965, TUSG); Shiogama (Jack, 27 Aug. 1905, GH). GUNMA. Adzuma-gun, Adzuma-machi, Haramachi (Masuda in 1940, MAK); Myogisan (Ishii 8, TNS)*; Tano-gun, Nakazako-mura, Kamigahara, Mt. Kanoosan, alt. 1100 m (Murata, Ohba, Kadota & Midorikawa 1072, TI). TOCHIGI. Shiobara, Shiogama–Shionoyu (Kanai, 27 May 1959, TI); Nikko (Matsumura, 29 July 1877, TI); Nikko-shi, Jakkou-no-taki (Ohba & Akiyama 1383 & 1384, TI); Mt. Akanagi-yama (Nakai, 5 Aug. 1931, TI). IBARAKI. Tsukuba (Oowatari, 9 July 1895, TI); Nishiibaraki-gun, Minamikawane-mura (Iwama-mura) (Okamoto, 4 Aug. 1904, MAK); Tsuchiura-shi, Hirookahara, Otto-numa (Makino, 15 Aug. 1913, MAK). CHIBA. Awakominato-machi, Mt. Uchiurayama, alt. less than 300 m (Ohba et al. 6808042, TI); Mt. Kiyosumi-yama (Kiyozumi-yama) (Okuyama 22724, TNS); Kimitsu-gun, (Akimoto-mura), Mt. Kanô-san, alt. 150 m (Hurusawa 109, TI); Mt. Mitsuishi (Okuyama

22877, TNS); Okitsu-machi (Furuse, 16 Oct. 1954, TNS). SAITAMA. Chichibu, Takashino, Sadamine (Moriya, 30 Sept. 1961, TNS)*; Chichibu-gun, Nogami-machi, Nagatoro (Murata, Tateishi, & Kadota 743, TI); Chichibu, Kagemori (Moriya, 7 Aug. 1961, TNS)*; Chichibu, Mt. Mitsumine-san (Sakurai, July 1892, TNS)*; Chichibu, Mt. Kumakura-yama (Moriya, 5 Sept. 1970, TNS); Chichibu-gun, Ryoogami-mura, Sakamoto, Mt. Futagoyama, alt. 1000 m (Murata, Tateishi & Midorikawa 2441, TUSG); Kamabuse-touge (Setsuda, 13 Oct. 1940, TNS)*; Hannou, Mt. Tenran-zan (Hattori, 14 May 1922, TI); Hanno-shi, Mt. Izugatake (Akiyama 338, TI). TOKYO. Mt. Tenso-zan (Takeuchi, 10 Aug. 1948, TI); Kawanoriyama (Kimura, Sept. 1951, TI); Okutama, Hatonosu (Kanai 720, TI); Mt. Takamizu-yama (Mt. Takamizu) (Mizushima, 4 Sept. 1954, TI); Mt. Gozen-yama (Mizushima, 7 July 1946, TI); Nishitama-gun, Hinohara-mura, Fukamata, alt. ca. 700 m (Ohba 1033, TUSG); Itsukaichi, Sengen-one (Ohba & Akiyama s.n., TI); Itsukaichi, Tokisaka (Akiyama 294, 297, TI); Takao-san, alt. 300 m (Wilson in 1914, A); Mt. Jinba-yama (Kazami, 14 Oct. 1962, TNS); Asakawa, Koshi-tasawa (Kanai 705, TI); Sekido (Mizushima 2493, TI). KANAGAWA. Yokohama (Hisauchi 409, TI); Kamakura (Momiyama 682, TI); Zushi-shi, Zinmu-ji (Ohba & Murata 669, TI); Kurihama (Makino in 1934, MAK); Miura-kaigan (Ito, 12 Nov. 1970, TI); O-iso (Momiyama 944, TI); Mt. Tanzawa-yama, Mt. Yake-yama (Yamazaki, 6 Oct. 1942, TI); Aikou-gun, Susu-gaya-mura, Teraie–Monomi-touge, alt. 400 m (Kanai, 10 Nov. 1957, TI); Ashigarakami-gun, Kamihatano-mura, Horinishi (Kanai, 27 Aug. 1956, TI); Yamakita (Komori 1975, TI); Ashigarakami-gun, Mt. Takamatsu-yama (Henmi 2304, TNS); Hakone, Mt. Kintoki-yama (Mt. Kintoki) (Momiyama, 7 Sept. 1951, TI); Hakone (Maximowicz in 1862, GH); Ashiga-rashimo-gun, Hakone-machi, Tounosawa (Ohba & Akiyama 614–615, TI); Manadzuru (Hen-mi, 20 July 1952, TNS). YAMANASHI. Yanakawa–Mt. Kuratake-yama (Mizushima, 26 June 1949, TI); Mt. Mitsutouge-yama (Sawada, 23 Aug. 1930, TI); Mts. Misaka, Mitsuke Pass (Kanai, 18 Sept. 1955, TI); Minamitsuru-gun, Funatsu-mura, on shore of Kawaguchi Lake, alt. 830 m (Furuse, 11 Sept. 1958, A); Nisiyatusiro-gun, Lake Shoji-ko (Kanai, 15 Sept. 1957, TI); Nisiyatusiro-gun, Kamikuisshiki-mura, Onnasaka–Koseki (Kanai, 15 Sept. 1957, TI); Nisiyatusiro-gun, Ookawachi-mura, Shionosawa (Kanai, 22 Sept. 1957, TI); Mt. Misho-tai-yama, Shimosha–Okusha (Kanai, 3 Nov. 1956, TI); at the foot of Mt. Fuji-san, Lake Motosu-ko, Ryugatake rindou (Kanai, 1 June 1958, TI); Mt. Fuji-san, Aoki-ga-hara, near Hyoketsu (Ohba 6707, TUSG); Mt. Minobu-san (Okuyama, 26 Sept. 1954, TNS); Kobuchi-zawa (Togashi 7390, TI); Nirasaki-shi, Anayama, alt. 550 m (Tateishi 4486, TI); Mt. Houou-zan, alt. 1300 m (Uematsu, 1 Oct. 1949, TI); Mt. Koma-ga-take (Satake 1556, TI); Mts. Tenshi, Okunoyu (Shimobe) (Kanai, 21, Aug. 1955, TI). NAGANO. (Tschonoski in 1864, GH); Karuizawa, Myozenji (Greatrex K25/30, TI); Minamisaku-gun, Nakagomi-machi, Arafune (Sato, 5 Aug. 1958, TI); Komoro, Mt. Nunobikiyama (Hasegawa, 12 July 1964, TI); Minami-saku-gun, Minamiaiki-mura, Sankawa–Mt. Sekibutsu, alt.140 0–1920 m (Murata & Ohba 5239, TI); Kawanakajima (Matsumura, 16 July 1884, TI); Chiisagata-gun, Shioda-chō, Bessho-onsen (Kitagawa 5398, KYO); Chiisagata-gun, Niuchi-mura (Momose, 8 Aug. 1931, TI); Mt. Hachibuse (Momose 664, TI); at the foot of Mt. Senjo-dake, Todai (Uematsu, 24 Sept. 1949, TI); Ontakesan (Koidzumi, Aug. 1910, TI); Mts. Akashi, Mt. Hijiri-dake, Nishizawa, alt. 1400 m (Yamazaki, Matsuda & Fujita, 6 Aug. 1954, TI); Nakasendou, Agematsu–Fukushima (Ito, 15 Aug. 1891, TNS); Kamiina-gun, Inasato-mura, Mitsumine-gawa, alt. 1000 m (Matsuda, 16 July 1954, TI); Shimoina-gun, Ueura (Koidzumi, 23 Sept. 1932, TNS); Shimoina-gun, Aiji-mura, Komaba–Sonohara (Kanai, 29 July 1957, TI); Shimoina-gun, Hiraoka-machi (Takei, 12 Aug. 1957, TNS); Shimoina-gun, Ooshikamura, Ochiai (Asano 9239, TI); Shimoina-gun, Iida-shi, Kazekoshi (Furusawa, Aug. 1940, TI); Shimoina-gun, Yamabukimura, Shimo-machi (Asano 11229, TI); Shimoina-gun, Minamiwadamura, Sokoike–Nagoyama (Asano 8120, TI); Shimoina-gun, Hiraokamura, Orochi (Asano 24425, TI). SHIZUOKA. Numadzu (Tako,

15 Aug. 1907, MAK); Mts. Ashitaka, Takayama (Kanai 6078, TI); Fuji, Subashiri-guchi (Muramatsu, 30 July 1924, TI); Fuji-gun, Kamiide-mura, Inokashira–Wakimizu-touge, alt. 1000 m (Kanai, 23 Sept. 1957, TI); Fuji-gun, Kamiide-mura, Mt. Kenashi-yama, alt. 1100 m (Kanai, 18 May 1958, TI); Ookouchi-mura, Mafuji-mura (Furuse 19885, TI); Ihara-gun, Mt. Kunou-zan (Mizushima 1640, TI); Umegashima, Ôtaki (Okuyama, 3 Oct. 1951, TNS); Abe-gun, Ikawa-mura, Tashiro-gawa (Sugimoto 20435, TI); Haibara-gun, Honkawame-cho, the top of Mt. Fudou–the foot of Mt. Morosawa, alt. ca. 1000 m (Koyama, Miki & Yahara 134, KYO); Shida-gun, Setonotani-mura (Shimizu 157, TI); Iwata-gun, Fukuroi-machi (Hashimoto, 24 July 1932, TNS); Kakegawa-shi, the foot of Mt. Ogasayama (Ohba 3496, TI). AICHI. Tomiyama-mura, Kawauchi (Torii 12404, TNS); Kita-shitara-gun, Tsugu-mura (Torii, 18 July 1954, TUSG); Kitashitara-gun, Miwa-mura, Uren-gawa (Yamazaki, 30 July 1957, TI); Minamishitara-gun, Hourai-machi, Mutsudaira, alt. 100–150 m (Mimoro, Tsugaru & Nishiyama 2004, TNS); Nagashino-mura (Torii 6108, TUSG); Shinshiro-shi (Torii 6135, TUSG); Mt. Ishimaki-yama (Torii 6162, TUSG); Moya (Makino, 30 Sept. 1905, MAK); Toyohashi, Ishimaki, Wada (Torii, 20 Aug. 1967, KYO). GIFU. Mt. Ena-san (Sakurai, July 1903, TNS); Toki-gun, Tajimi (Shiota, 17 July 1933, KYO); Toki-gun, Tsumaki-machi (Shiota, 17 Sept. 1933, KYO); Toki-gun, Ichinokura-mura (Shiota, 15 Aug. 1929, KYO); Kamo-gun, Higashi-shirakawa-mura (Shiota 8096, A); Mugi-gun, Suhara-mura (Shiota 6487, A); Mugi-gun, Mino-machi (Shiota 5390, A); Mugi-gun, Itadori-mura, NW foot of Mt. Kooka-san, Kabe–Taragadani-goe (Fukuoka & Yamashita 222, KYO); Kani-gun, Toyooka-machi, Ueno (Imadzue, 21 Aug. 1933, A); Gujo-gun, Yatomi-mura (Shiota, 27 July 1932, TI); Gujo-gun, Hachi-man-machi (Hiroe, 22 Aug. 1948, KYO); Gujo-gun, Kawai-mura (Shiota 7654, A); Gujo-gun, Yatomi-mura, Banba (Shiota 7253, A); Ishidzu-mura (Sugino, 20 Aug. 1952, TNS); Yourou-gun, Yourou-machi, Yourou-no-taki (Kanai, 1 Oct. 1957, TI). SHIGA. Mt. Ibuki-yama (Matsumura, 1 Aug. 1881, TI); Gamou-gun, Nishiooji (Hashimoto, 22 Aug. 1947, TNS); Sugi (Hashimoto, 8 Aug. 1945, TNS); Mt. Watamuki-yama (Senbongi–Kumano) (Araki 5744, SHO); Koga-gun, Tsuchiyama-cho, Tamura-jinja, alt. 260 m (Murata, 24 Oct. 1964, KYO); Mt. Hira-san, Kido (Hashimoto, 7 Nov. 1944, SHO); Katatatyo, Tochu–Sakashiba (Kitamura & Murata, 5 Nov. 1961, KYO); Ootsu-shi, Mt. Ousaka-yama (Hashimoto 11347, KYO). MIE. Komono, Yunoyama (Sono, 23 Aug. 1908, TNS); Mie-gun, NE of Mt. Gozaisho, Yuno-yama–Asake, alt. 600–800 m (Fukuoka, 4 Aug. 1963, KYO); Inabe-gun, Ishigure-mura, Mt. Ryûgadake (Fukuoka, 6 Aug. 1963, KYO); Suzuka-gun, at the foot of Mt. Nonobori, around so-called Byôbuiwa (Shimizu, 16 July 1959, KYO); Aki-gun, Mt. Kyogamine (Imai, 10 Oct. 1976, SHO); Ise Shrine forest, Kamijiyama (Nitta, 12–16 Aug. 1963, KYO); Watarai-gun, Mise-tani (Oomiya-machi) (Umemura, 11 Aug. 1895, MAK); Watarai-gun, Shimadzu-mura, Kowa-mura (Kuwana, 20 July 1905, MAK); Mt. Asama-yama (Momiyama 1037, TI); Ayama-gun, Ueno (Koreishi, July 1903, MAK); Isshi-gun, Tarou-mura, Mt. Kuroso (Kuruson) (Kanai, 20 Sept. 1962, TI); Naka-gun, Akame (Tagawa, 14 Aug. 1950, KYO); Osugidani Hinode-dake–Momonoki, alt. 600 m (Yamazaki, Yamazaki & Morita, 6 June 1973, TI); Minami-muro-gun, Gogou-mura, Momozaki–Yuya (Kanai, 28 April 1957, TI); Minamimuro-gun, Mihama-cho, Hūden-toge, alt. 250 m (Mimoro, Tsugaru & Deguchi 4285, KYO); Ootou-yama (Okamoto, 27 Oct. 1952, TNS); Mt. Odaigahara (Ohwi 9103, TNS). NARA. Oodai-ga-hara, Shinonoha (Ohwi, 3 Sept. 1936, KYO); Yoshino-gun, Kawakami-mura, Yamato (Koidzumi, Sept. 1924, TNS); Yoshino-gun, Tenkawa-mura, Mt. Ohmine, Mt. Chosendake–Tubonouchi, alt. ca. 700 m (Fukuoka & Hotta, 6 Aug. 1962, KYO); Yoshino-gun, Shimokita-yama-mura, Zenki–Zenkiguchi (Murata & Shimizu, 18 July 1954, KYO). WAKAYAMA. Kumano-shi, Isato-machi, Ooidani, Kitayama River, upstream (Furuse, 13 Oct. 1961, A); Higashimuro-gun (ut "Kitamuro-gun"), Kitayama-mura, Komatsu (Furuse, 6 Oct. 1962, A); Higashimuro-gun, Kumanogawa-cho, Koguchi–Ohara, along the River Wadagawa (Murata,

15 Oct. 1960 & 13587, TI); Mt. Nachi-san (Matsumura, 29 July 1883, TI); Shinguu (Chiho-mine) (Satake & Okuyama, 11 Oct. 1942, TNS); Higashimuro-gun, Takada-mura (Ito, 5 Oct. 1930, TI); Higashimuro-gun, Hongu-cho, Mt. Ooto-zan, interior of Shizukawa, Oosugi-dani (Naruhashi 1759, KYO); Higashimuro-gun, Koza-cho (Uno, 21 July 1952, A); Isl. Ooshima (Tsukamoto, 9 Oct. 1940, KYO); Nishimuro-gun, Nakaheji-cho, Fukusada–Nonaka (Mimo-ro, Tsugaru & Fujimoto 4062, KYO); Nishimuro-gun, Tanabe-machi (Nakajima 1310, TI); Namera-dani (Araki 5742, SHO). KYOTO. Sokaku-gun, Kasagi-cho, Kasagiyama, alt. ca. 150 m (Murata 20457, KYO); Kyoto-shi, Ukyo-ku, Mt. Atago-yama, alt. 100 m (Murata 20445, KYO); Hodzu-kyo (Murata 5950, TNS). HYOGO. Takarazuka-shi, Horai-kyo (Togashi, 23 June 1968, KYO); Miki-shi, Shimizu-cho, Mitsuda, Yamadagawa, alt. 100–200 m (Murata & Nishimura 201, KYO); Kobe-shi, Kita-ku, north slope of Mt. Rokkô, Gokuraku-chaya–Arima-cho, alt. 400–800 m (Shimizu & Kurosaki 14, SHO); Shikama-gun, Yumesaki-cho, Ko-ura (Kawara)-keikoku (Hashimoto 10061, SHO); Isl. Awaji, Inohana-dani (Tashiro, 23 Sept. 1927, KYO). OKAYAMA. Wake-gun, Saeki-machi, Ohara (Ookubo, 19 Aug. 1977, KYO)*; Ate-tsu-gun, Arato-mura, Kounoie (unknown collector, 1 July 1929, KYO); Akaiwa-gun, Taka-tsuki-mura, Mt. Takakura (Uno, 15 Aug. 1952, A).

SHIKOKU. KAGAWA. Shoozu-gun (Shoodo-shima), Kankakei, alt. 700 m (Ohashi, Ohba & Murata, 10 Nov. 1975, TI); Shouzu-gun, Kusakabe-mura (Honma, 5 Aug. 1905, MAK); Yashima (Nakai, 1 June 1941, TI); Kida-gun, Mt. Goken-zan (Matsui, 17 Aug. 1910, MAK). TOKUSHIMA. Naka-gun, Washiki-machi, Nakagawa (Yamazaki, 13 Aug. 1976, TI); Naka-gun, Hinotani-mura (Yoshinaga s.n., TI); Mima-gun, Handa, Okuyama (Kasai, 21 Sept. 1913, MAK); Oe-gun, Koyadaira-mura (Nikai 1295, TI). EHIME. Uma-gun, Oku-no-in (Araki 5743, SHO); Mt. Ishidzuchi-san (Makino, 22 July 1905, MAK); Nishiuwa-gun, Misaki-cho, Takaura, alt. 10 m (Fukuoka 8691, KYO). KOUCHI. Kouchi, Engyo-ji (Shibuya, 28 July 1958, TI); Nangoku-shi, Shiraki-dani, alt. ca. 100 m (Ohashi, 20 June 1973, TUSG); Tosa-gun, Kagami-mura, Sadanaga (Kikuchi et al., 16 Oct. 1968, TUSG); Takaoka-gun, Mt. Yokokura-yama (Kimura, 29 Oct. 1939, TI); Ochi (Watanabe, 16 Sept. 1889, GH); Takaoka-gun, Niyodo-mura, Mori (Fujita 341, KYO); Hata-gun, Isl. Oki-no-shima (Yamawaki, 14 Oct. 1944, TI).

KYUSHU. OITA. Shimoge-gun, Masaka-mura (Yamamoto, 3 Aug. 1921, KYO); Mt. Hiko-san (Honda 1055, TI); near Beppu (Ishidoya, Aug. 1917, A). NAGASAKI. (Maximowicz in 1863, A; Faurie 3069, A); in insul. Iwosima *leitum* (Siebold s.n., L 908.118-1370—Syn-type of *L. Buergeri* Miq.); Oomura-shi, Mt. Tara-dake, Housen-ji, alt. 650 m (Shimada, 30 July 1958, KYO); Kokuzou (Higashisonogi-gun) (Makino, 23 Oct. 1904, MAK); Isl. Hirato-jima (Hashimoto, 16–19 Aug. 1951, TI). MIYAZAKI. Takachiho (Momiyama, 8 Oct. 1958, TI); Nishiusuki-gun, Hinokage-machi, near Furuzono, alt. 150–250 m (Tateishi 3358, TI); Higashiusuki-gun, Toogoo-machi, Yamage (Murata 8590, TI); Koyu-gun, Kishiro-mura, at the foot of Mt. Osuzu-yama (Hurusawa 32, TI); Mt. Osuzu, alt. 900 m (Sako, 20 Oct. 1960, MAK). KUMAMOTO. Uto, Mt. Misumi-dake (Kôzuma, 23 June 1912, TNS); Kuma-gun, Itsu-ki-mura, Touji (Ohba & Akiyama 1031, TI). KAGOSHIMA. Mt. Kirishima, alt. 100–1000 m (Tashiro in 1917, A); Mt. Takakuma-yama (Yamazaki, 12 Aug. 1942, TI); Kimotsuki-gun, Nejime-cho, Mt. Nokubidake, alt. 300 m (Sako 2452, KYO); Kaseda-shi, Mt. Isoma-yama (Ohba & Akiyama 875, TI).

This species is greatly different from the other species of sect. *Macrolespedeza*. It is apparently a more woody plant than others and grows in shadier and/or rockier places. It blooms from (late June) July to September, earlier than others. One form with elliptic leaves with a round apex was named f. *Oldhamii* (Miq.) Sugimoto (=*L. Oldhamii* Miq.). The shape of leaves varies, however, so it is difficult to distinguish it based only on the shape of leaves.

Forma **albiflora** Honda in *Bot. Mag. Tokyo* **46**: 421 (1932), ut "*L. Buergeri* Miq. var. *Oldhamii* (Miq.) Maxim. f. *albiflora* Honda"—Ohwi, *Fl. Jap.*, 678 (1953); *Fl. Jap.* rev. ed., 789 (1965); *Fl. Jap.* Eng. ed., 559 (1965); *Fl. Jap.* new ed., 789 (1975).

*L. Buergeri* Miq. var. *albiflora* Honda in *Bot. Mag. Tokyo* **46**: 421 (1932).

*L. Buergeri* Miq. f. *leucantha* Sugimoto, *New Keys Jap. Trees*, 463 (1961), *nom. superfl.*

Japanese name. Shirobana-kihagi (Honda, 1932).

Corolla white.

Rarely distributed in the range of this species.

Specimens examined.

HONSHU  Shizuoka. Minamizaki (prov. Izu) (Sugimoto 21869, ti—Holotype of *L. Buergeri* var. *Oldhamii* f. *albiflora* Honda); Ikawa (prov. Sagami) (Sugimoto 20435, ti—Holotype of *L. Buergeri* var. *albiflora* Honda).

Forma **angustifolia** Makino in *Bot. Mag. Tokyo* **20**: 41 (1906)—Hatusima in *Mem. Fac. Agr. Kagoshima Univ.* **6**: 6 (1967)—Ohwi, *Fl. Jap.* new ed., 789 (1975).

*L. Buergeri* Miq. var. *Kinashii* Ohwi in *Journ. Jap. Bot.* **26**: 234 (1951); *Fl. Jap.*, 678 (1953); *Fl. Jap.* rev. ed., 789 (1965); *Fl. Jap.* Eng. ed., 559 (1965).

Japanese name. Tachige-kihagi (Ohwi, 1951), Hosoba-kihagi (Makino, 1906).

Stem and branches with patent hairs.

Distr. Sporadic throughout the range of the species.

Specimens examined.

SHIKOKU.  Tokushima. Kamo (prov. Awa), Yoshii-mura (Nishimoto, 15 Sept. 1904, ti—Holotype of *L. Buergeri* Miq. f. *angustifolia* Makino).

Among the specimens cited under *L. Buergeri*, those referable to f. *angustifolia* are marked with an asterisk.

This form has branches and inflorescence-axes with patent hairs. Some plants with branches and inflorescence-axes with ascending hairs are also found sporadically.

**Lespedeza Maximowiczii** Schneid.

*Ill. Handb. Laubholz.* **2**: 113, fig. 70i, 71h–i² (1907)—Momiyama in *Journ. Jap. Bot.* **9**: 129 (1933), ut "*Maximowicziana*"—Ohwi, *Fl. Jap.*, 678 (1953); *Fl. Jap.* rev. ed., 789 (1965); *Fl. Jap.* Eng. ed., 559 (1965); *Fl. Jap.* new ed., 789 (1975)—Kitamura & Murata, *Col. Ill. Herb. Pl. Jap.* **2**: 99 (1961)—Kung in *Chinese Journ. Bot.* **1**: 21 (1963)—Kitagawa in *Journ. Jap. Bot.* **45**: 123 (1970)—Lee in *Bull. Seoul Nat. Univ. For.* No. 2, 22 (1965); *Ill. Fl. Korea*, 468 (1979)—Kitagawa, *Neo-Lineamenta Fl. Mansh.*, 406 (1979)—Ohashi in Satake et al., *Wild Fl. Jap. Herb.* **2**: 205 (1982).

[Plate 23; Figs. 47 & 48]

*L. Sieboldii* Miq. in *Ann. Mus. Lugd.-Bat.* **3**: 47 (1867), *pro parte, quoad specim.* Oldham 333 et *Prol. Fl. Jap.*, 235 (1867), *pro parte.*

*L. Buergeri* (non Miq.) Maxim. in *Acta Hort. Petrop.* **2**: 353 (1873), *pro parte*—Palibin in *Acta Hort. Petrop.* **17**: 65 (1898)—Nakai in *Journ. Coll. Sci. Univ. Tokyo* **26** (*Fl. Koreana* I): 154 (1909).

*L. bicolor* (non Turcz.) Forbes et Hemsley in *Journ. Linn. Soc.* **23**: 179 (1887), *pro parte.*

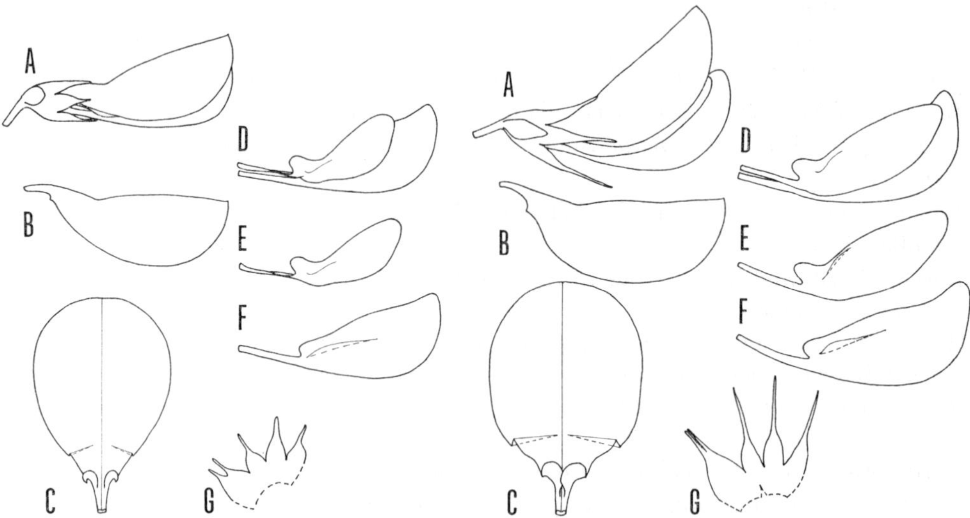

Fig. 47.  Flowers of *L. Maximowiczii*.
A, flowers, lateral view;   B, standards, lateral view;   C, standards, opened;
D, wings and keel-petals;   E, wings;   F, keel-petals;   G, calyces, dissected.
Left. Tsushima Isl. Toyosaki-mura (Nakashima, 14 Aug. 1939, KAG). Right.
Korea, Mt. Jiri-san (Yamazaki & Yamazaki, June 1982, TI).   All×3.3.

*L. bicolor* Turcz. var. *intermedia* (non Maxim.) Palibin in *Acta Hort. Petrop.* **17**: 64 (1898).

*L. Buergeri* Miq. var. *Oldhamii* (non Maxim.) Nakai in *Journ. Coll. Sci. Univ. Tokyo* **26** (*Fl. Koreana* I): 154 (1909).

*L. Buergeri* Miq. var. *praecox* Nakai in *Bot. Mag. Tokyo* **25**: 55 (1911); in *Journ. Coll. Sci. Univ. Tokyo* **31** (*Fl. Koreana* II): 467 (1911).

*L. Friebeana* Schindl. in Fedde, *Repert. Sp. Nov.* **9**: 514 (1911); in Sargent, *Pl. Wils.* **2**: 111 (1914), *typum non vidi*.

*L. praecox* (Nakai) Nakai [*Catalog. Sem. & Spor. Univ. Imp. Tokyo*, **20**, no. 566 (1914), *comb. nud.*] ex Koidz. in *Bot. Mag. Tokyo* **39**: 24 (1925).

*L. tomentella* Nakai, *Veg. Isl. Wangtō*, 16 (1914), in nota sub *L. Friebeana*, *nom. nud.*—Makino & Nemoto, *Fl. Jap.*, 735 (1925).

*L. Buergeri* Miq. var. *tricolor* Nakai, *Veg. Isl. Wangtō*, 9 (1914), *nom. nud.*

*L. praecox* (Nakai) Nakai var. *tomentella* Nakai, *Veg. Diamond Mts.*, 176 n.379b (1918), *nom. nud.*

*L. Oldhamii* (non Miq.) Nakai in *Bot. Mag. Tokyo* **33**: 202 (1919).

*L. Oldhamii* Miq. var. *tomentella* Nakai in *Bot. Mag. Tokyo* **33**: 203 (1919).

*L. Oldhamii* (non Miq.) Makino et Nemoto, *Fl. Jap.*, 735 (1925).

*L. Oldhamii* Miq. var. *tricolor* Nakai in *Bot. Mag. Tokyo* **37**: 78 (1927).

*L. Maximowiczii* Schneid. var. *tomentella* (Nakai) Nakai, *Lesp. Jap. Korea*, 39 (1927).

*L. Maximowiczii* Schneid. var. *elongata* Nakai, *Lesp. Jap. Korea*, 40 (1927).

*L. Maximowiczii* Schneid. var. *tricolor* (Nakai) Nakai, *Lesp. Jap. Korea*, 40 (1927).

*L. densiflora* Uyeki in *Bull. Nat. Hist. Soc. Corea* **20**: 17 (1935), *typum non vidi*.

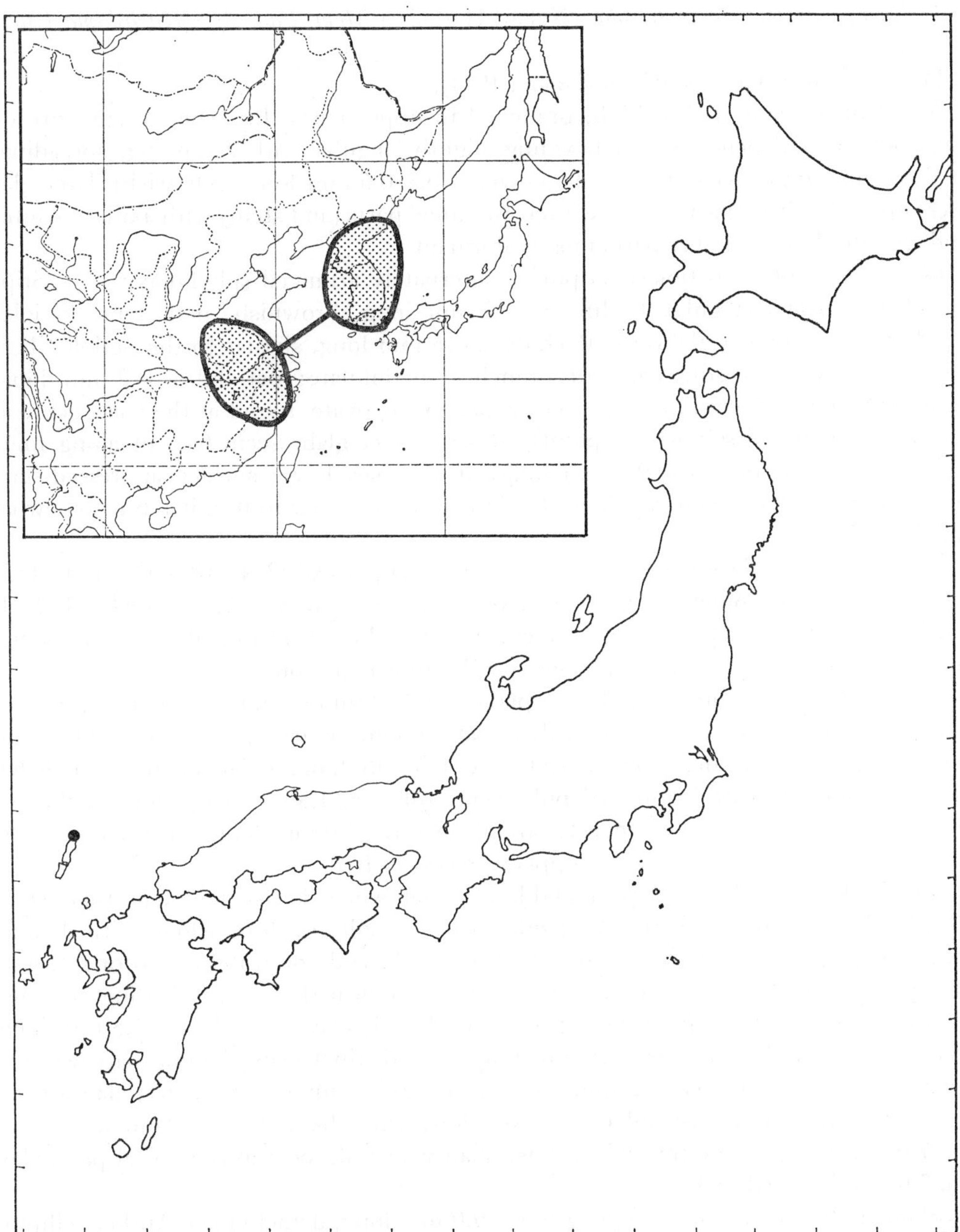

Fig. 48.   Distribution of *L. Maximowiczii*.

*L. Buergeri* Miq. subsp. *praecox* (Nakai) Hatus. in *Mem. Fac. Agr. Kagoshima Univ.* **6:** 6 (1967).

*L. Buergeri* Miq. subsp. *praecox* (Nakai) Hatus. f. *tomentella* (Nakai) Hatus. in *Mem. Fac. Agr. Kagoshima Univ.* **6:** 6 (1967).

*L. Buergeri* Miq. subsp. *tricolor* (Nakai) Hatus. in *Mem. Fac. Agr. Kagoshima Univ.* **6**: 6 (1967).

Japanese name. Chosen-kihagi (Nakai, 1914).

A perennial plant, 1–3 m high, branched in upper parts. Branches terete, spreading, sometimes dependent; when young slightly angular with ascending, spreading, or ascending appressed-sericeous hairs (hairs 0.3–0.6 mm long, yellowish), later glabrescent; most distal parts die in winter, proximal parts (and stem) with axillary winter buds (winter buds at lower parts mostly dormant).

Leaves trifoliolate, petiolate, stipulate, alternately arranged (phyllotaxy 1/2). Stipules free, linear-triangular to linear, 2–7 mm long, brownish, persistent. Petioles 0.5–4 cm long, almost glabrous. Rachides 5–15 mm long, similar to the petiole. Terminal leaflets (at middle parts of branches) petiolulate; petiolules 1–2 mm long, swollen; lamina 1–5 cm long, 0.5–3 cm wide, entire, ovate, round at the base, acute at the apex (the apex itself with a point), upper surface glabrescent (except along midrib) or sericeous (hairs ca. 0.5 mm long) at anthesis, lower surface appressed-sericeous (hairs ca. 0.5 mm long, yellowish); lateral ones similar to terminal but somewhat smaller.

Inflorescence axillary racemous, one (rarely 2–3) per axil, 2–4(–10) cm long including the peduncle, somewhat compactly to loosely 4–12-flowered; peduncles 1–2(–4) cm long, hairy like the young branch. Primary bracts ovate, about 1 mm long, pubescent, persistent; secondary bracts similar to primary ones.

Flowers 9–10 mm long at anthesis; pedicels 0.8–2 mm long, terete, curved, tomentose. Bracteoles at the base of the calyx ovate to elliptic, ca. 1.5 mm long, pubescent outside, glabrous inside, persistent. Calyx 4–5 mm long, campanulate, four-lobed above the middle part, appressed pubescent; tube ca. 1.5 mm long; lobes subequal in length or the lower one longest; lateral ones ca. 3.0 mm long, ca. 0.7 mm wide, triangular, acuminate at the apex; upper one two-clefted.

Standard shorter than keel-petal and longer than wings, keel-petal longer than wings (K>S>W). Standard red-purple inside, paler outside, ca. 9 mm long with distinct claw, ca. 6 mm wide; lamina round to elliptic or broadly obovate, round or retuse at the apex with or without mucro, inflexed-auriculate near the base, spreading in anthesis from the part near the base; auricles developed; claw whitish, ca. 1.5 mm long. Wings deeper red-purple, ca. 8.5 mm long with distinct claw; lamina narrowly oblong, ca. 6.5 mm long, ca. 2.5 mm wide, auriculate at upper basal part; claw ca. 2.5 mm long, whitish. Keel-petal ca. 9.5 mm long with distinct claw; lamina obovate, ca. 7.0 mm long, ca. 3.0 mm wide, paler than wings, deepest in the apical part; claw ca. 3.0 mm long, whitish.

Stamens 10, nearly the same length, ca. 9.0 mm long, diadelphous. Anthers elliptic, shallowly retuse at the apex, ca. 0.5 mm long; before anthesis yellow.

Pistils ca. 9.3 mm long; ovary elliptic, ca. 1.8 mm long at anthesis, pubescent, stalked; stalk ca. 1.3 mm long; style ca. 6.5 mm long, pubescent at the basal part, glabrescent at the apical part.

Fruits compressed, round or broadly elliptic, ca. 1.5 mm long, ca. 5 mm wide, subsessile, slightly to densely pubescent or nearly glabrous.

Seeds reniform, 3–4 mm long, 2–3 mm wide.

Distr. Japan (north of Tsushima Island), Korea, and China.

Specimens examined.
Japan or Korea (ut "Japan & Korean Arch. coll.") (Oldham 333, L 908.119-583—Syntype of *L. Sieboldii* Miq.).

**Japan.**
KYUSHU. NAGASAKI. Isl. Tsushima. Toyosaki-mura (Nakajima, 14 Aug. 1939, KAG); Toyosaki-mura, Toyo (Hara s.n., MAK). Cultivated in Tokyo (ex Isl. Tsushima, Kamiagata-gun, Toyo) (Akiyama, 5 June 1984, TI).

**Korea.**
Port Chusan (Wilford in 1895, CAL). Wangto (Nakai 592, TI—Lectotype of *L. Oldhamii* Miq. var. *tomentella* Nakai; Nakai 600, TI—Lectotype of *L. Oldhamii* Miq. var. *tricolor* Nakai). CHOLLANAM. Mt. Jiri San, Rokotan–Kegon-ji (Yamazaki & Yamazaki, 21 Aug. 1982, TI). KYONGSANGNAM (KEINAN). Toksan (Mori 183, TI—Holotype of *L. Maximowiczii* Schneider var. *elongata* Nakai). KYONGSANPUK. Mt. Woonmoon (Kim, 10 June 1984, A). KANGWON (KOGEN). Yangyang-gun, Seomyun, Osekoncheon–Sulak San, alt. 700–800 m (Murata & Ko 21084, TI); Kongo-san (Wilson, 30 June 1918, TI).

**China**
CHEKIANG. Yen Tang Shan, alt. 500–700 m (Chiao 14735, A).

This species is rare in Japan and only known from Toyosaki and Waniura, in the northernmost part of Tsushima Island (Kami-shima), Kyushu. It is abundantly distributed in Korea, especially in the southern parts, and also reported from China (Honan, Chekiang, and Anhwei (Kung, 1963)). *L. Buergeri* is not found in Tsushima Island and Korea. The distribution pattern of these two species is interesting. Hatusima (1967) recognized this species as a subspecies of *L. Buergeri*. But this is well distinguishable from the latter by calyx-lobes with acuminate apex and red-purple flowers. This species is woody like as *L. Buergeri*. The direction of hairs varies by branch; inflorescence-axes and young branches have ascending hairs. The degree of hairiness of the upper surface of leaves varies even in the same locality by population or by individual. A typical form with glabrous upper surfaces of leaves is common. A form with sparsely pubescent upper surfaces of leaves is called var. *tomentella* Nakai. Hatusima recognized this variety as a form and made a new combination under *L. Buergeri*, therefore, a new combination, *L. Maximowiczii* f. *tomentella*, is needed. It is difficult, however, to clearly distinguish this form from the typical glabrous form, because the degree of hairiness changes gradually.

Forma **albiflora** Uyeki in *Journ. Kor. For. Soc.*, No. 194 (1942), *typum non vidi*.
Corolla white.

This is a white-flowered form of *L. Maximowiczii*. As we cannot examine the type specimen nor any other specimens, the details about this form are uncertain.


## Natural Hybrids

In Japan natural hybrids of sect. *Macrolespedeza* have been recorded.
**Lespedeza × cyrto-Buergeri** S. Akiyama et H. Ohba in *Journ. Jap. Bot.* **57**: 238 (1982).

*L. Buergeri* Miq.×*L. cyrtobotrya* Miq.

Japanese name. Okutamahagi (S. Akiyama & H. Ohba, 1982).

Description and specimens were presented in S. Akiyama & H. Ohba (1982).

**Lespedeza×kagoshimensis** Hatus. in *Journ. Jap. Bot.* **38**: 155 (1963), *pro. sp.*—
Ohwi, *Fl. Jap.* rev. ed., 791 (1965); *Fl. Jap.* new ed., 791 (1975)—S. Akiyama & H.
Ohba in *Journ. Jap. Bot.* **58**: 251 (1983).

*L. formosa* (Vog.) Koehne var. *shiroyamensis* Hatus. in *Mem. Fac. Agr. Kagoshima
Univ.* **6**: 8 (1967), *nom. nud., ut nov. comb.*

*L. argyrophylla* Hatusima in *Mem. Fac. Agr. Kagoshima Univ.* **6**: 12 (1967).

*L. formosa* (Vogel) Koehne var. *kagoshimensis* (Hatus.) Hatus. in *Ann. Rep. Yokosuka City Mus.* No. 14, 4 (1969), *comb. nud.*

*L. Buergeri* Miq.×*L. formosa* (Vogel) Koehne subsp. *velutina* (Nakai) S. Akiyama
et H. Ohba var. *satsumensis* (Nakai) S. Akiyama et H. Ohba.

Japanese name. Shiroyamahagi (Hatusima, 1963).

Description and representative specimens were presented in S. Akiyama & H. Ohba
(1983c).

**Lespedeza homoloba** Nakai×**L. formosa** (Vogel) Koehne subsp. **velutina** (Nakai)
S. Akiyama et H. Ohba.

*L. homoloba* Nakai×*L. kiusiana* Nakai, S. Akiyama & H. Ohba in *Journ. Jap. Bot.*
**58**: 102 (1983).

Japanese name. Bitchu-tsukushihagi (S. Akiyama & H. Ohba, 1983).

Description and specimens were presented in S. Akiyama & H. Ohba (1983a).


## Horticultural Species

In Japan *Lespedeza* is cultivated as an ornamental plant; the origins of these ornamental *Lespedeza* are uncertain. In 1775 Yokoyama described many cultivated and
wild *Lespedeza* in Japanese, but it is difficult to detect to which species these belonged.
In this paper we treat cultivated species separately from wild species.


**Lespedeza japonica** L. H. Bailey

*Cyclop. Americ. Hort.* **2**: 904 (1901)—Nakai, *Lesp. Jap. Korea*, 23 (1927), *excl.* var.
*intermedia* Nakai, var. *angustifolia* (Nakai) Nakai, var. *gracilis* Nakai, var. *retusa* Na-
kai, et *syn. L. intermedia* (non Britt.) var. *alba* Nakai—Murata in *Acta Phytotax. Geobot.* **29**: 100 (1978), *excl.* f. *angustifolia* (Nakai) Murata, f. *sericea* (Hatus.) Murata,
var. *australis* (Hatus.) Murata et *L. Thunbergii* (DC.) Nakai f. *angustifolia* (Nakai) Ohwi
in *syn.*—Ohashi in Satake et al., *Wild Fl. Jap. Herb.* **2**: 205 (1982), *pro parte, excl.* f.
*angustifolia* (Nakai) Murata, var. *australis* (Hatus.) Murata et *syn. L. Thunbergii* (DC.)
Nakai f. *angustifolia* (Nakai) Ohwi. [Plates 24 & 25; Fig. 49]

*L. japonica* is here treated as a composite cultivated species including derivatives
from hybrid forms between *L. formosa* (especially subsp. *velutina*) and other wild
species. The type specimen of Cv. *Nipponica* is most similar to wild subsp. *velutina*.
Usually this cultivated species is characterized and distinguished from another cultivated species, *L. Thunbergii*, by the persistent short hairs on its upper leaf surface.
Nearly glabrous forms or sparsely pubescent forms are also sometimes found.

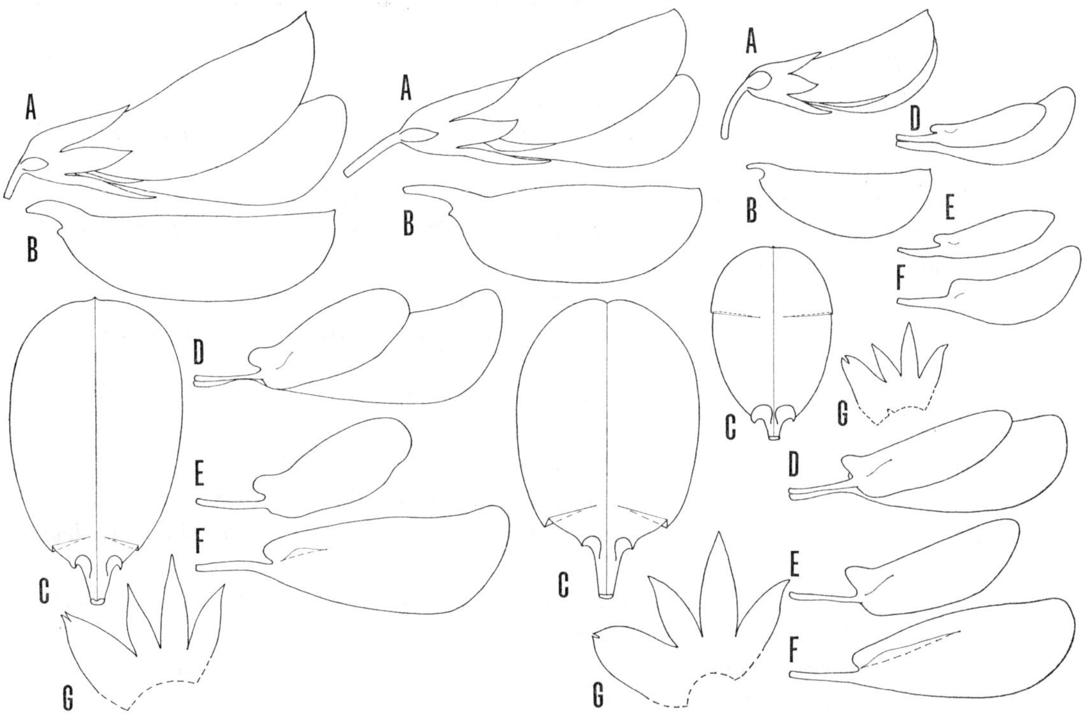


Fig. 49.　Flowers of *L. japonica*.
A, flowers, lateral view;　B, standards, lateral view;　C, standards, opened;
D, wings and keel-petals;　E, wings;　F, keel-petals;　G, calyces. Left.
Cv. *Nipponica*; cultivated in Bot. Gard. Nikko (Ohba & Akiyama 1042, TI).
Center: Cv. *Japonica*; cultivated in Tokushima (Kato 53, KYO—Holotype of
*L. Thunbergii* f. *leucantha* Murata). Right. Cv. *Hiratsukae*; cultivated in
Tottori (Hiratsuka, 10 Sept. 1973, TI—Holotype of *L. Hiratsukae* Nakai).
All×3.3.

Key to cultivars
1.　Flowers white ................................................................. cv. Japonica
1.　Flowers red-purple ............................................................. cv. Nipponica
1.　Flowers various, some flowers white with or without red-purple patches and others
　　red-purple with or without white patches ..................................... cv. Versicolor
1.　Flowers bicolored, standard white with purple base, wings whitish purple or purplish
　　red, keel-petals white ....................................................... cv. Hiratsukae

## Cv. **Japonica**

*Desmodium racemosum* (non DC.) Sieb. et Zucc. var. *albiflorum* Sieb. ex Miq. in
*Ann. Mus. Bot. Lugd.-Bat.* **3**: 48 (1867), *in nota sub L. cyrtobotrya* Miq. et *Prol. Fl.
Jap.* 236 (1867).

　　*L. bicolor* (non Turcz.) Maxim. α. *typica* Maxim. in *Acta Hort. Petrop.* **2**: 355
(1873), *pro parte, quoad specim.* "*Colitur circa templa Nagasaki floribus candidis.*"

　　*L. bicolor* Turcz. var. *intermedia* Maxim. f. *albiflora* Matsum. in *Bot. Mag. Tokyo*
**16**: 69 (1902).

*L. bicolor* Turcz. var. *albiflora* Maxim. *in sched.* ex Schneid., *Ill. Handb. Laubholz.* **2**: 113 (1907), *pro syn.*

*L. Sieboldii* Miq. var. *albiflora* (Maxim. *in sched.* ex Schneid.) Schneid., *Ill. Handb. Laubholz.* **2**: 113 (1907)—Makino & Nemoto, *Fl. Jap.*, 736 (1925).

*L. formosa* (Vogel) Koehne var. *albiflora* Schindl. in Engl., *Bot. Jahrb.* **49**: 582 (1913).

*L. bicolor* Turcz. var. *alba* Bean, *Trees Shrubs Brit. Isl.* **2**: 16 (1914).

*L. japonica* L. H. Bailey var. *albiflora* (Sieb. ex Miq.) Nakai, *Lesp. Jap. Korea*, 25 (1927) *excl. quoad specim.* Nakai 5576.

*L. inabensis* Nakai in *Journ. Jap. Bot.* **15**: 531 (1939).

*L. penduliflora* (Oudem.) Nakai var. *albiflora* (Sieb. ex Miq.) Ohwi in *Journ. Jap. Bot.* **26**: 234 (1951).

*L. bicolor* Turcz. f. *alba* (Bean) Ohwi in *Bull. Nat. Sci. Mus. Tokyo* **33**: 77 (1953).

*L. Thunbergii* (DC.) Nakai var. *albiflora* (Maxim. *in sched.* ex Schneid.) Ohwi f. *albiflora*, *Fl. Jap.* rev. ed., (790) 1438 (1965); *Fl. Jap.* new ed., 790 (1975).

*L. Thunbergii* (DC.) Nakai cv. Albiflora, *Fl. Jap.* Eng. ed., 559 (1965).

*L. formosa* (Vog.) Koehne f. *albiflora* (Sieb. *in sched.* ex Miq.) Hatusima in *Mem. Fac. Agr. Kagoshima Univ.* **6**: 8 (1967).

*L. Thunbergii* (DC.) Nakai var. *acutifolia* (Nakai) Hiyama ex Murata f. *leucantha* Murata in *Acta Phytotax. Geobot.* **29**: 105 (1978).

Japanese name. Shirobanahagi (Nakai, 1927), Shirahagi (Hort., Yokoyama, 1775).

Corolla white. Widely cultivated as an ornamental plant.

Representative specimens.
**U.S.A.** PHILADELPHIA. Meeban (Bailey, 26 Sept. 1899, BH—Holotype and isotypes of *L. japonica* L. H. Bailey). MASS. Northampton (Canning, 27 Sept. 1899, BH).

**Japan.** (Buerger s.n., L 998.118-1080 left).   TOTTORI. Tottori-shi, cultivated (Nakada 27, TI—Holotype of *L. inabensis* Nakai). TOKUSHIMA. Katsuura-gun, Kamikatsu-cho, Jigan-ji, cultivated (Kato 53, KYO—Holotype of *L. Thunbergii* (DC.) Nakai var. *acutifolia* (Nakai) Hiyama ex Murata f. *leucantha* Murata).

White-flowered *Lespedeza* was cultivated in gardens and many names were described under different species as varieties or forms. *L. inabensis* was distinguished from *L. japonica* by Nakai (1939) based on the glabrous upper surface of leaves. Because the hairiness of the upper surface of leaves is variable, it falls into the synonymy of *L. japonica*.

### Cv. **Nipponica**
*L. nipponica* Nakai, *Lesp. Jap. Korea*, 10 (1927)—Kitamura & Murata, *Col. Ill. Herb. Pl. Jap.* **2**: 99 (1961), *pro parte*, excl. var. *satsumensis* (Nakai) Murata et *L. kiusiana* Nakai, *L. japonica* L. H. Bailey var. *angustifolia* Nakai, var. *gracilis* Nakai, var. *retusa* Nakai in *syn.*

*L. bicolor* (non Turcz.) Hook. f. in *Bot. Mag.* **153**: t. 6602 (1882), *pro parte*, excl. *L. bicolor* Turcz. in *syn.* cit.

*L. liukiuensis* Hatusima in *Mem. Fac. Agr. Kagoshima Univ.* **6**: 12 (1967).

Japanese name. Nishikihagi (Nakai, 1927).

Corolla red-purple. Widely cultivated as an ornamental plant.

Representative specimens.

**Japan**    Cultivated in the Botanical Gardens Koishikawa, University of Tokyo, Tokyo (unknown collector, 25 Sept. 1879, TI—Holotype of *L. nipponica* Nakai); Cultivated in the Botanical Gardens, Nikko, University of Tokyo, Tochigi Pref. (Ohba & Akiyama 1042, TI). OKINAWA. Nago-machi, cultivated (Amano 7018, KAG—Holotype of *L. liukiuensis* Hatusima); Naturalized on open wayside near Yagachi (Hatusima 17963, KAG).

*L. nipponica* is described based on a cultivated stock in the Botanical Gardens, University of Tokyo, which is now lost, and white-flowered *L. japonica* is also described on a cultivated stock in the U.S.A. These two, now cultivated widely in Japan, are similar to each other. Putative hybrids between *L. nipponica* and other species, such as *L. homoloba* and *L. bicolor*, are sometimes cultivated, but morphologically are difficult to clearly distinguish from *L. nipponica*, because flowers change their forms transitionally. It is uncertain whether the type of *L. nipponica* is of hybrid origin or not. Although the variation of *L. formosa* is wide and *L. nipponica* and *L. japonica* can be found in its variation range, we would like to treat *L. nipponica* and *L. japonica* as cultivated species including these hybrids and *L. formosa* as native species.

Cv. **Versicolor**: Murata in *Acta Phytotax. Geobot.* **29**: 100 (1978)—Ohashi in Satake et al., *Wild. Fl. Jap. Herb.* **2**: 205 (1982).
   *L. japonica* L. H. Bailey var. *versicolor* Nakai in *Bot. Mag. Tokyo* **45**: 121 (1939).
   *L. penduliflora* (Oudem.) Nakai var. *albiflora* (Sieb. ex Miq.) Ohwi f. *versicolor* (Nakai) Ohwi, *Fl. Jap.*, 679 (1953), *nom. nud.*
   *L. Thunbergii* (DC.) Nakai var. *albiflora* (Sieb. ex Miq.) Ohwi f. *versicolor* (Nakai) Ohwi, *Fl. Jap.* rev. ed., 790 (1965), *nom. nud.*
   *L. Thunbergii* (DC.) Nakai cv. Versicolor: Ohui, *Fl. Jap.* Eng. ed., 559 (1965).
   *L. formosa* (Vog.) Koehne f. *versicolor* (Nakai) Hatusima in *Mem. Fac. Agr. Kagoshima Univ.* **6**: 8 (1967).
   Japanese name. Somewakehagi (Nakai, 1939), Sasahagi (Hort., Yokoyama, 1775).
   Corolla white with or without red-purple patches and red-purple with or without white patches in a plant. Cultivated as an ornamental plant.
   Representative specimen.
**Japan.**    HONSHU. Kanagawa Pref., Hakone, Sokokura (cult.) (Nakai, 5 Oct. 1930, TI—Holotype of *L. japonica* L. H. Bailey var. *versicolor* Nakai).

### Cv. **Hiratsukae**
   *L. Hiratsukae* Nakai in *Journ. Jap. Bot.* **14**: 637 (1938).
   Japanese name. Hime-nishikihagi (Nakai, 1938).
   Standard white with purple base, wings whitish purple or purplish red, keel-petals white. Rarely cultivated as an ornamental plant.
   Specimen examined.
**Japan.**    Cultivated in Tottori (Hiratsuka, 10 Sept. 1937, TI—Holotype of *L. Hiratsukae* Nakai).

This is considered to be a hybrid between *L. japonica* cv. Nipponica and *L. Buergeri*, because it has flowers similar to those of *L. Buergeri* although somewhat lighter in color and the shape of leaflets is also similar to that of *L. Buergeri*.

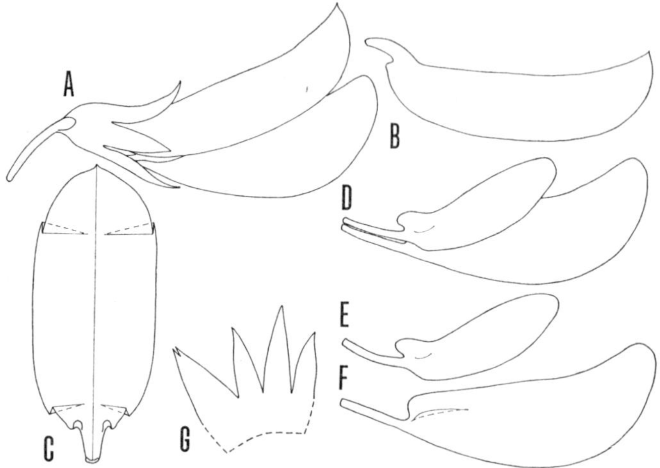

Fig. 50.   Flower of *L. Thunbergii* (cultivated in Bot. Gard. Koishikawa, Akiyama 170, TI).
A, flower, lateral view;   B, standard, lateral view;   C, standard, opened;
D, wing and keel-petal;   E, wing;   F, keel-petal;   G, calyx. All×3.3.

## Lespedeza Thunbergii (DC.) Nakai

*Lesp. Jap. Korea*, 15 (1927), *excl. syn. L. formosa* (Vog.) Koehne—Ohwi, *Fl. Jap.* rev. ed., 790 (1965); *Fl. Jap.* Eng. ed., 559 (1965); *Fl. Jap.* new ed., 790 (1975), *excl.* var. *obtusifolia* (Nakai) Ohwi, var. *albiflora* (Schneid.) Ohwi, var. *satsumensis* (Nakai) Ohwi—Hatusima in *Mem. Fac. Agr. Kagoshima Univ.* **6**: 11 (1967), *excl.* f. *sericea* Hatus.—Ohashi in Satake et al., *Wild. Fl. Jap. Herb.* **2**: 205 (1982), *excl.* f. *macrantha* (Honda) Murata, var. *satsumensis* (Nakai) Ohwi.          [Plates 26 & 27; Fig. 50]

*Hedysarum heterocarpon* (non L.) Thunb., *Fl. Jap.*, 287 (1784).

*Desmodium Thunbergii* DC., *Prodr.* **2**: 337 (1825).

*D. racemosum* (non DC.) Sieb. et Zucc. in *Abh. Muench. Akad.* **4**. Abt. 2: 121 (1845), *pro parte.*

*D. penduliflorum* Oudem. in *Neerlands Pl.* **2**: t. 2 (1866)—Carriere in *Rev. Hort.* **45**: 211, *cum. tab. col.* (1873).

*L. Sieboldii* Miquel in *Ann. Mus. Bot. Lugd.-Bat.* **3**: 47 (1867), *excl. quoad specim. cit.* Textor, Oldham 333 et *Prol. Fl. Jap.*, 235 (1867), *pro parte*—Franchet & Savatier, *Enum. Pl. Jap.* **1**: 101 (1875)—Bailey, *Cyclop. Americ. Hort.* **2**: 904, fig. 1264 (1901)—Schneider, *Ill. Handb. Laubholz.* **2**: 113, fig. 71e–g (1907)—Makino & Nemoto, *Fl. Jap.*, 736 (1925), *excl.* var. *albiflora* Schneid.

*L. bicolor* Turcz. *γ. Sieboldii* (Miq.) Maxim. in *Acta Hort. Petrop.* **2**: 356 (1873), *pro parte*—Matsumura in *Bot. Mag. Tokyo* **16**: 70 (1902), *excl.* f. *sericea* Matsumura; *Ind. Pl. Jap.* **2**: 267 (1912), *pro parte.*

*L. racemosa* Dippel, *Handb. Laubholz.* **3**: 720 (1893).

*L. penduliflora* (Oudem.) Nakai in *Bot. Mag. Tokyo* **37**: 79 (1923), *excl. quoad specim. cit.* Nakai 5578, Nakai 9823, Imai—Ohwi, *Fl. Jap.*, 679 (1957), *excl.* var. *sericea* (Matsum.) Ohwi, var. *albiflora* (Sieb. ex Miq.) Ohwi—Kitamura & Murata, *Col.*

*Ill. Herb. Pl. Jap.* **2**: 99 (1961), *excl.* var. *sericea* (Matsum.) Ohwi et *L. japonica* L. H. Bailey in *syn*.

Japanese name. Miyaginohagi (Hort.; Kaibara, 1698; Matsumura, 1902).

A perennial plant, 1–1.5 m high in late summer and early autumn. Stem dependent later, 0.3–0.5 m high in the first year, rarely branched, terete, slightly angular with sparsely ascending appressed hairs (hairs 0.3–0.6 mm long, whitish); aerial part (except the basal part ca. 5 cm from ground level) annual; basal part with axillary and adventitious winter buds. In the second year a few branches emerge from the basal part of the stem, 1–1.5(–2) m long, similar to the stem of the first year, but branched or rarely branched in upper part; most of the aerial part annual; the basal parts (near ground level) with axillary and adventitious winter buds. Branches in subsequent years (5–20 or more in number) coming from the basal parts of stem and branches every year and the number of branches increases according to the size of the stock.

Leaves trifoliolate, petiolate, stipulate, spirally arranged. Stipules free, linear-triangular to linear, 2–7 mm long, brownish, persistent. Petioles 1–6 cm long, hairy like the young branch. Rachides 5–30 mm long, similar to the petiole. Terminal leaflets (at middle parts of branches) petiolulate; petiolules 1–3 mm long, swollen; lamina 3.5–9 cm long, 2–5 cm wide, entire, narrowly elliptic, cuneate at the base, acute at the apex (the apex itself with a point), upper surface glabrous (except along the midrib) at anthesis, lower surface appressed-sericeous (hairs 0.5 mm long); lateral ones similar to the terminal ones but somewhat smaller.

Inflorescence axillary racemous, one per axil, 10–15 cm long including the peduncle, somewhat compactly to loosely 10–14-flowered; peduncles 1–2(–4) cm long, hairy like the young branch. Primary bracts ovate, about 1 mm long, pubescent, persistent; secondary bracts similar to the first ones.

Flowers 12.5–15.4 mm long at anthesis; pedicels 2–5 mm long, terete, curved, tomentose. Bracteoles at the base of the calyx narrowly elliptic to ovate, 1–1.5 mm long, pubescent outside, glabrous inside, persistent. Calyx 5.3–5.4 mm long, campanulate four-lobed above the middle part, densely appressed pubescent; tube 1.8–1.9 mm long; the lower lobe longest; lateral ones 3.4–3.6 mm long, 1.1–1.2 mm wide, triangular-lanceolate, acute at the apex but not acuminate; upper one can be slightly two-clefted.

Standard shorter than keel-petal and longer than wings, lamina narrowly elliptic, round at the apex with or without a point, slightly inflexed-auriculate near the base, reflexed in anthesis from near the base. Wings deeper red-purple than standard, lamina narrowly obovate, auriculate at upper basal part, tapering at lower basal part. Keel-petal paler than wings, lamina narrowly obovate. Stamens 10, nearly the same length, diadelphous. Anthers elliptic, shallowly retuse at the apex, ca. 0.7 mm long; before anthesis yellow, sometimes abortive. Ovary elliptic, pubescent, stalked; style pubescent at the basal part, glabrescent at the apical part.

Fruits rarely compressed, elliptic, slightly to densely pubescent or nearly glabrous, stalked. Seeds reniform.

Cultivated widely as an ornamental plant.

Representative specimens.

Cultivated in Japan. (Siebold s.n., l 908.118-2391—Lectotype of *L. Sieboldii* Miq.;

Thunberg? s.n., L 908.118-2353); Culta in H. B. (Siebold s.n., L 908.118-2356—Syntype of *L. Sieboldii* Miq.); Nagasaki (Oldham 329, L 908.118-2383); (Buerger s.n., L 908.118-2392 —Syntype of *L. Sieboldii* Miq.). Aomori. Naigyushi (cited as "Nadeshiko" by Nakai, 1927) (Kinashi s.n., TI). Cultivated in Botanical Gardens, Koishikawa, University of Tokyo, Tokyo (Akiyama 170, TI).

Usually this species is treated as conspecific with *L. patens*. It is, however, noticeable that the former is different from the latter; *L. Thunbergii* is very vigorous in the region where *L. patens* is not distributed; on the other hand, cultivated *L. patens* does not grow well. *L. Thunbergii* bears only a few seeds, branches are much dependent, the color of its leaves is dark green, the wing is not as short as that of *L. patens*. It should be noted that the origin of *L. Thunbergii* might be a hybrid between *L. patens* and other species. The flowering season of *L. Thunbergii* is earlier than that of other species except for *L. patens*, *L. Buergeri*, and *L. Maximowiczii*. As *L. Buergeri* has whitish yellow standard and keel-petals, a hybrid between *L. patens* and *L. Buergeri* is expected to have paler red-purple flowers like those of *L.×cyrto-Buergeri* and *L.× kagoshimensis*. But the flowers of *L. Thunbergii* are not paler red-purple. The shape and color of leaves of *L. Thunbergii* are similar to those of *L. Maximowiczii*. The possibility remains that *L. Thunbergii* has a hybrid origin between *L. patens* and *L. Maximowiczii*. However, this possibility is difficult to prove because there is no effective way to detect it at present. For example, the somatic chromosome number of *L. patens*, *L. Thunbergii*, and other species of sect. *Macrolespedeza* is 22 so far as has been examined. We attempted to create artificial hybrids but failed, because when the flowers were covered with fine nets to avoid pollinators, they lost vigor and fell down.

# Imperfectly Known Taxa

*Lespedeza albiflora* Ricker in *Amer. Journ. Bot.* **33**: 257 (1946).

*L.* × *angustifolioides* T. Lee in *Bull. Seoul Nat. Univ. For.* No. 2, 3 (1965).

*L. anhweiensis* Ricker in *Lingnan Sci. Journ.* **20**: 200 (1942).

*L. bicolor* Turcz. f. *patens* Nakai ex Hatus. in *Mem. Fac. Agr. Kagoshima Univ.* **6**: 14 (1967).

*L. bracteolata* Ricker in *Lingnan Sci. Journ.* **20**: 201 (1942).

*L. Buergeri* Miq. f. *purpurascence* H. Koidz. in *Journ. Pl. Iwateken* **2**: 77 (1937).

*L. Buergeri* Miq. f. *retusa* (Hatus.) Sugimoto, *New Keys Jap Trees*, 463 (1961).

*L. Buergeri* Miq. var. *retusa* Hatus. in *Journ. Jap. Bot.* **12**: 876 (1936).

*L. chekiangensis* Ricker in *Lingnan Sci. Journ.* **20**: 201 (1942).

*L.* × *chiisanensis* T. Lee in *Bull. Seonl Nat. Univ. For.* No. 2, 8 (1965).

*L. cyrtobotrya* Miq. f. *semialba* T. Lee in *Bvll. Seoul Nat. Univ. For.* No. 2, 11 (1965).

*L. Dunnii* Schindl. in Engl. *Bot. Jahrb.* **49**: 585 (1913).[1]

*L. japonica* L. H. Bailey var. *ovata* Ricker in *Amer. Journ. Bot.* **33**: 258 (1946).

*L. Merrillii* Ricker in *Lingnan Sci. Journ.* **20**: 202 (1942).

*L. Metcalfii* Ricker in *Amer. Journ. Bot.* **33**: 258 (1946).

*L. paradoxa* Ricker in *Amer. Journ. Bot.* **33**: 258 (1946).

*L.* × *patentibicolor* T. Lee in *Bull. Seoul Nat. Univ. For.* No. 2, 25 (1965).

*L.* × *patentielongata* T. Lee in *Bull. Seoul Nat. Univ. For.* No. 2, 26 (1965).

*L. penduliflora* (Oudem.) Nakai subsp. *cathayana* Hsu et al. in *Acto Phytotax. Sin.* **11**: 193 (1966).

*L. robusta* Nakai, *Lesp. Jap. Korea*, 72 (1927).

*L.* × *Schindleri* T. Lee in *Bull. Seoul Nat. Univ. For.* No. 2, 28 (1965).

*L. Thunbergii* (DC.) Nakai subsp. *cathayana* (Hsu et al.) Hsu et al. in *Acta Phytotax. Sin.* **21**: 306 (1983).

---

[1] We examined the type specimens in HK but no specimen, except the type, and no living material has been available. It is doubtful that this species belongs to sect. *Macrolespedeza.*

# Summary


The section *Macrolespedeza* of the genus *Lespedeza* is revised based on comparative studies.

Variation of floral characteristics within an individual (phenotypic plasticity), within a population (individual variation), and between populations (collective variation), and developmental changes of floral characters are described to clarify the characteristics by which species of sect. *Macrolespedeza* are circumscribed. The following are taxonomically significant: the shape of the lateral calyx-lobe, the shape of the apex of the lateral calyx-lobe, the ratio of the length of the lateral calyx-lobe to the length of the calyx-tube, the color of the corolla, the ratio of the length of the wing to the length of the keel-petal, the ratio of the length of the standard to the length of the keel-petal, the ratio of the length of the wing to the length of the standard, the shape of the standard, the shape of the auricle of the standard, the ratio of the length of the standard-claw to the length of the standard, the shape of the lamina of the wing, the ratio of the length of the wing-claw to the length of the wing, the shape of the lamina of the keel-petal, and the ratio of the length of the keel-petal-claw to the length of the keel-petal.

The section *Macrolespedeza* is divided into three series, ser. *Macrolespedeza*, ser. *Formosae* (*ser. nov.*), and ser. *Heterolespedeza* (*stat. nov.*). The section *Macrolespedeza* is represented by nine species, two subspecies, two varieties, ten forms, and two horticultural species. The species and infraspecific taxa of sect. *Macrolespedeza* are described in detail.

# Acknowledgments

This paper is the main part of a Ph.D. thesis accepted by the University of Tokyo (Akiyama, 1985), carried out under the direction of Associate Professor Hideaki Ohba, University Museum, University of Tokyo, to whom I am deeply indebted. Thanks are due to Professor Kunio Iwatsuki, University of Tokyo, for his encouragement. I am also grateful to the late Dr. Hiroshi Hara (then Professor Emeritus of the University of Tokyo) for his valuable suggestions and encouragement. I would like to thank Mr. Gen Murata, Kyoto University, for his valuable comments. Dr. Mitsuo Suzuki, Kanazawa University, Dr. Nobuhira Kurosaki, Shoei Junior College, Mr. Kenji Midorikawa, Tokyo, Mr. Sadao Tsutsui, Fukuoka, Mr. Toshio Sakata, Kagoshima, Mr. Hideo Nagase, Gifu, Mr. Nobuo Futamura, Gifu, and Mr. Noboru Kuramoto, Tokyo, were kind enough to supply many valuable materials for this study or accompany me on field surveys. Comments and constructive criticism were offered by the members of the University Museum and the Botanical Gardens, University of Tokyo.

I also express my appreciation to the curators of the following herbaria: A, CAL, GH, HK, K, KAG, KANA, KIEL, KYO, L, LE, MAK, NA, SAP, SHO, TI, TNS, TOFO, and TUSG, for permitting me to survey herbarium specimens and to borrow valuable collections.

# References


Akiyama, S. (1985) Taxonomic studies of the Japanese species of the section *Macrolespedeza* of the genus *Lespedeza* (Leguminosae). (Dr. thesis, University of Tokyo; unpublished).

—— & H. Ohba (1982) Studies on hybrids in the genus *Lespedeza* sect. *Macrolespedeza* (1) A putative hybrid between *L. Buergeri* Miq. and *L. cyrtobotrya* Miq. *Journ. Jap. Bot.*, **57**: 232–240.

—— & —— (1983a) Studies on the hybrids in the genus *Lespedeza* sect. *Macrolespedeza* (2) A hybrid swarm between *L. homoloba* Nakai and *L. kiusiana* Nakai. *Journ. Jap. Bot.*, **58**: 97–104.

—— & —— (1983b) Taxonomical notes on *Lespedeza satsumensis* Nakai and *L. formosa* var. *australis* Hatusima (Leguminosae). *Journ. Jap. Bot.*, **58**: 135–145.

—— & —— (1983c) Studies on hybrids in the genus *Lespedeza* sect. *Macrolespedeza* (3) A putative hybrid between *L. Buergeri* Miq. and *L. satsumensis* Nakai. *Journ. Jap. Bot.*, **58**: 248–252.

—— & —— (1985) The branching of the inflorescence and vegetative shoot and taxonomy of the genus *Kummerowia* (Leguminosae). *Bot. Mag. Tokyo*, **98**: 137–150.

—— & —— (1988) Taxonomical note on *Lespedeza formosa* (Vogel) Koehne. In Ohba, H. & S. B. Malla. (ed.), The Himalayan Plants, **1**: 217–229.

Bentham, G. (1865) Leguminosae. In Bentham, G. & J. D. Hooker, Genera Plantarum **1**: 434–600. Reeve, London.

Candolle, A. P. de (1825) De la germination des Légumineuses. Mémoires sur la famille des Légumineuses, 61–122.

Clewell, A. F. (1964) The biology of the common native lespedezas in southern Indiana. *Brittonia*, **16**: 208–219.

—— (1966) Natural history, cytology, and isolating mechanisms of the native American lespedezas. *Bull. Tall Timbers Research Stat.*, No. 6, 1–39.

—— (1971) In Löve, A., IOPB Chromosome number report XXXII. *Taxon* **20**: 349–356.

Cooper, D. C. (1936) Chromosome numbers in the Leguminosae. *Amer. Journ. Bot.*, **23**: 231–233.

Davis, P. H. & V. H. Heywood (1963) Principles of Angiosperm Taxonomy. 558 pp, Robert E. Krieger Publishing Co., New York.

Duke, J. A. & R. M. Polhill (1981) Seedlings of Leguminosae. In Polhill, R. M. & P. H. Raven, Advances in Legume Systematics, Part 2, 941–949. London.

Ferguson, I. K. & J. J. Skvarla (1981) The pollen morphology of the subfamily Papiliono-ideae (Leguminosae). In: Polhill, R. M. & P. H. Raven, Advances in legume systematics, Part 2, 859–896. London.

Hatusima, S. (1963) A new *Lespedeza* from Kyushu. *Journ. Jap. Bot.*, **38**: 155–157.

—— (1967) *Lespedeza*: Sect. *Macrolespedeza* and *Heterolespedeza* from Japan, Corea and Formosa. *Mem. Fac. Agr. Kagoshima Univ.*, **6**: 1–17.

—— (1969) *Lespedeza*: Sect. *Macrolespedeza* and *Heterolespedeza* from Japan, Corea and Formosa. *Ann. Rep. Yokosuka City Mus.*, No. 14, 1–7 (in Japanese).

Hochreutiner, B. P. G. (1934) Validity of the name *Lespedeza*. *Rhodora*, **36**: 390–392.

Huang, T.-C. & H. Ohashi (1977) Leguminosae. In H. L. Li et al. eds. Flora of Taiwan. **3**: 148–421. Epoch Publ. Co., Taipei.

Ikuse, M. (1956) Pollen grains of Japan. Hirokawa Publ. Co., Tokyo (in Japanese).

Kawakami, J. (1930) Chromosome numbers in Legminosae. *Bot. Mag. Tokyo*, **44**: 319–328. (in Japanese).

Kitamura, S. & G. Murata (1961) *Col. Ill. Herb. Pl. Jap.*, **2**: 96–100, Hoikusha, Osaka (in Japanese).

Kondo, K. et al. (1977) Interspecific variation of karyotypes in some species of *Lespedeza* and *Dismodium*. *La Chromosomo*, **2**(5): 123–137.

Lee, T. B. (1965) The *Lespedeza* of Korea (1). *Bull. Seoul Nat. Univ. For.*, No. 2, 1–34.

Lee, Y. N. (1969) Chromosome numbers of flowering plants in Korea. *Journ. Korean Res. Inst. Ewha Women's Univ.*, **11**: 455–478.

Maximowicz, C. J. (1873) Synopsis generis *Lespedezae*, Michaux. *Acta Hort. Petrop.*, **2**: 329–388.

Maekawa, F. (1938) Shirobanahagi no jiseichi ka. *Journ. Jap. Bot.*, **14**: 72 (in Japanese).

Momiyama, Y. (1933) On the bud of *Lespedeza Buergeri* Miq. *Journ. Jap. Bot.*, **9**: 129–130 (in Japanese).

Murata, G. (1978) Taxonomical notes 12. *Acta Phytotax. Geobot.*, **29**: 95–105 (in Japanese).

—— (1983) On the lectotype of *Lespedeza bicolor* f. *sericea* Matsum. *Acta Phytotax. Geobot.*, **34**: 126–129 (in Japanese).

Nakai, T. (1925) Who is really the author of Michaux: Flora Boreali-Americana? *Bot. Mag. Tokyo*, **39**: 215–219 (in Japanese).

—— (1927) *Lespedeza* of Japan & Korea. *Bull. For. Exp. Stat. Chosen*, **6**: 1–101.

—— (1939) Notulae ad Plantas Asiae Orientalis IX. *Journ. Jap. Bot.*, **15**: 523–541.

Ohashi, H. (1971) A taxonomical study of the tribe Coronilleae (Leguminosae), with a special reference to pollen morphology. *Journ. Fac. Sci. Univ. Tokyo*, Sect. III, **11**: 25–92.

—— (1981) Notes on *Lespedeza thunbergii* (DC.) Nakai. *Journ. Jap. Bot.*, 239–244 (in Japanese with English summary).

—— (1982a) Nomenclatural changes in Leguminosae of Japan. *Journ. Jap. Bot.*, **57**: 29–30 (in Japanese).

—— (1982b) *Lespedeza*. In: Satake, Y. et al. eds., Wild flowers in Japan, herbaceous plants (including dwarf subshrubs). **2**: 204–206, Heibonsha Ltd., Tokyo (in Japanese).

—— (1986) Notes on *Lespedeza thunbergii* and its wild ancestor, var. *patens* (Leguminosae). *Journ. Jap. Bot.*, **61**: 114–126 (in Japanese with English summary).

——, R. M. Polhill, and B. G. Schubert (1981) Tribe 9. Desmodieae (Benth.) Hutch. (1964). In: Polhill, R. M. & P. H. Raven, Advances in legume systematics, Part 1, 292–300.

Ohwi, J. (1953) *Lespedeza*. In: Flora of Japan, 677–680, Shibundo, Tokyo (in Japanese).

—— (1965a) *Lespedeza*. In: Flora of Japan, rev. ed., 788–792, Shibundo, Tokyo (in Japanese).

—— (1965b) *Lespedeza*. In: Flora of Japan, Eng. ed., 557–560, Smithsonian Institution, Washington, D.C.

—— (1975) *Lespedeza*. In: Flora of Japan, new ed., 788–792, Shibundo, Tokyo (in Japanese).

Pierce, W. P. (1939) Cytology of the genus *Lespedeza*. *Amer. Journ. Bot.*, **26**: 736–744.

Ricker, P. L. (1934) The origin of the name *Lespedeza*. *Rhodora*, **36**: 130–132.

Schindler, A. K. (1912) *Kummerowia* Schindler novum genus Leguminosarum. *Fedde, Repert.*, **10**: 403–404.

—— (1913) Einige Bemerkungen über *Lespedeza* Michx. und ihre nächsten Verwandten. Engler, *Bot. Japhrb.*, **49**: 570–658.

Vassilczenko, I. T. (1937) Morphologie de Keimung der Leguminosen in Zusammenhang mit ihrer Systematik und Phylogenie. *Act. Inst. Bot. Acad. Sci. USSR*, Ser. 1, **4**: 347–425.

Young, J. O. (1940) Cytological investigations in *Desmodium* and *Lespedeza*. *Bot. Gaz.*, **101**: 839–850.

# List of New Names

*Lespedeza* series *Formosae* S. Akiyama et H. Ohba.

*Lespedeza* series *Heterolespedeza* (Nakai) S. Akiyama et H. Ohba.

*L. bicolor* Turcz. f. *niveoflora* S. Akiyama et H. Ohba.

*L. formosa* (Vogel) Koehne subsp. *velutina* (Nakai) S. Akiyama et H. Ohba f. *alba* (Nakai) S. Akiyama et H. Ohba.

*L. patens* Nakai f. *sericea* (Matsum.) S. Akiama et H. Ohba.

# Index of Latin Names

Names in italics are synonyms and those in bold are new names.

# Index of Japanese Names